PENGUI

VOY

Stephen J. Pyne is a Regents Professo... School of Life Sciences, Arizona State University. An award-winning environmental historian, he is the author of *Year of the Fires*, *The Ice*, and *How the Canyon Became Grand*. He is a recipient of the Robert Kirsch Award from the *Los Angeles Times*.

---

## Praise for *Voyager*

"*Voyager*, a meditation on the nature and meaning of exploration itself, disguised as a chronicle of the life and times of a space mission."

—*The New York Times Book Review*

"For space geeks, it's a sweet read; for everyone else, it's an eye-opener."

—*Time*

"Even the most passionate aficionado, who devoured every digital bit sent back by the Voyagers, will find this overview enriching."

—*The Washington Post*

"The Voyager spacecraft have not only clocked up a far better understanding of the outer planets, they also illustrate mankind's third great age of discovery, according to Stephen Pyne in a fascinating new book."

—*The Economist*

"The Voyager story itself is an amazing one, and Mr. Pyne tells it skillfully. . . . Mr. Pyne deftly shows how the development of rocketry, of orbital science, and of computer technology all came together just in time to take advantage of a once-every-176-years planetary alignment that would allow a spacecraft to make close passes of outer planets— Jupiter, Saturn, Uranus, and Neptune—all in one long trip." —*The Wall Street Journal*

"Pyne's book isn't just an overview of the Voyager program; it's a sweeping history of what Pyne calls the 'third age of discovery,' beginning with the first sputterings of Sputnik and reaching all the way to our recent space shuttle disasters. Along the way, we're treated to a dense but intriguing sweep of the eras of exploration past." —Salon.com

"Today both Voyagers are still in operation and are passing beyond the edge of the solar system, serving as distant ambassadors for humankind. In this book, Pyne puts that quest in grand perspective." —*Science News*

"A challenging but immensely rewarding read."

—*Kirkus Reviews* (starred review)

"Pyne manages to set alight the story of the Voyagers as few space writers have ever done. In a marvelous twist on the usually self-important chronicles of space missions, he places the Voyagers in their historical and sociological context."

—*The Dallas Morning News*

"This is not just a history of the remarkable Voyager program. Instead, it is an attempt to analyze it as part of the broad sweep of exploration stretching back to the likes of Christopher Columbus, focusing more on the politics and culture behind these ventures than on their scientific returns."

—*New Scientist*

"Pyne captures the Western passion for exploration and the lure of the unknown, while relating the fascinating story of two fragile spacecraft continuing after three decades their brave quest across space and time."

—*Publishers Weekly* (starred review)

"If NASA's historic achievement in manned spaceflight is a source of wonder and pride for all Americans, the agency's epic triumphs with satellites should be just as widely known and celebrated. Stephen Pyne's masterful *Voyager* shows us just how the United States continues to command the forefront of a third age of discovery in exploring our beautiful and astonishing universe."

—Craig Nelson, author of *Rocket Men: The Epic Story of the First Men on the Moon*

"Stephen Pyne's *Voyager* is perhaps the greatest of his many books, much as the Voyager voyages are perhaps the greatest achievement of what Pyne terms 'A Third Great Age of Discovery.' Pyne chronicles these space expeditions in brilliant and detailed prose that is almost a new and poetic scientific language. It is a must-read for every intelligent human."

—William H. Goetzmann, Jack Blanton Professor of American Studies and History

"There is an argument that what Americans do best is send extraordinary missions into space. Stephen Pyne, with a swift eye to the history of great exploration, positions the Voyager probe perfectly as it exits our solar system, beyond which are regions that exceed our imagination."

—George Butler, producer and director of the documentary films *The Endurance: Shackleton's Legendary Antarctic Expedition* and *Roving Mars*

"Like the phenomenal missions it evokes, *Voyager* is destined for enduring eminence. A mind-boggling synthesis of science, history, and poetry."

—Laurence Bergreen, author of *Marco Polo: From Venice to Xanadu* and *Over the Edge of the World: Magellan's Terrifying Circumnavigation of the Globe*

# Voyager

## EXPLORATION, SPACE, AND THE THIRD GREAT AGE OF DISCOVERY

Stephen J. Pyne

PENGUIN BOOKS

PENGUIN BOOKS
Published by the Penguin Group
Penguin Group (USA) Inc., 375 Hudson Street, New York, New York 10014, U.S.A. • Penguin Group (Canada), 90 Eglinton Avenue East, Suite 700, Toronto, Ontario, Canada M4P 2Y3 (a division of Pearson Penguin Canada Inc.) • Penguin Books Ltd, 80 Strand, London WC2R 0RL, England • Penguin Ireland, 25 St. Stephen's Green, Dublin 2, Ireland (a division of Penguin Books Ltd) • Penguin Books Australia Ltd, 250 Camberwell Road, Camberwell, Victoria 3124, Australia (a division of Pearson Australia Group Pty Ltd) • Penguin Books India Pvt Ltd, 11 Community Centre, Panchsheel Park, New Delhi - 110 017, India • Penguin Group (NZ), 67 Apollo Drive, Rosedale, Auckland, 0632, New Zealand (a division of Pearson New Zealand Ltd) • Penguin Books (South Africa) (Pty) Ltd, 24 Sturdee Avenue, Rosebank, Johannesburg 2196, South Africa

Penguin Books Ltd, Registered Offices:
80 Strand, London WC2R 0RL, England

First published in the United States of America by Viking Penguin,
a member of Penguin Group (USA) Inc. 2010
Published in Penguin Books 2011

1 3 5 7 9 10 8 6 4 2



THE LIBRARY OF CONGRESS HAS CATALOGED THE HARDCOVER EDITION AS FOLLOWS:
Pyne, Stephen J., 1949-
Voyager : seeking newer worlds in the third great age of discovery / Stephen J. Pyne.
p. cm.
Includes bibliographical references and index.
ISBN 978-0-670-02183-3 (hc.)
ISBN 978-0-14-311959-3 (pbk.)
1. Voyager Project. 2. Astronautics—United States—History.
3. Aeronautics—United States—History. 4. Planets—Exploration. I. Title.
TL789.8.U6V5275 2010
919.9'204—dc22 2009046305

Printed in the United States of America
Set in ITC Legacy
Designed by Daniel Lagin

*To Sonja*
*a polestar, as always*
*and Lydia, Molly, Karlie, Ashley, Lindsey, Colten, Julie, and Ivy*
*a new constellation in the heavens*

An age will come after many years when the Ocean will loose the chains of things, and a huge land lie revealed; when Tethys will disclose new worlds and Thule no more be the ultimate.

**—Seneca, *Medea* (ca. AD 40)**

Throw back the portals which have been closed since the world's beginning at the dawn of time. There yet remain for you new lands, ample realms, unknown peoples; they wait yet, I say, to be discovered.

**—Richard Hakluyt, *The Principal Navigations, Voyages, Traffiques, and Discoveries of the English Nation* (1587)**

Come, my friends,
'Tis not too late to seek a newer world.

**—Alfred Tennyson, "Ulysses" (1842)**

There are no more unknown lands; the new frontiers for adventure are the ocean floors and limitless space.

**—J. Tuzo Wilson, *I.G.Y.: The Year of the New Moons* (1961)**

# Contents

# Illustrations

**IMAGES (IN INSERT)**

## FIGURES (ALL IN APPENDIX)

### THE GRAND TOUR AND ITS ENCOUNTERS



### THE COLDEST WAR

# Mission Statement: Voyager of Discovery

On August 20 and September 5, 1977, two spacecraft, Voyager 2 and Voyager 1, respectively, lifted off atop Titan/Centaur rockets from Cape Canaveral, Florida, to begin a Grand Tour of the outer planets. Some 33 years and 21 billion kilometers later, having surveyed Jupiter, Saturn, Uranus, Neptune, their moons, and the interplanetary medium, and sent back enough digital information to fill a Library of Congress, they find themselves within the diaphanous heliosheath that divides the Sun from the stars. They have sufficient power for another ten years of operation, and racing at 440 million kilometers a year, that should grant them enough stamina to sail beyond the reach of the solar gases altogether and enter the interstellar winds before they expire.

The Voyager mission culminated what many of its contemporaries have come to regard as a golden age of American planetary exploration and what the future may well identify as the grand gesture of a Third Great Age of Discovery, the most recent revival of geographic journeying and questing in a chronicle that, for Western civilization, traces back to the fifteenth century. Yet what it has done for geography, the Voyager mission also does for history: the saga of the Voyagers' trek is carrying the inherited narrative of exploration to its outer limits, and perhaps beyond.

---

The countdown for the Grand Tour began twenty years earlier when, within the context of the International Geophysical Year (IGY), the Soviet Union lofted Sputnik 1 into orbit on October 4, 1957. The feat caused a sensation. Orbiting every ninety minutes, it sent out a beep that broadcast both triumph and taunt. The United States failed in its counter-launch of Vanguard 1 before succeeding with Explorer 1, on January 31, 1958. Thereafter, the two rival superpowers of the cold war began a long volley of launches that extended their geopolitical rivalry beyond the confines of scientific competition and the gravitational reach of Earth, and that, whether they intended the outcome or not, helped birth a new epoch of geographic exploration. Modern rockets had breached an ancient barrier to beyond.

From 1962 to 1978 America launched a flotilla of spacecraft—Pioneer, Mariner, Viking, and Voyager—to orbit the inner planets and survey the outer ones. In an outburst comparable to that of the sixteenth-century Portuguese *marinheiros* into and across the Indian Ocean or the eruption of eighteenth-century circumnavigators into the Pacific, American spacecraft would visit all the major bodies of the solar system save Pluto. But even as Voyager was wending through the asteroid belt, the institutional apparatus that launched it was collapsing. Less than a year after Voyager 1 launched, the Pioneer Venus Orbiter fired off, and then, for eleven years, no other American spacecraft trekked beyond the bonds of Earth. Throughout those years the Voyagers defined the American planetary program.

For some participants, those years bracket the acme of American space exploration. For others, the borders of a golden age are more elastic, reaching back to Explorer 1 or even to the lonely experiments of Robert Goddard and extending onward to the recent journey of Cassini to Saturn and the brave sojourn of New Horizons to extra-planetary Pluto and the Kuiper Belt. Others scorn so American an emphasis and insist that the circle include such visionaries as Hermann Oberth and Konstantin Tsiolkovsky, the relentless outpouring of Soviet spacecraft, the proliferating efforts of the European Space Agency, and the satellites of Japan, India, and China, all of which expand the scope of the undertaking. For a few observers—the most visionary or fanciful—the epic begins in evolutionary mists, when marine life first crawled onto land, or when early *Homo* of an unquenchable curiosity acquired fire; and for such prophets, the tale can end only when humanity, perhaps genetically self-engineered beyond recognition, wanders outside the solar system altogether.[1]

But all seem to agree that what lifted off from launchpads at Tyuratam and Cape Canaveral was something both novel and inevitable, and that it portended a fabulous new era of discovery, a journey beyond anything humanity had known before. In the past, geopolitics had repeatedly projected rivalries internal to Europe outward onto a wider world. With the launches of Sputnik and Explorer the cold war prepared to do the same, and this time the competition would transcend Earth itself. It would reach beyond sordid politics and the blinkered ambitions of its originating time and place. Those who meditated on the launches believed they were witnessing a transit of history that would define the dimensions of human time, like the transit of a planet across the Sun that demarked the dimensions of space. They did not know what lay beyond, or how they might reach it, only that they were determined to go.

After the launch of Explorer 1 in 1958, a press conference led to one of the canonical images of the ensuing space age, as three men—Wernher von Braun, James Van Allen, and William Pickering—hefted a mock-up of Explorer 1 over their heads. That tableau made visible the competition that drove the launches, a cold war that refused to be bound by Earth and carried its geopolitical jousting into space. But there, too, in the personality of each man and the cultural ambitions he represented, were the internal rivalries, both institutional and intellectual, behind the launch. Each man stood for a competing vision of what space might represent.

Wernher von Braun was in the tableau because he had overseen the development of the rocket that launched Explorer 1, and was the most prominent advocate for space as an arena for extraterrestrial colonization. He represented a tradition that identified far journeying with human settlement; he promised, in particular, to project the American experience outward to the Moon, then to Mars, and ultimately beyond. The von Braun narrative was one in which explorers were simply the vanguard of colonizers. It built on a heritage of European expansion since the Crusades.

James Van Allen was there because he had developed the instrument package for the satellite, and he became a public voice for space as a new laboratory for science. He stood for a tradition of systematic inquiry that had happily looked beyond Earth since the pre-Socratics, one that Galileo had refurbished and bonded to modern science with his invention of the telescope and, through that instrument, his questioning of the Ptolemaic model of the solar system. Rockets promised to take proxies of scientists—robots with instruments—beyond the limits of earthbound telescopes. The Van Allen narrative was one of an ever-searching science, always pushing against the limits of what was possible to know, an inquiry that had easily allied itself with geographic exploration.

The third man was William Pickering, then director of the Jet Propulsion Laboratory, the institution responsible for getting rocket and satellite together and on trajectory, and he would argue that Explorer 1, as its name indicated, was a vehicle for interplanetary discovery. Its larger purpose was to continue an erstwhile legacy of exploration—one certainly outfitted with the instruments of modern science, and one that would likely lead someday to outposts staffed with humans, if not to outright colonization. But sending robotic spacecraft was in and of itself a worthy exercise in an honored tradition of exploring. Of course that sentiment was never alone sufficient to justify such costly enterprises, which had to acquire other ends as well; but ultimately America should explore because that is what America had always done and what made America what it is. The Pickering narrative was one organized around half a millennium of geographic discovery by Western civilization and its rambunctious offspring, not least the United States.

All three visions shaped the evolving American space program, although each sought to point it in a different direction. Von Braun looked to the Moon and Mars, both potential sites for settlement. Van Allen sought to place instruments beyond the obscuring atmosphere, whether in wide Earth orbit or through interplanetary space.

Pickering (JPL) wished to send spacecraft to other planetary worlds; the journey itself was a medium for the message; the enterprise was, literally, a voyage of discovery. When money was flush, all three ambitions could claim more or less what they wanted. But when money became tight, the program had to choose among them; and while some tasks could overlap—the visions were not mutually exclusive—not every project or launch could satisfy each ambition equally, and consensus could fray and unravel.

This internal competition, separate from any geopolitical rivalry, was not solely about money. It was also about meaning; it was about what the Voyagers might do, what they signified, and why we should care. Explorer 1 could hold the space clans together, for each party could project into future launches what it hoped would happen. But that was possible because Explorer 1 first raised the curtain; twenty years later it was not possible to stage from a collective script. The forced unity among contestants could no longer heft a common project. In reality, one purpose or another would tend to dominate.

Voyager chose the journey; or rather, it passed through all of the purposes, as it did regions of the solar system. It began as part of cold war saber rattling little distinguishable from geopolitics, or at least the squabbling for spheres of influence if not outright imperialism that had so shaped the history of both the United States and the USSR. But in bypassing the Moon and Mars, Voyager left that purpose behind. It then became more purely science as its Grand Tour subjected the miniature systems of the outer planets to instrumented scans, one stunning encounter after another. Still, it continued, and as it has persisted it has become more and more the expression of an earlier genre of exploration, the quest, for which the narrative itself is a purpose and product. As the Voyager twins pass outside the solar system, pushing well beyond their engineered limits, they have seemingly transcended their origins and the received story by which to explain their mission.

The Voyagers are among exploration's purest expressions, and among both its strangest and its most revelatory. For more than five hundred years the West has relied on exploration to shape its encounter with a wider world, and to seek newer ones. That process has alloyed with adventure, curiosity and colonization, wanderlust, greed, war, pilgrimage, slavery, trade and missionizing, technological innovation, sheer animal instincts for survival, and moral imperatives, a vast historical cavalcade that has morphed and trod across, below, and beyond Earth, yet has organized itself in ways that align with the ambitions, the understandings, and the hopes of its sustaining society.

That chronicle has been neither random nor unbroken. It tacks and veers with changes in the technologies of travel and in modes of inquiry, with the opening of previously unknown lands, with the onset of fresh competitors and aspirations—in brief, with the core values and understandings of the culture that propels it. There are periods of quickening and of slackening; times when enthusiasm flames, and times when it smolders; eras when aggressive expansion seems irresistible, and eras when it appears to intellectuals as quaint, repugnant, or laughable. Twice in the past, geographic discovery as a project had been so reorganized in its fundamentals as to constitute an identifiable phase; call it a Great Age. This happened with the Great Voyages of the Renaissance, and it happened again as the Enlightenment dispatched Corps of Discovery to resurvey the old lands and to inventory whole continents with the sharpened eyes of science. Such eras rise from the general chronicle like mountain ranges, between which lie valleys of exhaustion or indifference. Beginning with the International Geophysical Year, another such Great Age has come into definition, drawn to new domains of geographic discovery, equipped with robots and remote sensors, and outfitted with a very different cultural syndrome, what might be termed a Greater Modernism. The Voyager mission nests within this narrative, and it may serve for this latest phase as a defining gesture of what it is about.

Yet a certain uneasiness hovers over this latest phase. It's dehumanized in sometimes unsettling ways. It has dispatched expeditions to new worlds, yes, but worlds inoculated against life, where no natives can guide and enlighten explorers, where no explorer can possibly live off the land, and where colonization is a fantasy. And perhaps most fundamentally the Voyager mission and its kind do not rely on human discoverers. The mantle of explorer rests on robots. How is this exploration? In what respects is this age continuous with and distinct from those that went before? In what ways does this new era of discovery recapitulate old ones, and in what ways does it exhibit new conceptions as well as technologies? How might its own enterprise be a novelty as spectacular as anything its journeys have unveiled? What has the era meant to the half-millennial saga of geographic exploration by Europe and its cognate civilizations? And why might that matter?

All this, Voyager has gathered from history, as its instruments have collected cosmic rays from deep space. The Voyager mission amassed the pieces of this new era, miniaturized and assembled them into working machines, and dispatched them on an immense journey. Those two spacecraft look back across five centuries of looking outward. Their trek is a complex fugue of past and future, of tradition and novelty, a narrative that pushes onward to newer worlds by constantly realigning with a legacy of exploration to older ones. The Voyagers' journey is an apt symbol for what might be considered a Third Great Age of Discovery, which is our own.

# part 1

## THE BEGINNING OF BEYOND: JOURNEY OF AN IDEA

I cannot rest from travel; I will drink
Life to the lees . . .
There lies the port; the vessel puffs her sail;
There gloom the dark, broad seas.

**—Alfred Tennyson, "Ulysses"**

# 1. Escape Velocity

A Great Age of Discovery, like the epic voyages it encompasses, requires places to visit, the means to get there, and the will to go. For decades, after sledging over Antarctica and suffering through world wars and a global depression, exploration had sunk into a deep trough. But by the late 1950s all those errant requirements, separately evolving, had come into auspicious alignment. With Sputnik, a new era of planetary exploration achieved escape velocity.

There now existed technologies to carry instruments and people into places implacably hostile to life, and once there, to measure, map, and inventory by remote sensing. Some were mechanical devices such as rockets and submersibles; some, intellectual inventions such as maneuvers to boost spacecraft through gravitational assists. Together, machines and minds made possible forays over the ice fields of Antarctica and Greenland, across and down into the oceans' abysses, and through the solar system, the primary geographic arenas for new discovery. And there were motives aplenty from science, ever competitive, always pushy; from communities of believers and enthusiasts, explorers and space cultists, eager to exploit a congealing alloy of interests; from cultural longings, especially traditions of adventuring and questing, however projected afar; from businesses keen to supply the products of desire, particularly when refracted through governments; military ambitions, either for defense or forward strategies to forestall moves by antagonists.

This bundle of new motives, means, and opportunities made the exploration of the solar system possible, beginning with planet Earth. By themselves, however, they could not guarantee a new great age. Left to its own momentum, scientific inquiry and technological inventiveness would have slowly opened more of Antarctica, here and there, as coastal bases appeared and sledging parties converted from dogs to Sno-Cats and Bell 212 helicopters. Instrumented vessels would have probed the continental shelves and the occasional abyssal plain. Rockets would have supplanted balloons for sampling the upper atmosphere and the fringes of interplanetary space. Geographic adventuring would have merged with extreme sports and exotic tourism. The process would likely have been slow and sporadic, not unlike the Portuguese coasting that mapped the shores of Africa in fits and starts over the course of the fifteenth century or that slowly seeped across South America during the sixteenth.

The cold war, however, whipped those long swells into the whitecaps of a cultural storm. Golden ages involve concentrated outbursts, contained frenzies of discovery, typically within a single generation or two. The dynamics of the latter twentieth century offered a close paraphrase of those that powered previous outbursts such as the Iberian reconnaissance of the late fifteenth and early sixteenth centuries, the Russian leap across Eurasia in the mid-seventeenth century, the British and French circumnavigations of the late eighteenth, the North American blitz from the Appalachians to the Pacific in the early nineteenth century, the three-decade scramble across Africa that commenced in the late nineteenth, and the twentieth century's spectacular if brief heroic age in Antarctica. Now, as another fever of discovery spread, new expeditions looked to their predecessors not only for navigational aid but for models by which to define their character.

## THE COLDEST WAR

In 1599 Vargas Machuca asserted the valence between geopolitics and exploration by declaring simply, "*a la espada y el compas, mas y mas y mas y mas.*" By the sword and compass, more and more and more and more.

Some 350 years later the cold war reconfirmed that alliance. Rockets and remote sensing owed their rapid development to military sponsorship and the perceived need to control the new high ground of near-Earth space. Submersibles plumbed the oceans to map the terrain for nuclear-armed submarines. Both kinds of vehicles, moreover, whether beyond the atmosphere or beneath the seas, required secure communications, always an incentive to investigate new media and the paths of technology. The bonding of science with a national security state, begun during World War II, stabilized amid the early cold war and was then bolstered by the political panic that followed Sputnik. The primary institutions of contemporary exploration, even those nominally civilian, as often as not had Defense Department funding or acted as civilian surrogates to the same geopolitical ends–NASA doing for the space program, for example, what it was hoped the Peace Corps might do to forestall insurgencies. The superpower rivalry played out amid gestures of cultural superiority, from counting Olympic medals to technological triumphalism in rocketry, and it played out on the coldest of terrains from Antarctic ice to deep-ocean abyss to interplanetary space.

The prime mover of exploration remained what it had historically always been: the competition not between Europe and others but among the Europeans themselves, or among their former colonies and empires. However removed from overt confrontation, however benign compared to hot-war battlefields, space launches were a form of saber rattling. Whatever they had to say about distant planets, they spoke first to and about Earth. Just as Europe's internal competitions had pushed it into colonial conflicts, so the cold war projected Earth's dominant rivalry into the heavens. This appeal to exploration for geopolitical ambitions had a long pedigree. The monarchy that England could not assault in Madrid it could hobble on the Spanish Main. What France failed to achieve in Alsace, it could relocate to the Atlas Mountains. What Holland could not seize in North America, it might claim in the Spice Islands. The containment of communism that the United States was unable to achieve in the Mekong Delta it might overcome on the Sea of Tranquility. If it trailed the USSR in rockets, it could best its rival in satellites and science. "Space exploration," conceded Bruce Murray, co-founder of the Planetary Society and director of JPL during Voyager's passage through Saturn, "burst forth amid open belligerence and armed confrontation between the United States and the Soviet Union."[1]

Curiosity, a passion to explore, economic spinoffs, intellectual sparks cast from the whetstone of political necessity, cultural rejuvenation, government investment as a fiscal stimulant, renewed frontiers—whatever reason *could* be associated with the enterprise likely *would* be. After World War II, American oceanographers, for example, seemingly appealed to any and every cause that might fill their coffers and advance their standing as scientists. If the U.S. Navy had technical needs, then it could meet them. If foreign policy required international "cooperation," then that became the sustaining rationale. If assistance in "development" for emerging nations drove national interest, then that served. But in the postwar era, all these jostled under the covering umbrella of the cold war. National defense, broadly defined, was the overarching conceit, both unanswerable and sufficiently elastic to justify and disguise whatever real purpose a proponent intended.[2]

Yet those selfish pursuits emanated from a common culture, and it is precisely such amalgams—the sloppier and less precise, the better—that allow scattered acts of discovery and exploring expeditions to congeal into a Great Age of Discovery. The power of exploration derives from the power it shares with its sustaining society. The more deeply it can draft from that culture, the more interests it can tap, the more robust will be its support and the richer its impact. Still, something has to hold those oft-disparate pieces together to reach a goal: a colossal rivalry does just that, as internal dissensions shrink in comparison with the distance to a common foe. That was the catalytic effect of the cold war. Paradoxically, while they emanated from very different societies, the two superpowers displayed vital commonalities. Both identified themselves as expansionist nations founded on an ethos of exploration. Both committed themselves to similar state-sponsored technologies and came to mirror each other in their exploring styles; even civilian institutions morphed into looking-glass versions of their military cognates.[3]

The core issue for those interested in transforming stunning gestures into a program of space exploration and colonization was not whether the cold war could spark a golden age—it clearly did. The issue was whether such an era could sustain itself once that catalytic competition was gone.

## EXPLORATION'S NATION

What this era of discovery means—why it happened when it did, what precedents it taps, even whether it constitutes a special age at all—depends on how those framing issues are placed and to what purposes. It configures one way within the cold war; another within sagas of imperial expansion, folk wanderlust, or colonization; still others within chronicles of technological innovation and scientific inquiry. Most advocates are eager to seize whatever justifications can be mustered, gathering any and all auxiliaries under the banner of their master purpose.[4]

But every proponent of a space program has instinctively appealed to exploration as a cause and consequence. Whatever else planetary spacecraft do, they explore, and whatever else they might mean, they belong necessarily in that illustrious pantheon of humans and societies that have pushed beyond frontiers to reveal a wider world. Against such motives, boosters assert, there can be no appeal. Exploration is politically deserving, culturally enriching, and genetically obligatory. It just is, and it must be.

In July 1969 William Pickering, director of JPL, responded to a request from Thomas O. Paine, head of NASA, for thoughts about the future direction of agency programs. Over the previous decade, Pickering noted, "the contest with the Soviet Union" provided the "necessary incentive" for Apollo. After Apollo 11 that external stimulus no longer existed, and Pickering, perhaps surprisingly, argued against an explicit replacement such as "a scheduled goal to land men on Mars." But how otherwise to "describe and justify the NASA purpose" remained tricky.[5]

He ticked through the usual roster. The advancement of scientific knowledge was "not worth $4 billion for a relatively narrow area of knowledge having no obvious relevance to everyday problems." The spinoffs from new technologies were "not good enough." (The public had already waited in vain for ten years for some "dramatic application of space technology" and had gotten Tang and the DustBuster.) The sense of "human adventure" was "interesting" but awkward to justify at these expenditures. National security was important, but it was hard to see how the present array of programs contributed. No single factor was, in truth, sufficient.[6]

Rather, the answer lay in "combining the several factors." That required an "integrating factor," which in Pickering's mind should be "exploration." Exploration could rally public sentiment. The country had always celebrated its "explorers and pioneers who tamed a continent." Now that the entire Earth had been "thoroughly explored from pole to pole, from mountain top to ocean depth," it fell to NASA to project that saga across the solar system. This, he concluded, was "a fitting task for the U.S."[7]

Exploration was, in brief, the final frontier of justification that potentially absorbed all the others and whose claims, in some respects, seemed unanswerable. Yet the appeal the space community made to the actual history of exploration was often both banal and irrelevant. That chronicle existed in the minds of many proponents only to motivate. The worst offenders were the colonizers, who saw themselves as true cosmopolites—literally citizens of the cosmos, who were willing to tap any nationalist chauvinism that would advance their cause. In particular, they became adept at phrasing the arguments for space travel in ways that would make critics seem to question the significance of the New World and especially the success of the American experiment.

In writing a history of space travel, with the help of two associates, Wernher von Braun concluded with a panegyric:

> During the Renaissance, Prince Henry the Navigator of Portugal established in his seaside castle of Sagres the closest precedent

to what the space community is trying to accomplish in our time. He systematically collected maps, ship designs, and navigational instruments from all over the world. He attracted Portugal's most experienced mariners. He laid out a step-by-step program aimed at the exploration of Africa's Atlantic coast as well as the discovery of the continent's southernmost tip, which he knew had to be circumnavigated if India were to be reached by the sea. With equal determination he pushed for the possibly shorter westbound route to the Far East. Prince Henry trained the astronauts of his time—men such as Bartolomeu Diaz, Ferdinand Magellan, and Vasco de Gama, and he created the exploratory environment that launched Christopher Columbus from neighboring Spain on his historic voyage.[8]

Not a word of that valedictory screed is true. This is an engineer's history, redesigned as one might rework a faulty engine to make it run better. It is a prophet's history, selectively culled and shaped to anticipate a premonitory future. And it is a rationalizer's history, gliding over the character of early rocketeers—the freelance pilots of the twentieth century—who offered their services to whatever power might advance their millennial ends. It is, too, a manifesto that justifies exploration as a scouting party for colonization. Discovery nests within a narrative of Western imperialism. It does not exist within a narrative of exploration as an act and institution in its own right.

To the believers, historical distortions mattered less than the value of historical appeal. "Henry the Navigator," von Braun continued, "would have been hard put had he been requested to justify his actions on a rational basis, or to predict the payoff or cost-effectiveness of his program of exploration. He committed an act of faith and the world became richer for it. Exploration of space is the challenge of our day. If we continue to put our faith in it and pursue it, it will reward us handsomely." Again, not a word rings true, save the appeal to faith, that history might justify what reason could not. What matters to the prophet is motive: if old faiths could move mountains, new ones could move to Mars.[9]

Still, whatever else it was, the space program *was* at least partly exploration, and its historical position remains a triangulation between past and future. If there is a new great age of discovery aborning, it will help to know its parentage, for like all progeny, the offspring will share some traits with its forebears and show new ones. The Voyagers tapped into that heritage and took it in directions never before attempted, so much so that they displayed the ideal expression not only for a new era of exploration but also for what might be considered a new species of explorer.

# 2. Grand Tour

In its origins, Voyager was both a mission in search of an opportunity and an opportunity in search of a mission.

Although the means to send spacecraft beyond Earth's gravity had become possible, and the rivalry with the USSR, rekindled by hysteria over Sputnik, made some undertaking obligatory, it was not obvious where such vehicles should go. The "new ocean" of space—a Black Sea of Darkness, as it were—beckoned powerfully, if vaguely. Beyond the Moon, an obvious first port of call, destinations depended on personalities, institutional preferences, ideas in the wind, real or anticipated moves by rivals, and that element of chance that is both randomness and opportunity.

## FROM THE EARTH TO THE MOON, AND BEYOND

The clamor of voices after Sputnik was furious. Whatever the gloss of science applied to Sputnik, journalists and the public saw the satellite as a surrogate for nuclear-armed intercontinental ballistic missiles (ICBMs). Congress immediately commenced inquiries to restore American prestige, principally under the Preparedness Investigating Subcommittee of the Senate Armed Services Committee, which for its

Texan chair, Lyndon Johnson, meant being first and biggest. The Pentagon made its own claims early and loudly, offering compelling reasons why space belonged under its aegis, particularly the near-Earth environs of orbiting satellites for weather, communication, and espionage, and of course the powerful rockets necessary to launch payloads for any purpose; each service had its own claims, and collectively the Department of Defense funded an Advanced Research Projects Agency. The Rocket and Satellite Research Panel, a scientific advisory group under the Naval Research Laboratory, issued proposals within six weeks after Sputnik. The National Academy of Sciences evolved a Space Science Board out of its International Geophysical Year panel. The President's Science Advisory Committee staked a claim. The National Advisory Committee for Aeronautics (NACA) insisted that space best belonged within its bailiwick. And of course there were professional societies, commercial vendors, and citizen prophets. Each proposed programs tailored to its own purposes.[10]

What emerged was a consensus that the United States needed to demonstrate political will by besting the Soviet technological triumphs, that this might be achieved most blatantly by a high-visibility mission, and that however powerful the military substrate, the program ought to be ostensibly civilian. The institutional compromise, urged by a cautious President Eisenhower, was to recharter the National Advisory Committee for Aeronautics, originally established in 1915 to coordinate among academia, industry, and government, into a National Aeronautics and Space Administration (NASA) on July 29, 1958. The technological compromise was to have the civilians adapt military rockets to nominally scientific goals. And the political compromise was to leave NASA's founding purposes ambiguous, allowing each competing group to see in the NASA charter the realization of its own ambitions. Beyond an understood imperative to produce a publicity event equivalent to Sputnik, everyone could project onto NASA what they wished to see. The agency thus had a difficult birth that led to a troubled adolescence.

As long as money sloshed through the system and a clearly defined space race was on, the fissures within NASA were small and easily spanned. But the conflicts were present at the creation: military control versus civilian purposes, the need to ensure engineering success versus the desire to expand the frontiers of science, the preference for many small vehicles versus a few opulent ones, the desire—as the Rocket and Satellite Research Panel proposed in November 1958—to conduct "scientific exploration" and, at the same time, to carry out "the eventual habitation of outer space." One group envisioned a specific political reply to the Soviet challenge; another, an agenda for scientific discovery through prosthetic vehicles that could carry soundings from Earth's upper atmosphere to those of the other planets; still others, an implied mandate to colonize the solar system for which robotic spacecraft had meaning only as reconnaissance parties for subsequent human settlers; and most, perhaps all, hoped that such ambitions complemented rather than competed. The quarrels over means and ends, they wanted to believe, were merely squabbles over timing.[11]

NASA's charter did not prescribe particular programs, only an institution and a process. Conflicting interests would meet where a democracy best handled them, in open politics. Significantly, NASA's first administrator, T. Keith Glennan, was at once an academic (Case Institute of Technology), a military researcher (Navy's New London Underwater Sound Laboratory), and a member of a government agency (Atomic Energy Commission). NASA simply consolidated existing programs, which basically meant continuing NACA research and extending projects begun under the International Geophysical Year, discipline by discipline. At its organization meeting in June 1958, the National Academy of Sciences' Space Science Board concluded that "the immediate program would integrate results" of the IGY Committee's study, "which is now or has recently been completed." In brief, the emerging space establishment rounded up the usual suspects and prepared to give them money and rockets to go "beyond the atmosphere."

From the outset planetary exploration was deemed an essential part of a space program. The inspiration and prototype came from the International Geophysical Year (1957–58). As with so much of the

Third Age, IGY was catalyst, announcement, and model. It first melded science, exploration, new technologies, and geographies of discovery into a style that was recognizably different from earlier epochs of exploration, and so helped kindle a Third Age, much as the eighteenth century's endeavors to survey the transits of Venus had galvanized a Second Age. This was evident not only in its themes but in how it had morphed beyond them. IGY had begun more modestly, as a Third International Polar Year that intended to focus on Antarctica. But the opportunities seemed too fabulous not to scale up into a full-body scan of planet Earth, including its upper atmosphere; and it was within this context that Sputnik and Explorer launched. It seemed unavoidable, then, that IGY's successor institutions such as NASA would likewise leap beyond their founding conceptions, in this case to push beyond Earth to the solar system overall. The real questions were not whether to go but when, where, and how.

The Moon was an obvious first target, and as soon as the early Sputniks had racked up their orbital triumphs, the Soviets sent Lunik I past the Moon and Lunik II to its surface (both while the IGY was still in progress); and Lunik II sent gasps around the world when it broadcast a photo of the Moon's dark side. So even before NASA was founded, America planned to launch counter-probes to the Moon under the auspices of the Defense Department's Advanced Research Projects Agency—three Pioneer spacecraft by the air force and two through a collaboration between the army and the Jet Propulsion Laboratory.

JPL, in particular, was already imagining a future of planetary exploration far beyond the Moon. It had long ties with the army, not only with the "Propulsion" of its moniker, but more significantly for interplanetary travel, with guidance systems, telemetry, and communications—an early recognition that hardware was only as good as its accompanying software. A mere seventeen days after Sputnik I, the Lab urged that America "regain its stature in the eyes of the world by producing a significant technological advance over the Soviet Union." Director William Pickering recommended going to the Moon "instead of just going into orbit." But the institution aspired to much more. In January 1959 NASA accepted JPL's concept for a deep-space communication network, and in March it approved Project VEGA, an upper-stage rocket and spacecraft, thus partially substantiating JPL's bid to become the prime center for extra-lunar space flights. Ideas swelled like a supernova.[12]

Still, when congressional hearings in March reviewed "The Next Ten Years in Space, 1959–1969," the focus was on near-Earth sites that were achievable with expected engineering (and likely military) payoffs: orbital settings, the Moon, the neighboring Earth-like planets. Exploration for its own sake had little standing. What was done not only had to be doable. It also had to satisfy a political purpose.

Meanwhile, the Soviets had essentially claimed the Moon. A true triumph to counter the endless string of Soviet firsts required a still untouched target. That pointed to Venus and Mars. In February 1961 the Soviet Union made two attempts at Venus. Sputnik IV failed on launch, and Venera I missed its rendezvous with the planet. The opportunity for an American coup remained open. In November JPL engineer Allan Hazard submitted "A Plan for Manned Lunar and Planetary Exploration" that would place astronauts permanently on the Moon by the early 1970s, on Mars by mid-decade, and on Mercury, Venus, and "the outer Planets and their satellites" at some unspecified time thereafter. Though officially disavowed, the scheme showed where Lab thinking was headed. Then NASA pulled the plug in December 1959 by reversing itself and canceling Project VEGA.[13]

The Moon was far closer and more essential. Whatever might go to the planets would be field-tested on the Moon. NASA's founding interplanetary mission, in fact, began as a Moon shot before being diverted into orbit between Earth and Venus. While a first for America, Pioneer 5, jointly assembled by the air force, the Space Technology Laboratories, and Goddard Space Flight Center, had the appearance of a consolation prize after the Soviet spectaculars on both sides of the Moon. There seemed little more possible. The United States lacked a rocket capable of launching interplanetary spacecraft, and there was little incentive to push to the planets when the Moon still proved elusive. Nonetheless, Pioneer 5 sparked a need for an administrative structure—what became NASA's Office of Lunar and

Planetary Programs, which included an Office of Space Science and Applications.[14]

An office, however, needs missions, and missions need both a vehicle and a purpose. What emerged was Ranger, a program for lunar exploration based on a prototype for a true interplanetary spacecraft, one that was not simply an instrumented hunk of metal but an automated system. A JPL team sketched the basics in February 1960 with a study, "Spacecraft Design Criteria and Considerations; General Concepts, Spacecraft S-1." This established the foundational framework for JPL vehicles: a hexagonal frame, modular electronics and subsystem compartments, and a "bus-and-passenger" design that could accommodate a variety of payloads in what became a "hallmark of lunar and planetary missions." Ranger was its prototype, the progenitor for what became Mariner and through Mariner, Voyager.[15]

Yet Ranger's record was ghastly—six failures, finally followed by three redemptive successes. Its successor, the soft-landing Surveyor, hit the Moon five times out of seven. With the announcement by President Kennedy that the United States would seek to put a man on the Moon and return him safely by the end of the decade, the gravitational pull of Earth became less than the political attraction of the Moon.

Still, the planets beckoned. With a new upper-stage rocket, the Centaur, under development, the idea was floated to use some of its developmental test flights to send a spacecraft to Venus or Mars. The launch would occur anyway; the planetary probe would free-ride. Although the needs of Centaur determined the launch window, here was an opportunity to go beyond the Moon, and with NASA's acquiescence, JPL projected a new class of spacecraft, Mariner. The hope was that Mariner A would go to Venus, and Mariner B to Mars, probably in 1962, when planners projected a happy coincidence of developmental timetables and favorable trajectories. Planetary exploration was officially on the books.

Then Centaur encountered troubles, and scrambling for an alternative, JPL proposed in August 1961 that instead of a Titan/Centaur rocket, the launch could rely on an Atlas/Agena, and instead of a new spacecraft, they could construct one out of Ranger, what became Mariner R. So, despite having gone no farther than orbits around Earth, NASA approved two launches to Venus for July–August 1962. With barely a year to prepare, "not knowing that the proposed mission was almost impossible," as Oran Nicks recalled, "we laid out a plan, reprogrammed funding and hardware, and went ahead and did it." Mariner 1 failed on launch; Mariner 2 made the first planetary flyby, coming within 34,400 kilometers of Venus on December 14, 1962. A month before closest approach, NASA approved two Mariner-class spacecraft to go to Mars in 1964.[16]

Going to other planets carried the rivalries advertised in Explorer 1 farther out. Escaping Earth did not, for one, mean escaping the cold war, whose rivals soon sought allies among the planets. Mars and Venus were the prizes, and while both superpowers sent probes to each, the United States held a particular fascination with Mars, while the USSR came to regard Venus as its sphere of influence. But neither did leaving Earth dissolve the rivalries within the American space community. In particular, scientists, colonizers, and explorers could squabble over where and when to go.

Early on, a consensus emerged for Mars. After the success of Mariner 4's flyby in 1964, proposals for further unmanned probes flashed like meteor showers. The American Astronomical Society sponsored a symposium in 1965 on "Unmanned Exploration of the Solar System." The NAS Space Science Board declared that Martian exploration, in particular, including a search for life, should be a "National Goal in Space," a planetary counterpart to the Apollo enterprise. At the same time JPL constituted a Mars Study Committee. Headed by Bruce Murray, the committee stimulated a furious discussion that led to proposals for a dazzling full-bore program of exploration that would include flybys, orbiters, and landers. The last would involve a new state-of-the-art spacecraft called Voyager.[17]

But going to Mars in a colossal way was less politically compelling than getting out of Vietnam, and being tied neither to military necessity nor to Great Society programs, the ballooning costs made the project both visible and vulnerable. In August 1967 Congress canceled funding. By various juggling, and by reverting to a plain vanilla Mariner spacecraft, NASA salvaged enough money to keep the planetary program flying. There were two Venus flights scheduled for 1967 and two Mars flights for 1971. The grandiose Voyager program, once killed, was subsequently resurrected as Viking, a mission to Mars timed for (and justified by) the American bicentennial.[18]

Much as the planetary program had to insinuate itself, almost by accident, into the dominant Moon agenda, so a scheme to visit the outer planets had to finesse its way past Mars.

## THE GRAND TOUR CONCEIVED

All the parts came together. Motive: a surrogate cold war played out in space. Means: the rapidly revolving technologies of rockets, spacecraft, communications, and guidance systems. These were trickier, because in the mid-1960s the capabilities for travel beyond the inner planets did not exist. Opportunity: a Grand Tour to the outer planets so compelling it could move the fantastic into the realm of the hypothetical. Although projects were only as good as the capability to encode them into metal and missions, a fabulous idea could—just might—impose a reality of its own.[19]

The practical range of space travel remained limited to the combustion that engineers could ram through thrusters, and Mariner 2 was already pushing those limits. While bigger rockets, with bigger payloads, were on the drawing boards, travel beyond the Earth-like inner planets demanded a far greater propellant. Even the most powerful rocket, the Saturn V, would require thirty years to send a probe to Neptune solely on its own impulse. Such propulsion alone could never make the trek quickly enough. The payload spacecraft would expire from natural causes before it reached the farthest planets, and politics would never commit to projects that imposed immediate costs for such remote payback. For scientists, too, the horizons were dim; and for public politics, they were as invisible to the naked eye as Uranus.

When a solution was found, it came from software rather than hardware. It appeared in the form of a suggestion that it was possible to outflank the superpower sparring around the inner planets and go directly to the outer ones, and that the propulsion to do so was latent within the very purpose of the mission. Robert Frost once explained that a poem, like a block of ice, should "ride on its own melt." Far planetary exploration needed a mission that could likewise ride on the melt of its design.

The two vital insights emerged not from machine shops or government offices but from densely mathematical studies of hypothetical trajectories. One involved finding a way to get spacecraft to the outer planets, and the other, a time and reason to do so.

The first took shape in 1961, from work by a mathematics graduate student, Michael Minovitch, hired for the summer by JPL's mission-design program to explore trajectories to carry spacecraft from Earth to Venus and back again. Over the next couple of years Minovitch went further and came to realize that a spacecraft behaved like a small planetary body, subject to the same gravitational accelerations and decelerations as asteroids and comets. By approaching a larger body in the same direction as its orbital motion, a spacecraft would accelerate, and thereby achieve additional momentum—a "slingshot" effect. Relative to the planet, it would lose what it gained as it sped away, but relative to the Sun, it would have gained overall, and where the planet producing acceleration was massive, the spacecraft would acquire far more propulsion than it ever could from prospective launch vehicles on Earth, which were soon to approach their upper limits. Moreover, the gravity-assist maneuver could be used more than once in a single mission.[20]

Minovitch organized his thoughts in a 1963 technical report to JPL's trajectory group. His insight did not instantly galvanize his colleagues, however, and they continued to experiment with mixes of propulsion systems, trajectories, and potential projects. Whatever its mathematical elegance, the concept had little engineering relevance until it got coded into a machine and a mission. The defining event came the next year, when Gary Flandro, a postdoc at JPL, distilled the "gravitational perturbation technique" to its essential equations and applied the results to a select suite of trajectories for the outer planets. The "great challenge," Flandro appreciated, was "to try to make exploration of the outer planets practical."[21]

His research took him to Walter Hohmann, and then to Gaetano Arturo Crocco, an Italian who in 1956 had published a scheme for repeated close flybys of Earth, Mars, and Venus in what he termed a "grand tour." Mostly Flandro fixed on Krafft Ehricke's *Space Flight*, which expressed in general language what might be interpreted as gravity assistance. Then Joe Cutting, the group's supervisor, recommended Minovitch's work, particularly that which targeted Jupiter. Here, Flandro thought, was "the key to the outer solar system." To his mind, however, the work done so far had been "elementary" and abstract. What was needed were calculations for "realistic mission profiles so that estimates of actual flight times, payloads, and planetary approach distances and speeds could be made." Especially critical was an identification of "launch windows."[22]

What emerged by the spring of 1965 was a stunning recognition that a once-in-176-years alignment of planets meant that a single spacecraft could fling itself from one to another within a period comparable to the life of a probe. The required conjunction would occur in the early 1980s, which meant that a spacecraft launched in the late 1970s could reach Jupiter, the critical accelerant, just in time to ride a gravitational wave train to Saturn, Uranus, and Neptune. It was, Flandro recalled, "a rare moment of great exhilaration." But as always the inspiration was only as good as its expression. He found that the "trajectory computer programs" crafted by Minovitch were "not truly adequate," and he replaced them with a "hand method using tabulations and graphs." Since Minovitch worked at night, he and Flandro did not meet. It mattered little, since it was not Minovitch's general solution that made Voyager possible, or his computational methods, but the providential alignment of the outer planets that tipped the scale of possibilities. Flandro circulated his results internally in 1965, even as the American Astronomical Society sponsored a symposium on "Unmanned Exploration of the Solar System." The next year, he published the scheme for the space community in *Acta Astronautica*, identifying ideal launch dates and sketching prospective trajectories.[23]

Initially, there was ample skepticism, even at JPL. At issue were engineering concerns: guidance, communications, and particularly the durability of spacecraft, for even an accelerated journey to Uranus would take ten years, while existing mechanical and electronic devices could barely survive a nine-month trip to Mars. But the idea itself soon accelerated, flung from one study to another much as the hypothetical spacecraft it imagined might careen from planet to planet. In December 1966 Homer Joe Stewart, director of JPL's Advanced Studies Office, instantly saw the significance of the idea and published a prospectus in *Astronautics and Aeronautics*, and it was he who proposed to transfer Crocco's term "grand tour" from the inner planets to the outer. Meanwhile, Bruce Murray, who had previously headed the Mars Study Committee, was busy outlining the particulars–what seemed to most partisans to be veritable axioms–for the mission, or rather a suite of missions. The scientific harvest was stupendous, the engineering challenge magnificent, the potential cultural impact "great and enduring," and its value as cold war propaganda immeasurable. Here was a space spectacular that America was especially equipped to win. The Grand Tour was a noble complement to Apollo.[24]

By early 1969 JPL had sketched the practical requirements of spacecraft design and launch trajectories for variants of an outer-planet survey, and was even using its dazzling vision as a recruiting device. James Long summarized the options for a multiple-year, multiple-mission reconnaissance, concluding that such a scheme was feasible as well as "a timely–virtually unique–opportunity for exploration." Within the year William Pickering was promoting the Grand Tour through *American Scientist*. Throughout, there were two selling points: one, that the cost of going to the outer planets was essentially the same as going to Jupiter alone, since the critical added velocities came from Jupiter's gravitational assist; and two, that the alignment of the planets was, from a programmatic perspective, providential. There was every incentive not only to seize the "super-window" offered but to make as complex a mission as possible in order to magnify that opportunity, perhaps including orbiters and landers. What the Viking mission would do on Mars, the Grand Tour could do across the solar system.[25]

In this, the first of its journeys, Voyager had achieved a breakthrough by means of a shift in reference frames, and that is exactly what Voyager has continued to do throughout its long trek. The process started at its conception, by seeing a hypothetical spacecraft differently, and it will end by forcing us to see ourselves differently. But as William James once put it, "truth *happens* to an idea." Gravity assist happened when Mariner 10, launched in 1973, swung around Venus and Mercury exactly as predicted. The scheme worked, shaving rocket fuel loads as much as 70 percent. A spacecraft could reach the outer planets and still be sentient, if not fully ambulatory.

The primary propulsion system for Voyager was, in the end, intellectual. It was the idea of the Grand Tour.

It seemed irresistible. Both the scientific community and the American media seized on the scheme. By 1971 the Space Science Board had endorsed the proposal; NASA lent approval for further studies, and JPL had plunged into a flurry of inquiries. Here was the scotched Mars Voyager resurrected. Here was a dazzling prospect for technological challenge and first-order exploration. Here was promised a marker in the centuries-long saga of exploration. The last time this planetary conjunction had appeared, Alexander von Humboldt was halfway through his monumental trek across South America; before that, Jan Carstensz had just navigated around Australia's Gulf of Carpentaria, finding nothing of worth, and Plymouth Plantation had settled in New England, still looking eastward, and more intent on the reformation of the Old World than westward to the colonization of a New; and before that, Portugal had barely passed Cape Verde, still stutter-stepping toward the Renaissance's Great Voyages. By the next conjunction, 2153, the planets would have been explored singly. The exploration of the solar system for the first time could happen only once. A Grand Tour could do that.

Enthusiasm was keen: opportunity seemed like inevitability. How could anyone oppose such a fabulous conception? By 1977, when the ideal launch window would appear, the technology, the managerial skills, the instrumentation, and the hands-on experience of navigating probes through interplanetary space would be sufficient. Were the launch window to come five years earlier, those means would not have existed, and had they arrived five years later, the opportunity might have been lost. It was inconceivable that anyone might deny the Grand Tour its call to destiny.

Studies boomed, not least because the options for surveying the outer planets, especially if two spacecraft were launched, were surprisingly rich. Three combinations particularly intrigued planners. One was a three-planet mission to Jupiter, Uranus, and Neptune. Another three-planet mission targeted Jupiter, Saturn, and Pluto. The last version was a four-planet mission to Jupiter, Saturn, Uranus, and Neptune. Each mission would involve multiple launches. It was even possible, if barely, to imagine a flotilla of missions that would collectively sample all of the outer planets with reasonable completeness.[26]

Meanwhile, on July 31, 1969, only fifteen days after Apollo 11 landed on the Moon, Mariner 6 entered orbit around Mars, followed five days later by Mariner 7, offering for the first time an effort to synchronize two simultaneously exploring spacecraft. To observers it seemed that a Grand Tour had to follow. Nothing else could claim anything like its cachet. It remained only to select the most savory of the potential offerings and then render it into a machine to cross the solar system, even though neither the technology nor launch vehicle for such an enterprise existed, nor the political determination to go. JPL's preference was for a program built around a new vehicle, a spacecraft more durable, expansive, and autonomous than the Mariner series, and one that could survive a hostile decade in space. This required, among many features, a computer that could undertake on its own some routine testing and repairs, and an energy system not dependent on sunlight. The new vehicle, called Thermoelectric Outer Planet Spacecraft (TOPS), promised to be as expensive as it was daring. The Grand Tour deserved nothing less.

By 1970 NASA had formally assigned an Outer Planets Grand Tours program to JPL, which endowed a Grand Tour Project Office, which commenced to sort through the preliminary studies and announced a call for proposals for the scientific package. Estimated costs ranged from $750 million to $900 million, a prince's ransom at the time. Still, the Office of Management and Budget (OMB) liked it as an expression of national prestige, and approved some funding for planning. On March 7 President Nixon declared unequivocally for a Grand Tour, with preparations to begin in 1972. By summer some five hundred scientists had submitted proposals for what would be a dozen experimental slots. But in some respects, the winnowing hardly mattered. Where so little was known, everything was there to be learned. JPL petitioned NASA headquarters for a formal authorization, and NASA approved, contingent on congressional funding.[27]

## THE GRAND TOUR OPPOSED

Yet what excites some can cause in others an allergic reaction. The Grand Tour unexpectedly rallied a grand alliance to oppose it. The usual motives came into play, from squabbling over money to disciplinary jealousies, some reasons honorable, some petty. The bottom line was, what the cold war gave, it could also take away, by proposing competitors eager for the same government funds and stature. Nor was the space community united. What seemed to true believers a juggernaut providentially aligned with the stars appeared to others more like Oliver Wendell Holmes's rickety "one-hoss shay."

The most universal concern was cost, which was serious and might go ballistic despite close attention. Even before Apollo 11, NASA's budget was fast ebbing, and the commitment to the space shuttle threatened to become (and did become) a fathomless fiscal sinkhole. Apollo 16 and 17 were canceled. President Nixon decided, along with NASA strategists, that the future belonged with the space shuttle, an American Concorde, which could also quell industry unrest over the cancellation of a supersonic transport (SST). There was not enough in the Treasury to support both robotic Grand Tours and human-steered SSTs, much less fight foreign wars and finance a Great Society. Besides, the political calculations were overwhelming. The shuttle meant thousands of jobs; an outer-planets spacecraft, prime work for a thousandth as many voters. The reality was what Harris "Bud" Schurmeier, a key JPL manager and prime mover behind the embryonic Voyager mission, blandly said it was, that "planetary exploration has inevitably been insignificant on a national scale, low on the agenda, small in the budget." Start-up funds for the Grand Tour were slashed to $10 million from a requested $30 million.[28]

The funding might be finessed. A single mission, even if expensive, might be easier to sell than five, each of which offered tempting political targets. But what threatened to fatally compromise the Grand Tour was intramural fighting among space scientists. A Grand Tour was big science that risked alienating academic little science, and it was a big-budget item that competed directly with a big-budget item craved by astronomers: a space telescope. Those groups whose research did not extend to the outer planets or who were not slated to join the eleven science teams on the spacecraft offered tepid support, or became hostile, viewing the funds lavished on the Grand Tour as siphoning precious monies that might go to their own projects. There was already a big-science mission on the boards, the reincarnation of Mars Voyager into Viking; and the search for life on Mars seemed likely to muster broader enthusiasm from both the scientific community and public than mapping the magnetosphere of Uranus.

Even enthusiasts for outer-planet exploration could legitimately doubt whether now was the proper time, or a Grand Tour the proper means. The NAS Space Science Board had conducted a major study in 1965. While urging NASA to think beyond the Moon, the board had concluded that an intensive study of Jupiter was a better bet than far-ranging flybys; and that if missions to the other gaseous planets were feasible, single shots were a more conservative hedge than a romantic Grand Tour. NASA then constituted an Outer Planets Working Group, with representatives from all its field centers that had an interest in the scheme, and again the consensus favored parsing the Grand Tour into more manageable excursions or crafting a compromise in which two missions would each tour three of the outer planets—all this on the grounds that the engineering was more likely to succeed and the richer returns would be more widely distributed among planetary scientists (a narrower version of the jobs argument).[29]

In June 1969 NASA invited the Space Science Board to reconsider.

The Board reconfirmed its belief that "study of the outer solar system" was a "major objective of space science" and that the community was "eager and excited." It urged that NASA increase the fraction of its budget devoted to planetary exploration, that Pioneer-class spacecraft lead the endeavor, and that Jupiter be the first object of outer-planet missions, with a mix of Grand Tour projects to follow if funding proved adequate. But it also reasserted its original position that the Grand Tour should be not one but several missions. Specifically, it argued for five, all of which would either target Jupiter or use it for gravity propulsion. The JPL scheme was deemed at or beyond the limits of engineering expertise. It was risky, its success uncertain, and its costs extreme.[30]

The Grand Tour's promised payoff, while glamorous, seemed meager compared to what might be done by other means, and worse, its unstanched costs threatened to bleed space science dry. (By now Voyager had become a surrogate for criticism of Viking, which was further along in its development and much more costly.) The SSB nearly recommended outright termination. After discussions with the OMB, NASA returned to what was becoming the default setting: two missions launched over two or three years to reach two or three planets. Had the Grand Tour been a launched missile, with its wobbly path so badly out of trajectory, it would have been blown up. It was kept, although its woes seemed an unsavory augury for what became Voyager. Meanwhile, the SSB, while sanctioning planetary exploration as a "major objective," recommended halving the funding level, thus taking away with one hand what it gave with the other. The SSB urged more near-Earth activity and, led by astronomers, wanted an orbiting telescope. If robotic spacecraft could do science that humans couldn't, then a telescope could do much of that science without the robots. The proposal in effect sought to sever science from exploration.[31]

Other critics, led most visibly by James Van Allen of the University of Iowa, protested the magnitude of the NASA investment in crewed programs to the detriment of robotic missions; the former did no science worth mentioning, and given a fixed (or inflation-ablated) budget, a sharp tilt toward manned space flights, in a kind of bureaucratic sheet erosion, would drain funds away from all the rest. (With no specific appropriation for the shuttle, NASA had to absorb internally its development costs, which escalated from $12.5 million in 1970 to $78.5 million the next year.) This promised to scrap everything that could not fit into the shuttle's bay. The opposition to crewed programs spurred a counter-protest against robots, and gave NASA critics overall additional arguments against both programs.[32]

The deeper concern, though only implied, was to question whether science needed spacecraft at all. For the present it did, as the SSB observed. Several fields of interest—the kind that the International Geophysical Year had pursued in the upper atmosphere of Earth—had "no known alternative to the techniques of direct observation near, or within, the object of investigation." The only way to put an instrument in such proximity was with a spacecraft. But if other means became available, they might well, from the perspective of science, be superior. One could dispense with spacecraft as robotic spacecraft did with astronauts.[33]

Even within NASA, competition was keen. The agency was a forced merger of institutions that had their own traditions, goals, and styles, and that were avid to contest against one another. So long as money had been ample, every research group and every facility could get something that it wanted; but when choices had to be made among them, a competitive scramble resulted among institutions, among projects, among personalities, among ideas, among visions of what "space," "science," and "exploration" meant. This held even as NASA's head, James Fletcher, insisted that no such internal trade-offs existed (no one believed him) and that the space science community had to unite behind all programs. "Science" by itself had "very little political support," he noted, but "space" meant "technology, applications, and political prestige" (all of which was marginally more acceptable). Regardless, the space science community fissured, and that chasm split the larger space lobby.[34]

The crunch came in 1971. The NAS Space Science Board met for an intensive study seminar in early August, argued again to bolster NASA's planetary program, reaffirmed a 1970 study that established priorities for space research, and concluded that regular or even intermediate funding was not sufficient for Grand Tour missions "without jeopardizing the Planetary Explorers [Pioneer] and key programs of the other scientific disciplines." The 1971 session also folded in recommendations from the President's Science Advisory Committee. For the outer planets, it urged as a priority a Jupiter study by "Pioneer-level technology." The JPL ambition for a Grand Tour with a new spacecraft found support only if NASA received its "HIGHER budget program"—an improbable outcome. Still, the SSB supported development of TOPS-class spacecraft and an enhanced Titan rocket to launch it; and it introduced the prospects for "satellite imaging," which it thought might "constitute the most solid justification for [a] Grand Tour." The intellectual returns from remote moons, however, could not outflank the staggering expenses and the aroused alliance of rivals within NASA and the space-science community.[35]

A body in shock pools blood from its outer limbs to its vital organs; so, too, with bureaucracies. Congress had not backed Nixon's rhetoric with real money. NASA redirected the flow of funds from the outer planets to near-Earth activities and to that most Earth-like planet, Mars. Like the tall poppy that gets cut down, the high-visibility Grand Tour was a tempting target—something almost everyone liked in principle but few were willing to sacrifice for. NASA decided that its institutional future lay with humans in space. In December 1971 Administrator James Fletcher wrote the Office of Management and Budget that NASA would cancel the Grand Tour. In January 1972 that announcement became public.[36]

## THE GRAND TOUR REVIVED

The Grand Tour was dead. Its spirit, however, proved harder to kill, and it quickly transmigrated into another avatar.

Within ten days, JPL proposed an alternative. It would scale back the full-bore Grand Tour into paired visits to Jupiter and Saturn. It would adapt the proven Mariner spacecraft rather than, Viking-like, invent a new one from scratch such as TOPS. It would pare costs from \$900 million to \$360 million. It would co-opt critics from the start, particularly enlisting support from the SSB and OMB. And it would, with sly ambiguity, allow for the option of tweaking the second flight to swing past Saturn and rendezvous with Uranus and perhaps Neptune. The metempsychosis was complete. The Grand Tour had become Mariner Jupiter/Saturn 1977 (MJS 77).[37]

NASA accepted the proposal in June and signed a formal project agreement in December, assigning Bud Schurmeier as project manager and assembling a Science Steering Group. JPL began reorganizing immediately. Even this furled-sail mission went far beyond anything attempted to date, not only geographically but technologically and administratively. (Pickering insisted that "something more than 'organized arm waving'" was needed to ensure confidence that a spacecraft could even survive to a "far planet encounter.") Just getting to Saturn was tricky enough: the projected three-and-a-half-year voyage was six times longer than any that had flown. The most advanced power source would last only a year; sensitive equipment had to endure the hazards of interplanetary and near-planetary space, ranging from micrometeorite bombardment to saturated radiation; and navigating a spacecraft through rings, radiation belts, and moons while adjusting sensor platforms demanded an unprecedented choreography of commands. Communications could barely handle a trek to Venus or Mars, and while improvements were expected, the distances were too great to rely on human judgment during crises, which meant a spacecraft would have to analyze and repair itself. Some technology had to be new, not simply adapted. Some could be expected to need correction based on the Pioneer 10's encounter with Jupiter in 1973 and Pioneer 11's encounters with Jupiter and Saturn in 1974 and 1979, respectively. The whole science selection process had to be renewed, which was certain to revive ill feelings. And while everything had to be done on schedule and on budget, nothing done should inherently prevent a reincarnated full-bore Grand Tour.[38]

The pressures, both internal and external, got worse with each year. The internal pressures were many: to reengineer the Mariner spacecraft to accommodate the best features of TOPS; to create sufficient reservoirs of power and propulsion for the vehicle; to upgrade the means of communication through the Deep Space Network with its receivers in

Spain, Australia, and California; to harden electronics after Pioneer 10 had its circuits fried by the gargantuan radiation around Jupiter; to create an onboard computer at a time when even simple personal computers did not exist; to absorb administrative regime changes as Bruce Murray replaced William Pickering and John Casani succeeded Bud Schurmeier; and to affix some kind of permanent message to the spacecraft beyond the plaques created for Pioneers 10 and 11. In October 1974 some 116 "concerns" remained before the Final Spacecraft System Design Review scheduled for March 1975 could determine if the spacecraft could meet launch dates.[39]

The external pressures were no less daunting: the constant downsizing of a post-Apollo NASA determined to retain Skylab, Apollo-Soyuz, and the shuttle, all major tributaries to the budget drain; the cost-accounting for work done by other NASA labs and the Defense Department; the need for contractors, though overseen by JPL; the awkward siting of JPL within the NASA bureaucracy as both an agency branch and a part of Caltech; and not least, the potential crisis over a launch vehicle. Planners had assumed from the onset that it would launch with a Titan IV (of which NASA still had two); but NASA's subservience to the shuttle had decreed that it be the agency's primary launcher, a decision that threatened to eliminate any prospects for outer-planet exploration. Yet throughout, the hope for a resuscitated Grand Tour remained embedded in the design. As John Casani remarked, "We knew what the strategy was; we knew what we were going to do; we knew what the decision points were."[40]

In February 1976 NASA granted final approval for "Mariner Jupiter/Saturn 1977 Planetary Exploration (Outer Planets Missions)." The mission called for two spacecraft that would journey for a scientific reconnaissance of the nearest of the outer planets and their moons. Launch would occur in 1977. But the enabling authorization included two prospects for an extended mission, provided the designated goals at Jupiter and Saturn were satisfied and the spacecraft were still performing well beyond their warranty ("a miracle," as Casani himself put it). One allowed for "retargeting" the second spacecraft to go to Uranus. The other imagined the trek of the spacecraft to the edge of the solar system, to the shores of interstellar space, the Third Age's Sea of Darkness.[41]

By month's end NASA had accepted in principle an extended mission, first to Uranus and, ultimately, beyond the solar system. The official Voyager Project Plan allotted one spacecraft, "as long as it continues to function," to break through the heliopause and into the interstellar medium. It allotted the other, if it, too, was still working, to "permit an encounter with Uranus," with objectives similar to those for Jupiter and Saturn. And implied, but left unsaid, was the prospect for an encounter with Neptune.[42]

Again, software—the dominance of guidance over hardwiring, the power of ideas and culture, the self-capacity to adapt—had proved critical. The will had found a way.

As the parts assembled, the official name, Mariner Jupiter/Saturn 1977 Planetary Exploration (Outer Planets Mission), or, in shorthand, MJS 77, seemed inadequate to the reality of the spacecraft and the half-submerged ambitions of the mission. Or at least it did to John Casani, who announced a competition for a new name. He had recently come on board as project manager, thought the label lame, awkward, and overly bureaucratic, and despite the fact that a contest had recently settled on an MJS 77 logo, he reopened the issue, with a case of champagne promised to the winner. It appeared that names no less than visions might be reborn.[43]

Candidates bubbled up: Nomad, Pilgrim, Pioneer, Antares, and a name that had surfaced for missions twice before but had never stuck, Voyager. It had most recently attached to the ambitious Mars mission that got scrapped before being resurrected as Viking. After a superstitious hesitation—the previous project had, after all, been canceled—a counter-consensus emerged. Voyager Mars had failed because of costs, not because it was a doomed idea, and it had spawned a marvelous replacement. "Voyager" thus seems to have careened around the bureaucracy and the imaginations of JPL designers in a kind of noumenal Grand Tour of its own. A general vote approved it. In March 1977, five months before liftoff, NASA agreed.[44]

---

There were three spacecraft constructed. The first (VGR 77-1) was trucked to Florida in March, feeling its way over highways and through the perils of overland traffic. The others arrived on April 21 and May 19.

There they underwent rigorous prelaunch checks. When the second (VGR 77-2) revealed fatal flaws, it was left behind as a bed for spares and as a model on which engineers could test reported glitches. It could make explicable in a lab what signals from the others might indicate were problems in space. VGR 77-3 took its place. The differences between the two launched spacecraft were minor. Voyager 2 included, for example, a slightly greater power source and several more robust sensors (and camera), since it might, just might, pivot around Saturn and go to Uranus. The Voyager triplets had become twins, and the greatest tag team in planetary discovery.

## EXPLORING POLITICS

The Voyagers' journey from vision to launch—through the vacuum and hazards of mind, institutions, engineering, and politics to the materialization of ideas and ambitions—was itself a daunting, extraordinary trek. It took as long to move from Flandro's 1965 insights to the actual launch in 1977 as it took the Voyagers to travel from Florida to Neptune. Political hazards could be as fatal as radiation, and Congress as dangerous as asteroid belts. But it had always been so.

Few private entities could afford the cost of exploring over the Ocean Sea or across new worlds, and as soon as they appealed for public support, they became subject to public control. When private companies dispatched exploring parties, they did so under the political cover of letters patent or charters, and even private persons could not evade the political context of their travels and required letters of transit or risk charges that they were spies. John Ledyard's daffy ambition to walk from Europe to America ended in 1788 when the Russian empress Catherine had him arrested and deported. Alexander von Humboldt had to seek permission from Carlos IV in 1798 before he could tour New Spain.

To fanatics aflame with their vision quest, the goal is so self-evident, so tantalizingly palpable, so urgent, that they can imagine any question or impediment only as crude harassment. To those less addled by such schemes, the expenditure of public money must satisfy public interest, and having foreign nationals wandering about the world can loose political conflicts that others will have to clean up. A traveler seized or killed becomes a traveler politicized, a hostage that demands rescue, a national honor besmirched and crying out for retribution, a cause célèbre. Voyager's long, troubled gestation is more norm than exception.

Conviction is not knowledge, resolve not rightness; and vision is only as good as the times it lives in. Five centuries before Voyager, Portugal waited ten years after the return of Bartolomeu Dias before it sent Vasco da Gama to India; yet the route was there, needing only the will to send vessels; and no other prize so dominated the imagination of the Great Voyages as the passage to India. Still, Portugal procrastinated, obsessed with more immediate crises and opportunities, not least a change in monarch and court. Exploration existed to promote Portugal, not Portugal exploration.

Today's partisans for the colonization of space had their counterpart in proponents for the development of the spice trade. The latter were loud, pesky, persistent, and they enjoyed an alliance of convenience with a claimant to the throne, Manuel I, who finally assumed the crown in 1495. All this makes reasonable the suggestion that da Gama's first voyage was a political bone tossed to appease a noisy distraction while the court attended to the important affairs of state, notably its enduring tensions with Castile and endless wars with Morocco. "The low priority given to this expedition, the appointment of a minor fidalgo to the command, and the fact that the mission carried with it so little in the way of diplomatic gifts or trade goods," suggests Malyn Newitt, indicate it was a "minimal gesture" to "silence" a marginal cabal, while providing the young monarch with a symbolic means of maintaining what had become a nearly century-long tradition.[45]

The politics could go both ways. While the obvious might lie rotting at anchor, the idiotic might set sail. Captain Thomas James found sponsors in his quest to discern a route to the South Seas through Hudson Bay. John Cleves Symmes inspired a republic skeptical of intellectuals to outfit ships to explore a "hole at the pole" that he argued was a mathematical certainty and would lead to the center of the Earth. Christopher Columbus and his brother hawked his ideas for a western voyage to the Indies—flawed, as it turns out, but lucky—for years. In the 1830s the United States could find no senior officer willing to command its great Exploring Expedition to the South Seas and around the world. It all depended on tide and time.

To those enthralled by the Grand Tour, the politics of authorization was tedious, perverse, and needless, an encumbrance and an embarrassment, a monarch-in-waiting begging alms in the sordid corridors of Washington. But over the centuries, while politics has erred in both omission and commission, it has remained the preferred medium for discourse. Occasionally it has financed the foolish, and not infrequently it has dismissed the savvy, but whatever else exploration was or aspired to be, it could not divorce itself from the politics of court and congress and especially not from the geopolitics of competing empires. The quirky trek of Voyager through NASA bureaucracy, scientific commissions, congressional committees, partisan critics, and prophetic seers places the mission squarely within a tradition harking back to the very origins of exploration as a systematic enterprise.

Because of politics, the expedition almost didn't happen, and because of politics, it finally did.

# 3. Great Ages of Discovery

The Voyagers left Earth atop Titan/Centaur rockets. The Titan had two stages, to which the Centaur, with its own propulsion system, added a third. Titan put Centaur into rough orbit; then the Centaur's rocket fired twice to bring it into cruise orbit. Finally the Voyagers' own propulsion modules sent them on their prescribed trajectories.

As they moved beyond the limits of Earth-orbiting satellites, the Voyagers also moved beyond the limits of earthbound exploration. Yet that past lifted them as surely as their Titan/Centaur rockets, whose stages of propellants might well stand for the three great ages of discovery of which the Voyagers were a culminating payload.

## EXPLORATION, LUMPED AND SPLIT

Why three ages? There are those who see many more, and those who see none at all, for exploration history, too, has its lumpers and splitters.

The lumpers view the long saga of geographic exploration by Western civilization as continuous and thematically indivisible. The Viking landers on Mars are but an iteration of the longships that colonized Greenland. The Eagle, the Command Module orbiter, and the

Saturn V rocket that propelled the Apollo 11 mission to the Moon are avatars of Columbus's *Niña, Pinta,* and *Santa Maria.* The "new ocean" of interplanetary space is simply extending the bounds of the old. The *ur*-lumpers would go further. The origins of all exploration, including Europe's, reside in the genetic code of humanity's inextinguishable curiosity. Even more, space exploration, they insist, shares an evolutionary impulse. Through humanity, life will clamber out of its home planet much as pioneering species crawled out of the salty seas and onto land. The impulse to explore is providential; the chain of discovery, unbroken; the drivers behind it, as full of evolutionary inevitability as the linkage between DNA and proteins. The urge, the motivating imperative, resides indelibly within our character as *Homo sapiens sapiens.*[46]

The splitters see it differently. Exploration pulses, expanding and contracting. Ming China launched seven dazzling voyages of discovery, and then outlawed all foreign travel and prohibited the construction of multimasted boats. Medieval Islam sponsored great travelers before shrinking into the ritual pilgrimage of the haj. The Norse spanned the Atlantic, then withered on the fjords of Greenland. Plenty of peoples have stayed where they were: they lacked the technological means, the fiery incentives and desperate insecurities, or the compelling circumstances to push themselves to explore beyond their homeland. Like Australia's Aborigines, they were content to cycle through their ancestral Dreamtime, and felt little urgency to search beyond the daunting seas or looming peaks. A walkabout was world enough.

To the splitters, what determines the cadences of exploration are the cultural particulars—the social conditions that prompt and sustain discovery. What is commonly called "geographic exploration" has been, in truth, a highly ethnocentric enterprise. It will thrive or shrivel as particular peoples choose. There is nothing predestined about geographic discovery, any more than there is about a Renaissance, a tradition of Gothic cathedrals, or the invention of the electric lightbulb. From such a perspective, the European era of exploration that has dominated the past five centuries is simply another in a constellation of cultural inventions that have shaped how peoples have encountered a world beyond themselves. It is an institution, and it derives much of its power because it bonds geographic travel to cultural movements, because it taps into deep rivalries, and because its narrative conveys a moral message. It can accordingly be parsed into historical eras.

For Western civilization, these fall most easily into three grand eras. Each had its primary geographic domain, each bonded with its prevailing intellectual syndrome, each tapped a moral energy. Each had its own peculiar dynamic of geopolitical rivals and cultural enthusiasms. Each found a gesture that came to express its character. And each stage had to be rekindled. A successful launch only appears continuous in broad-brush retrospect; on closer inspection, it shows a rhythm of spark and extinction.

## GREAT VOYAGES: THE RENAISSANCE EXPLORES

The Great Ages of Discovery opened with centuries of false dawns. Part of the difficulty is disentangling exploration from other forms of travel—from migration, walkabout, exile, wars of conquest, enslavement, trading expeditions, reconnaissance, long hunts, great treks, missionizing, pilgrimage, tourism, and just plan wanderlust. Roman merchants had contact with the Canaries and Cathay. European pilgrims trekked from Hibernia to the Holy Land. Franciscan scholars trudged to the court of the Great Khan. Each age of expansion, every expansionist people, experienced a burst of discovery about a larger world.[47]

What made events of the fifteenth century special was that these exploring contacts did not end in a rapid contraction. They became welded to a revived expansion of Europe that would stretch over half a millennium; they bonded with revolutionary epochs of learning and political reform. Exploring became institutionalized. Exploration became the outward projection of internal unrest that would not let the momentum long languish.

That Portugal pioneered the Great Voyages should alert us to the process's uncertain origins and its often desperate character. There was little in Portuguese history from which someone might predict, in 1450, that the nation would leap across whole seas and over unknown continents, establish the world's first global empire, and create the raw template for European expansion, whose outposts would survive until the twenty-first century. Yet that is precisely what happened. For several hundred years, exploring nations sought to emulate the Portuguese paradigm. Within a generation, it came to be said that it was the fate of a Portuguese to be born in a small country but to have the whole world to die in.

Why Portugal led remains an exercise in historical alchemy. One can find reasons for its ingredients, but not a simple explanation for why they mixed as they did. There is a certain logic embedded in Portugal's geographic setting. Here, at Europe's land's end, the two major traditions of boat construction converged, the Mediterranean with the Baltic. It was a place on the edge. Its isolation forced it to take to the sea; its smallness compelled it to find nimble ways to outflank rivals and enemies; its precarious politics surrounded it with competitors. In particular, it waged a ceaseless dynastic war with Castile that left those two states as the drivers of European expansionism during the late Renaissance; and it fought endlessly with Morocco. Under Henry the Navigator it had discovered and colonized Atlantic isles from 1420, a cameo of what it would attempt with its passage to India. Yet if the causes seem feeble, the outcome was unmistakable. More than anyone, Portuguese sailed the Great Voyages, whether as sponsors or pilots in the service of others, and they plotted out the terms of European imperialism. Columbus learned his trade on the Portuguese circuit. Magellan sailed for Spain only after his native Portugal had rejected his scheme.

Exploration became—directly, or indirectly through charters—an organ of the state, and because no single state dominated Europe, many joined the rush. Geographical exploration became a means of knowing, of creating commercial empires, of outmaneuvering political, economic, religious, and military competitors—it was war, diplomacy, proselytizing, scholarship, and trade by other means. For this reason, it could not cease. For every champion, there existed a handful of challengers. This competitive dynamic—embedded in a squabbling Europe's very fabric—helps explain why European exploration did not crumble as quickly as it congealed.

On the contrary, many Europeans absorbed discovery into their understanding of who they were, even in some cases writing explorers into a founding mythology, a cultural creation story. In short, where exploring became a force, something beyond buccaneering, it interbred with the rest of its sustaining society. The broader those cultural kinship ties, the deeper the commitment. Societies dispatched explorers; explorers reshaped society. Exploration became an institution. The explorer became a role.

The fabled Great Voyages announced a First Age of Discovery. Its particular domain was the exploration of the world ocean: it ultimately proved that all the world's seas were one, that it was possible to sail from any shore and reach any other. Of course there were some grand entradas in the Americas, and missionaries, Jesuits especially, penetrated into the vast interiors of the Americas, Africa, and Asia. But as J. H. Parry observes, it was the world sea that defined the scope and achievements of the First Age. Mapping its littoral was the era's finest intellectual achievement.[48]

The map reminds us that the First Age coincided with a Renaissance. The era unveiled two new worlds, one of geography, another of learning. Francis Bacon conveyed this sense perfectly when he used as a frontispiece to his *Instauratio Magna* the image of a sailing ship pushing beyond the Pillars of Hercules. The voyage of discovery became a metaphor for an age of inquiry that would venture far beyond the dominion of the Mediterranean and the inherited wisdom of the ancients. The discoveries overwhelmed a text-based scholarship. Scholasticism, that arid discourse that resulted from too many scholars and not enough texts, collapsed as new information poured into Europe like New World bullion into Spain, and like it, caused an inflationary spiral of knowledge.

The nature of learning differed, too. It came not from recovered texts but from newly discovered lands and peoples, and not from the ancients but from encountered living cultures. Very little of the terrestrial world Europe discovered was uninhabited, which is to say, unknown to humanity. It was unknown to Europe, and Europe learned about it through its indigenes. Interpreters, guides, cultural brokers—all assisted in the transfer of learning from various enclaves to Europe, which proceeded to sew them together into a global quilt. Voyages were the stitches; seas, straits, and societies, the patches. Over and again, explorers succeeded by relying upon (or seizing) local pilots, and by learning the language of, and emulating the dress and mores of, the native peoples. This meant that the central act of discovery, the encounter, was almost always an encounter between peoples.

An age of discovery thus demands more than curiosity and craft and yields more than data points or lore hoarded like bullion. Acquired knowledge has to be minted into useful currency; and exploring has to speak to deeper longings and fears and folk identities than science and scholarship. An expedition voyages into a moral universe that explains who a people are and how they should behave, that criticizes and justifies both the sustaining society and those it encounters. The Great Voyages provided that moral shock: they forced Europe to confront beliefs and mores far beyond the common understanding of Western civilization. The Renaissance expansion of Europe profoundly altered Europe's understanding of itself and its place in the world. There was plenty of hollow triumphalism, of course, but those contacts also inspired Montaigne's celebrated preference for the cannibalism of Brazil's noble savages to that of Versailles's courtiers, and Las Casas's excoriating denunciation of the conquistadors. The contacts also compelled a reexamination of the political and ethical principles underlying Christendom and its secular principalities.

While all peoples are ethnocentric, Europe was distinctive in that its *mappa mundi* placed the cartographic center of creation in the Holy Land, leaving Europe to the margins. In the early fifteenth century that displacement accurately depicted Europe's standing in the world. A century later, however, Europe could relocate itself to the center.

## CORPS OF DISCOVERY: THE ENLIGHTENMENT EXPLORES

By the early eighteenth century, exploration had found itself becalmed, even moribund. Discovery had achieved its purpose. It had found serviceable routes—the only ones, really—to the wealth of the East. Its sponsors felt little need to search for more, or to probe remote regions of the globe without prospects for commerce or plunder. Mariners did more poaching and piracy than original questing; the explorer blurred into the fantasist and the fraud, a promoter of Mississippi and South Seas bubbles. Expeditions of adventurers persisted largely because interlopers tried to outflank established competitors.

By almost any index, exploration sagged. Although missionaries and *bandeirantes* had worked through the rivers of South America, and fur traders had done likewise for the main lakes and rivers of northern Asia and North America, few new islands were unveiled, and nothing of the interiors of Africa, and nothing of the outlines of Australia and Antarctica; and what was learned was often hoarded. They had not, of course, surveyed the Earth in its fullness; until the late eighteenth century, even the world's coastlines still had unmapped gaps. But the implacable will (or the internal furies) that had driven explorers now flagged. Exploration seemed destined to be left marooned on the shore of a fast-ebbing historical tide. As with a Titan/Centaur launch, the saga of exploration also had its coasting periods.

Then the historical dynamics changed. A period of coasting and consolidating ceased. The elements for a revival of exploration positioned themselves. Cultural engines again burned and boosted a new stage of exploration upward.

The long rivalry between Britain and France, the penetration of high culture by the Enlightenment, and a hunger for new markets, all combined to move Europe again out of dry dock and onto the high seas of commerce and conquest. The grand tour became a global excursion around the Earth. Perhaps most extraordinarily, the missionary emerged out of a secularizing chrysalis as the naturalist. Increasingly, scientists replaced priests as the chroniclers and observers of expeditions—Linnaeus's apostles supplanted Saint Francis Xavier's Jesuits—and scientific inquiry substituted for and justified the proselytizing that had helped sanctify an often violent and tragic collision of cultures.

The era's annunciatory events were two sets of expeditions. The first was a paired undertaking sponsored by the Paris Academy of Sciences in 1735 to measure an arc of the meridian. One expedition went to Lapland under Pierre Maupertuis and one to Ecuador under Charles-Marie de la Condamine. For the first time abstruse questions of natural philosophy, in this case involving the shape of Earth and competing theories of gravity, drove expeditions. The second cluster proved significantly more impressive as it mounted an international campaign to measure the transit of Venus, first in 1761 and again in 1769.

Here was a scientific campaign urged by scientists, to be conducted by scientists, aimed at simultaneous global surveys that would measure a critical value needed to understand models of the solar system, whose calculated working had become the exemplar of Enlightenment science. It was, to advocates, a unique opportunity, a passage of astronomical alignments across a suitable civilizational setting. In its request for funding, the Royal Society of London appealed to two principal "Motives": the "Improvement of Astronomy and the Honour of this Nation." There was national glory to be gained from success, and national shame to be endured from failure, and of course one could necessarily expect economic spinoffs from the inevitable improvements in scientific knowledge that would result. The 1761 transit featured 120 observations, of which 106 were in Europe; the 1769 transit, 150 from European outposts around the Earth.[49]

The swarm of expeditions helped rouse geographic discovery from its long slumber; they defined the terms by which exploration, empire, and Enlightenment might find common causes; they midwifed a transition from an exploring science welded to natural philosophy to one bonded to natural history. The expeditions of Baptiste Chappe d'Auteroche to Tobolsk and then to Baja California, of Legentil de la Galaissière to the Indies and Alexandre-Guy

Pingrè to Rodrigues and Haiti, and especially the voyage of the HMS *Endeavour* under Captain James Cook to the Pacific galvanized public opinion and helped spark a revolution in scientific discovery—a model less for measuring the distance of Earth from the Sun than for inventorying the splendor of Earth. Here in cameo was demonstrated the ambition and means to inspire a new age of discovery.

Over the next century every aspiring great power dispatched fleets to seek out new wealth and knowledge, to loudly go where others had not yet staked claims. Once again, the rivalries among the Europeans were as great as anything between Europeans and other peoples. In 1769 James Bruce reached Lake Tana, the traditional source of the Blue Nile, while James Cook arrived at Tahiti to measure the transit of Venus. One journey represented rediscovery, a reconnection by a new sensibility with classical lore; the other, a new discovery, a barely known place subject to vision.

Circumnavigation revived, but ships proved mostly a means to reposition explorers, who promptly moved inland. The world's continents replaced the world sea as a primary arena for discovery, and the cross-continental traverse substituted for circumnavigation as its boldest expression. The voyaging conquistador metamorphosed into the Romantic naturalist. The transition matters because as the nineteenth century ripened, Europe was no longer content to remain as a trafficker on the beaches of the world sea. Like its exploring emissaries, it shoved and swarmed inland. Trading ventures became imperial institutions, coastal colonies evolved into continental nations, and the politics of commerce gave way to outright conquest. Exploration as a reconnaissance for trade segued into surveys for settlement; imperialism moved from founding coastal trading factories to establishing states over which they would rule, some of which they would populate with émigré Europeans.

The outcome was a fabulous era for exploring scientists. New intellectual disciplines bubbled up out of the slush of specimens shipped home. The returns from the earliest explorers to a particular place were often phenomenal—the scholarly equivalent to placer mining or, in the First Age, to the sacking of Tenochtitlán or Malacca. A revolution in geographic discovery again accompanied a revolution in learning, aptly symbolized by the simultaneous recognition by two exploring naturalists, Charles Darwin and Alfred Wallace, of evolution by natural selection.

The moral drama changed accordingly. Secularization and science translated Vasco da Gama's famous declaration that he had come to the Indies for "Christians and spices" into a cry for civilization and commerce. The deeper drama concerned that fraction of Europe's imperium colonized by European emigrants. These settler societies tended to look upon discovery as part of a national epic, and to honor explorers as vital protagonists—a Moses, an Aeneas—of those founding events. Their subsequent folk expansions proceeded hand in glove with formal exploration, such that Daniel Boone, not George Washington, became America's folk-epic hero. These were new worlds, premised on the prospects for a new order of society. America truly was, in William Goetzmann's words, "exploration's nation"; but so were Russia, Australia, Canada, and others.[50]

Discovery metastasized. As measured by the number of exploring expeditions, a slight increase appears in the latter eighteenth century and then erupts into a supernova of discovery that spans the globe. In 1859 the last unknown Pacific island, Midway, was discovered; by the 1870s, explorers had managed comprehensive traverses—cross sections of natural history—for every continent save Antarctica. With the partition of Africa, expeditions proliferated to assess what the lines drawn on maps in Berlin libraries actually meant on the ground. Exploration had become an index of national prestige and power. The first International Polar Year (1882) had turned attention to the Arctic. An announcement by the Sixth International Geographical Congress in 1896 that Antarctica remained the last continent for untrammeled geographic discovery inspired a swarm of explorers to head to its icy shores; even Belgium and Japan sponsored expeditions. (America's attention remained fixated on the North Pole and that other stampede to the Klondike.) Ernest Shackleton's celebrated 1914 Imperial Trans-Antarctic Expedition was, after all, an attempt to complete for that continent the grand gesture that had crowned every other.[51]

But Antarctica was the last. There were no more unvisited lands to traverse, other than such backwaters as the Red Centre of Australia, the crenulated valleys and highlands of New Guinea, and the windswept Gobi. The enthusiasm for boundary surveys and natural history excursions—for imperialism itself—waned with the slaughter of the Great War. Plotting the number of exploring expeditions reveals the Second Age as a kind of historical monadnock, rising like a chronological volcano above a level terrain. The peak crests in the last decades of the nineteenth century, as exploration crossed the summit of the Second Age. Then it began a descent down the other side.

Like a cycle of economic boom and bust, what had ramped up now ramped down. The process went into reverse—exploration's equivalent of deleveraging. The reasons are many. One is simply that Europe completed its swarm over the (to it) unknown surfaces of the planet. There was nowhere else for the Humboldtian explorer to go, and there were no more lands to meaningfully settle. Antarctica, the deep oceans, interplanetary space—these arenas for geographic discovery might be claimed, but they would not be colonized.

No less important, the dynamic behind exploration changed. The Second Age had kindled with a rivalry between Britain and France, much as the contest between Portugal and Spain had powered the First Age. Thereafter virtually every competition featured Britain, which is why its explorers so dominate the age. Britain and France clashed in India, the Pacific, and Africa; Britain and the United States in North America; Britain and Russia, the Great Game, across central Asia; Britain and all comers in Antarctica. But after the Great War, Britain and France could no longer afford the enterprise. Russia turned inward with revolution. The United States had few places other than Antarctica in which discovery had geopolitical meaning. The Second Great Age of Discovery, like the First before it, deflated. Moreover, the old rivalries, once projected outward, now turned inward, and Europe brought its colonial wars home in what ended with near self-immolation.

By the middle twentieth century, after two world wars, a global depression, and the sudden shedding of colonies, Kipling's "Recessional" had become prophetic. The Great Powers were exhausted, Europe sought to quench its internecine wars by severing its colonial ties, and the resulting decolonization accompanied an implosion of exploration. Europe turned inward, quelling the ancient quarrels that had restlessly and violently propelled it around the globe, pulling itself together rather than projecting itself outward.

And as in the past, there were cultural factors also at work. The Second Age had served as the exploring instrument of the Enlightenment. Geographic discovery had bonded with modern science such that no serious expedition could claim public interest without a complement of naturalists, while some of the most robust new sciences, such as geology and biology, relied on exploration to cart back the data that fueled them. Science, particularly natural history, had shown itself as implacably aggressive as politics, full of national rivalries and conceptual competitions, and through exploration, it appeared to answer, or at least could address, questions of keen interest to the culture. It could exhume the age of the Earth, reveal the evolution of life, celebrate scenic monuments to nationalism and Nature's God. Artists such as Thomas Baines and Thomas Moran joined expeditions, or as John James Audubon did, mounted their own surveys. General intellectuals eagerly studied narratives of discovery. Exploring accounts and traveler narratives became best sellers; explorers were cultural heroes; exploration was part and parcel of national epics; exploration was a means to fame and sometimes fortune. The Second Age, in brief, braided together many of the dominant cultural strands of its time.

By the early twentieth century, however, this splendid tapestry was unraveling. A Greater Enlightenment found itself challenged by a Greater Modernism. One consequence was that, in field after field, intellectuals turned to subjects that no longer lent themselves to explication by exploration. Natural scientists looked to the very large and the very small, to redshifting nebulae and subatomic particles or molecular genes. Artists turned inward, probing themselves and the foundations of art, not outward to representational landscapes. High culture was more inclined to follow Sigmund Freud into the symbol-laden depths of the unconscious, or Joseph Conrad into a heart of imperial darkness, than to ascend Chimborazo with Humboldt or to paddle with John Wesley Powell through the gorges of the Grand Canyon. The Second Age sagged not simply from the exhaustion of closed frontiers but from a more profound weariness and ironic dismay with the entire enterprise of Enlightenment and empire.

Once more Western exploration began to coast. In the early nineteenth century an intellectual could claim international acclaim by exploring new lands. By the early twentieth he could not, if he could even find suitable lands. There were a few spectacular exceptions. The gold-prospecting Leahy brothers trooped into the unknown highlands of New Guinea. Richard Byrd wistfully erected Little America on the Ross Ice Shelf. Roy Chapman Andrews, with carbine and Model T, whisked across the Gobi in search of dinosaur eggs, the very model of a Hollywood action hero (and inspiration for Indiana Jones). But there was overall a rueful, forlorn quality to the striving, aptly expressed when the American Museum of Natural History, with Andrews in command, dispatched an expedition to Shiva Temple, an isolated mesa within the Grand Canyon, to look for exotic creatures, as though it stood somewhere between the Galapagos and Shangri-La. Sixty years before, the Canyon had claimed center stage not only for geographical discovery but also for the answers it offered to fundamental questions about the Earth's age and organic evolution. Now the press boosted a minor foray into a journey to Arthur Conan Doyle's *Lost World*. Lost world, indeed.

## INTO THE ABYSS: MODERNISM EXPLORES

Then it happened again. New lands became available; new technologies and ideas made them enticing; and rivalries between institutions, companies, nations, and personalities mixed combustibly.

Of a sudden, unexpected realms had appeared. Places previously off-limits because of hostile environs, rivals of comparable power, technological feebleness, or ignorance or timidity had opened up. These were the ice sheets (and sub-ice terrains) of Greenland and especially Antarctica, the deep oceans with their abyssal plains and immense trenches, and of course a solar system, full of worlds that beckoned beyond the vision of earthbound observatories. As powerful instruments and remote-sensing technologies emerged, as manned vehicles and unmanned probes plummeted to the depths and beyond the atmosphere, the prospects for a revival of exploration became possible. Suddenly, a Pacific Ocean that had seemingly yielded its last island to discovery a century before revealed hundreds of new islands in the form of submerged seamounts.

Antarctica was the transition. It was an abiotic landscape not much accessible to Enlightenment art and science, with no prospects for colonizing settlement, the last of the continental frontiers and the one where the Second Age exhausted itself. Twice before—in 1882 and 1932—a global science had rallied for "polar years." The earlier versions had focused heavily on the Arctic; this time, after a half century of world wars, proponents hoped to concentrate on the Antarctic.

The scheme soon snowballed into a call for a more general eighteen-month scientific scan of the Earth, what became known as the International Geophysical Year (1957–1958). The roster of participants was a veritable United Nations of science—some sixty-eight nations in all. It was here, for the first time, that the contours of a new age of discovery came together. IGY's explorers would visit places inimical not only to humans but to life itself. They would rely on remote-sensing instruments, tracked vehicles, rockets, and robots. They would inventory a planet whole, of which Earth would be the prototype: the home planet became, intellectually, a new world, the first of a dawning age of discovery that would propagate to the fringe of the solar winds. The voyages that followed to planets such as Venus, Jupiter, and Neptune would carry essentially the same instruments and ask the same questions of them as IGY did for Earth.

Through the infrastructure provided by IGY, noted J. Tuzo Wilson, the "science of the solid earth" was "absorbed into the broader framework of a new planetary science." Yet Earth's fluids interested the founders as much as its solids. They peered with special fascination into the upper atmosphere—geophysics, after all, was embedded in the project's very name. Auroras in particular had pointed to Antarctica as an insufficiently exploited platform for earthly observation, which was where a third polar year had been headed. But the fast-morphing capabilities of rocketry made it possible to send instruments directly into the auroral belts. Both the United States and the Soviet Union had long-extant, if semi-dormant, plans to launch instrumented missiles beyond the realm of high-altitude balloons, leaving rockets as a new means to ask inherited questions.[52]

IGY escalated those scientific yearnings into a political probability. It did for the Third Age what the voyages of Columbus and da Gama did for the First, and the transits of Venus did for the Second. But where the successors to the transits belonged with Enlightenment and empire, the successors to IGY looked, however uneasily, to a Greater Modernism and a postcolonial age. The character of exploration morphed, for IGY did not simply revive the Second Age but assembled the pieces for a Third. The International Geophysical Year was barely three months along when, under its auspices, the Soviet Union launched Sputnik 1. Twenty years after Sputnik leaped into earthly orbit, the Voyagers were flying toward a rendezvous with the moons of Neptune.

Still, dazzling technologies and an invigorated curiosity are not enough to spontaneously combust into an era of exploration: cultural engagement also demands a sharp rivalry. Those competitive energies flourished with the cold war.

In retrospect, the great game between the United States and the Soviet Union lasted far less than those between Spain and Portugal, or Britain and France, but the era is young, and if it does in fact mark a Third Age, some other competitors, keen to secure national advantage or prestige through sponsored discovery, may emerge. Without the cold war, however, there would have been scant incentive to erect bases on the Antarctic ice, scour the oceans for submerged mountain ranges and trenches, or launch spacecraft. Two geopolitical rivals, both with active exploring traditions, chose to divert some of their contest away from battlefields and onto untrodden landscapes. The cold war was the final propulsion module that boosted planetary exploration out of Earth orbit.

Most observers assumed that technology drove discovery. Surely without rockets and remote-sensing devices, the age could not have unfurled. But inventions followed ideas. The pioneering rocketeers had envisioned migrations off Earth and experimented with rocketry as a technological means to move their enthusiasm into space. Nor was the apparatus of the Enlightenment designed to cope with the realms of the Third Age. It broke down on ice, abyss, and space as it did with atoms, relativistic quasars, and self-referential logic. The Second Age had neither the technology nor the software to plunge into those uninhabitable domains; but the intellectual revolution that we might lump together as Modernism could. The most successful explorers of the Third Age would be modernists, whether they willed it or not.

Yet there was a paradox in its pith. Modernism could deal with such realms, but most modernists lacked the incentive to do so. Outside the sciences they were more inclined to look inward than outward. The culture's software lagged behind its hardware. The ability to voyage anew appeared before the capacity arose among elites to wish to do it. Pragmatism was a philosophy suited for pioneering; existentialism was not. Those sciences that were most moribund were precisely those most closely bound with the Second Age. Geology was seemingly done more in libraries than in the field; certainly personalities such as Wilson, so despairing of earth science, hoped IGY would spark a reformation. A new era of exploration might ignite, as it had in the eighteenth century, a paradigm shift. (Interestingly, the anticipated revolution occurred exactly between the dates of the two editions of Thomas Kuhn's *The Structure of Scientific Revolutions*, 1962 and 1970, which introduced the concept of scientific paradigms, and thus furnished a kind of manifesto.)[53]

What bridged the gap, meanwhile, was popular culture. Exploration did not wither away, because the culture had not only institutionalized but also internalized discovery. This was a civilization that could hardly imagine itself as other than exploring. Accordingly, it forged new institutions, of which the International Geophysical Year is an apt annunciation; and with spacecraft dispatched across the solar system, it recapitulated the entire half-millennium saga of Western exploration. Those vessels crossed what enthusiasts were pleased to call "this new ocean," and in the outer planets they discovered new worlds, miniature solar systems, full of unknown seas and isles. Out of that encounter came a reformed earth science and a comparative planetary science.

In the Voyager mission, the Third Age stripped exploration down to its essentials. The twin spacecraft went into Earth orbit atop a long heritage of experience and expectation, but they had their own boosters as well, and they used them to point outward and fire into a distinctive trajectory that traced what may well be the age's most spectacular and defining gesture. If the Voyagers departed Earth full of inertia from the past, they also added a momentum of their own that sent them into the future. They looked back even as they looked beyond. Their trajectory was a constant triangulation of both.

# 4. Voyager

Exploration has always been something done by explorers. Of course they used the technology of their times. They traveled by ships, wagons, rail, balloons, and bathyspheres; they relied on astrolabes, compasses, barometers, and sextants. But it was the person of the explorer, or of an exploring corps, that propelled the expedition, did the necessary tasks, and embodied its purpose. To invoke the name of the explorer was to evoke the mission overall.

The Voyagers challenged this traditional formulation. They were robots: the technology *was* the exploring agent. They were something more than dumb machines, because they had a degree of autonomy and had to act, within limits, independently of their handlers. They were more than simple ganglia of instruments, because they carried their own motive power, which granted them the means to journey. Like any artifact of human contrivance, they exhibited a style that embodied the imagination and values of their creators, which granted them a simulacrum of personality. They could be anthropomorphized in ways that the *Santa Maria* or the HMS *Resolution* could not, yet they were far from human, or even alive. In sailing beyond the realm of previous exploration, they also journeyed beyond the realm of expectation about how discovery might be done and what constituted an explorer.

## VOYAGER AS SPACECRAFT

Like its mission, the Voyager spacecraft was an alloy of ambition and practice, a compromise between the high-end hopes of TOPS and the low-end practicality of the Pioneers that in the end left them as a modified Mariner. The Voyagers were large for their time, built around a hexagonal bus that could be modified to suit particular packets of instrumentation and propulsion. With their ultimate dimensions set by Titan/Centaur, they nonetheless boosted the Mariner frame into another weight class, like a caravel reworked into a small carrack.

Each spacecraft weighed 1,817 pounds, of which 232 pounds were its scientific payload. There were eleven instruments in all: cosmic ray and plasma detectors, narrow- and wide-angle imaging cameras, ultraviolet and infrared spectrometers, a photopolarimeter, a radiometer, high- and low-field magnetometers, a low-energy charged-particle detector, and antennas for planetary radio astronomy and plasma waves. Bolted against the bus frame was the vessel's most distinctive visual feature, a high-gain dish antenna. Bristling outward were booms to hold assorted instruments, two antennas, the low-field magnetometer, and the spacecraft's nuclear power source, a radioisotope thermoelectric generator. (The RTG meant Voyager lacked perhaps the most distinctive visual feature of Mariner: its fan of solar panels.) Adding to its weight were sixteen hydrazine thrusters, a suite of onboard computers, and a propulsion module. Reflective blankets wrapped those portions that required some temperature regulation. The core spacecraft was twelve feet high (the diameter of the high-gain dish antenna), but with its booms fully extended, its width swelled to fifty-seven feet, to its which the propulsion module added another nine. With its reach spanning three dimensions, Voyager claimed an impressive volume.[54]

They were larger than most American spacecraft and smaller than Soviet counterparts. Pioneers 10 and 11, which also had to fit into Centaur, were more compact, with a 9-foot main antenna, booms that reached out 10 feet, and a total weight of 570 pounds. They had shed weight by eliminating the onboard computer; by shrinking the size of the antennas, since communication would not have to extend to the outer planets; and by having the spacecraft spin, making it a gyroscope, rather than stabilizing it by multiple thrusters, as Voyager did. Soviet spacecraft were larger and more ponderous, since heavy-lift rockets were the norm and, from the onset, the Soviet space community intended to send people. At least at JPL, Americans were content with robots, and could miniaturize electronic prostheses, and in any event, granted the reality of their lighter-lift launch vehicles, they had to accept the smaller machines.[55]

The Voyager spacecraft was within the range of smaller exploring vessels. A nineteenth-century keelboat of the type used by Lewis and Clark and later by exploring fur trappers on the Missouri River, had a length of fifteen feet and a width of ten, comparable to Voyager in its Centaur shroud. Columbus's *Niña,* his favorite, which he used on his second voyage as well, was a caravel approximately fifty-five to sixty-seven feet long and twenty-one feet wide. The unfurled Voyager, including masts and booms, might claim a rudely comparable volume. For his imaging during the Great Surveys of the American West, relying on glass-plate photography, Timothy O'Sullivan required a small boat on the river and a mule-drawn covered wagon overland. Both vehicles were smaller than the cocooned Voyager.[56]

More interesting is the comparative scientific punch. The Great Voyages had almost none, save the captain's log and charts, and the overland entradas such as Coronado's or De Soto's left nothing more than letters and the occasional journal. The Second Age had science as a prominent purpose, but still could allocate only a fraction of space to its practice. Captain Cook's HMS *Endeavour* was exceptional in its commitment to the "scientifics" (not least because Joseph Banks paid for them), but they were five out of a crew of ninety-seven, and commanded a similar fraction of the working vessel, even allowing for use of the "great cabin." Charles Darwin shared a small room aft on the HMS *Beagle,* in which hammock, table, instruments, chest of drawers, and shelves constituted his residence, laboratory, and library; probably less than 2 percent of the available floor space of the ship. Where the science ratio was high, it was because the treks were short and did not require transport and sustenance over long distances. John Wesley Powell took three specially constructed dories and a pilot boat, nine men (none of them scientists), and a handful of instruments (sextant, compass, barometer) on his descent through the gorges of the Colorado River; science weighed literally less than his ration of beans. G. K. Gilbert took four men (one of them an assistant) and a complement of mules to the Henry Mountains, the last mountain range to be explored in the United States. Of the 138 species of goods carried, from machine oil and postcards to salt and rice, 41 (30 percent) had some direct bearing on doing science, though they composed probably less than 5 percent by weight; to this, one should add the mules, packs, blankets, and saddles, all of which reduced the scientific load to little more than a backpack. Even a dedicated vessel like the converted steam corvette HMS Challenger, surveying the world's oceans for over four years and 127,500 kilometers, with 15 of its 17 guns and spare spars removed, had only a scintilla of its space allotted to laboratories. Of three decks and the hold, scientific work claimed perhaps 2 percent, with an equivalent amount devoted to quarters for its practitioners. The reason of course is that most of the enterprise went to caring for the crew, and the larger the crew, the larger the proportion of space and time committed to their sustenance.[57]

The space program has investments both greater and lesser. For crewed vessels, the fraction is minimal. Apollo 11 carried small tools and returned 21.7 kilograms of moon rock. Of the twelve astronauts the Apollo program put on the Moon, only one, Harrison Schmitt, was a formal scientist. The reason for the consistently low proportion devoted to inquiry, not only in space but over the long arc of exploration, is of course that the transporting vessel must support not only itself but people, and can do so only by establishing an artificial habitat. It's as though Mariner 4 or Pioneer 11 had to launch encased in the assembly labs at JPL or Ames.

Robotic spacecraft could manage a far higher ratio of science to bulk. Of Voyager's total weight, 1,817 pounds, some 232 (almost 13 percent) came directly from scientific instruments, and a goodly fraction of the rest came from the thrusters and antennas required to position those instruments, not to transport or sustain the spacecraft overall. It was a powered high-tech lab. It needed to carry no flour or salt, no water, no rum, no saddles or pack frames; it housed no mess, no kitchen, no bunks, no dispensary; it held neither library nor ballast. And not least, the spacecraft carried no weapons. It had no armaments: no cannon, shot, or powder. It had shields to protect against micrometeorites and radiation and reflective wrapping to ward off sunlight, but no thick-beamed oaken sides to slow shot. Vacuum had its own peculiar hazards, but they were not such that an onboard arquebus or a guard of marines were needed to repel them.[58]

The chief check on Voyager's size was the properties of the launch vehicle—the thrust of the Titan rocket and the dimensions of the Centaur shroud. Once in space it would be weightless, in need only of internal power for electricity and thrusters. It was exploration stripped to its essences.

Actual fabrication started in 1975. Since each spacecraft was custom-built, testing was continual, and since not all hazards were known, engineers had to work without the prospects for repairs. That required close attention, constant testing, and redundancy.

Each NASA lab had its own protocols to validate design and parts. For JPL, since the early days of Ranger, this had meant running individual components through trials before assembly, then constructing a mock-up model to test the system's mechanical features, a thermal control model to verify its ability to withstand interplanetary space, notably heat and vacuum, and finally a proof test model that replicated the spacecraft as fully as possible. The standards were severe, not only because of the known hazards but also because of the uncertainties. Whether or not Voyager went on to Uranus, it was expected that a subsequent vehicle based on its pattern would, so the spacecraft had to survive the equivalent of several mechanical lifetimes beyond what had been attempted.[59]

The remaining hazards resided within the spacecraft and were inherent to long treks through interplanetary space. Even robots needed protection, and space-sensing instruments, a habitat. Their demands might command far less space and shielding than those for humans, but they required shelter surrogates nonetheless. To survive extremes of heating and cooling, the spacecraft needed some shielding; and to survive the gusts of Jovian (and other) radiation, it needed hardened electronics, just as vessels in tropical seas were sheathed with copper, or oaken vessels in polar seas strengthened against ice. For the rest it would have to depend on high-reliability components and redundancy. It could not return to have a faulty boom replaced or a hydrazine-powered thruster cleaned, as the vessels of the Great Voyages could replace a broken rudder or a torn sail. It could not replace scurvied or dead crew with new recruits.[60]

Over the course of construction the validation procedures racked up 3,500 reports of problems or failures—this out of a machine that had 65,000 individual parts, and complicated interactions among all of them, guided by first-time software. Some of those "parts," moreover, were themselves electronic composites. (Each Voyager was estimated to have the "electronic circuit complexity" of 2,000 color TV sets.) As launch approached, the pace of testing quickened—provisioning at port has always seemed a frenzy. Program manager John Casani recalled a "lot of crisis activity" during those last few months—"changing radio systems and switching things around," many of these difficulties "unexpected" and "quite challenging." In that final rush a major loss of detectors occurred, because the building in which the spacecraft was being tested got painted, and spray ("some hydrocarbons") was sucked inside the assembly room through the air handler and contaminated the sensors.[61]

Not all the testing failures resulted from instruments. Engineers had not fully appreciated the actual stresses from launch, and had not tested Voyager for them, and at one point during launch thought they "might have lost the spacecraft." What had actually happened was that Voyager's own self-protection systems had kicked in, confusing controllers, who thought they were witnessing outright failure instead of fail-safe software working as it should. The breakdown lay in not devising the right tests. Before Voyager 1 launched, some minor mechanical adaptations were made to dampen the effects the second time around. The reliance on software, however, suggested an alternative to simple redundancy: Voyager's onboard computers could be reprogrammed. That meant that not every task had to be replicated in metal, or potential failure anticipated; adjustments could be improvised as needed. Voyager could substitute flexibility and redundancy for size.[62]

Raymond Heacock observed that the level of redundancy was such that each Voyager contained "almost two spacecrafts within a single structure." In truth, the issue went well beyond that, since the Voyagers were triplets, and each did work. The proof test model proved critical when, during final assembly, problems cropped up in VGR-2, and the nominally spare spacecraft had to launch in its place. The on-ground model subsequently helped engineers understand glitches and breakdowns when Voyager 1 had its scan platform stick and then had a capacitor short out, and again when Voyager 2 lost its primary receiver. By experimenting with the grounded spacecraft, it was possible to replicate in the lab what telemetry was reporting from space.[63]

Despite breakdowns in some critical components and problems with "scientific instruments," Heacock noted that there was "essentially no mission loss due to failures." The prime objectives were met—and more. Redundancy once again proved an almost indispensable index of successful exploration.[64]

For that reason as well there were two spacecraft. Launch windows were unforgiving, interplanetary space hostile, and distances impossibly remote. If something went wrong, there was no opportunity to return, or to sail to a nearby port or to temporarily beach on a remote moon for careening or repairs. The simplest solution was to send multiple vessels.

Almost all the Great Voyages were fleets. Columbus sailed with three vessels, then seventeen (a hapless experiment in colonizing), six, and four; da Gama, Cabral, and Magellan each launched with five. John Davis sailed first with two, then four, and then three. James Cook had only the *Endeavour* on his first voyage (and nearly foundered on the Great Barrier Reef), then sailed with two vessels on his subsequent circumnavigations. There were of course solitary ships, mostly to the better known North Atlantic. On his first voyage, John Cabot discovered "New Found Land" in the *Matthew*, and returned to tell the tale. That reconnaissance inspired plans to establish a trading colony and a five-ship voyage, of which one ship crept back to Bristol while the other four were lost at sea.

Even major expeditions had to accept whatever ships they could claim, very few of which were new or prime. (Da Gama's voyage mattered sufficiently to warrant the construction of two new ships, but that was an anomaly.) While Magellan was assembling his fleet, the Portuguese consul in Seville reported gleefully that his ships were "very old and patched up," and that he personally "would not care to sail to the Canaries in such old crates; their ribs are soft as butter." In the end only one patched-up ship and eighteen men survived (although one, the *San Antonio*, had deserted earlier). That was far from exceptional. Even repairs required extra materiél, and unlike with Voyager, no spare vessel would be left at port to edify engineers. The Great Voyages went in groups.[65]

On the continents, save in extreme environments such as Antarctica, the Sahara, and Greenland, it was possible to live off the land or trade with indigenes, so the demands for redundancy could be reduced. That gain was frequently offset by the need in hostile or truly unknown lands for a military escort; the same was true for fleets. A sensible fraction of exploring armadas was devoted to firepower. The more a mission sought to go beyond simple reconnaissance, the more complex its arrangements, the greater the tendency to replace lone travelers with a corps. The Long Expedition of 1819–21 (the first of five by Stephen Long) took nineteen men up the Platte to the Rocky Mountains, beginning with the first steamboat, the *Western Explorer*, on the Missouri River. It was, one observer exalted, a veritable cavalcade of the era. "Botanists, mineralogists, chemists, artisans, cultivators, scholars, soldiers; the love of peace, the capacity for war: philosophical apparatus and military supplies; telescopes and cannon, garden seeds and gunpowder; the arts of civil life and the force to defend them—all are seen aboard."[66]

That was the norm. The magnitude of the investment, the uncertainties of the hazards, and the expectations of losses, all trended toward small numbers of costly expeditions rather than swarms of inexpensive ones. There were plenty of exceptions, especially by first-voyaging nations. John Cabot sailed to the New World in the *Matthew*, a navicula about the size of Columbus's *Niña.* Some single ships even ventured around the world, as Francis Drake did in the *Golden Hind* and William Dampier in the *Cygnet*, but they were privateers, and many of the famous scientific expeditions during the Second Age involved single vessels—think Thomas Huxley in the *Rattlesnake* and Charles Darwin in the *Beagle.* But the latter had well-established ports of call at which to provision, and they were surveying points of navigational interest such as the Strait of Magellan or the Torres Strait rather than blue-water voyaging into the unknown.

The geography of launch windows also argued for multiple or paired launches. Since the ideal Hohmann trajectories to Mars came every twenty-six months, it became the norm wherever possible to launch two spacecraft at each cycle. Some failed on launch; some failed at Mars; but few cycles passed altogether without some contact. Moreover, even paired spacecraft on the same mission might have incommensurable objectives: it was not possible, for example, to fully survey both Titan and Saturn's rings in one traverse. In this respect, as in many others, planetary exploration more resembles the voyages of the First Age than it does the Second. Voyager, however, was the last of the paired missions. Double-launching spacecraft had become, in John Casani's words, "an insurance policy that now we can't afford to take." In this, too, it resembles its Iberian predecessors who over time risked more and more with less and less.[67]

So there were two Voyagers, as there were two captains in the Corps of Discovery, and two ships on Cook's second and third voyages. But unlike those others, they would not communicate between themselves. They had separate tasks, and once separated, they would not rejoin.

## VOYAGER AS EXPLORER

Voyager was an autonomous spacecraft; by most standards, the first. "Autonomous," however, is a relative concept, and its operational meaning is specific to particular tasks.

There were things the Voyager spacecraft could do—had to

do—that previous spacecraft could not. It had to self-regulate. It had to monitor its machinery, take remedial steps in case of sudden ruptures or emergencies, restart, and from time to time accept new orders. Those needs increased over time. So great was the distance that even at the speed of light, messages to Earth and back would take too long, and so extended was the journey that constant text-messaging with, and micro-managing by, JPL would exhaust the spacecraft's reserves of energy. Voyager had to tend to itself—identify problems, shut down troubled parts, restart anew. It had to assist ground controllers with adjusting trajectories and transmitting scientific data. It had to execute the split-second maneuvers of close encounters through preprogrammed codes, each rewritten based on past experiences. It had to recognize when its onboard machinery for overseeing such tasks had itself failed or become corrupt. It had, within limits, to exercise self-control.

The technology that made this possible was the digital computer, which coevolved with the space program. Gordon Moore announced his eponymous law in 1965, the same year that Gary Flandro outlined the prospects for a Grand Tour. The second would depend on the first. Moore's Law states that the number of transistors that can be placed inexpensively on an integrated circuit increases exponentially—specifically, it doubles roughly every eighteen to twenty-four months. From this derives nearly every metric of computing power. By the time Voyager launched, twelve years after those dual announcements, six doubling periods had passed, or a sixty-four-fold increase in computing capabilities. Voyager was on the cutting edge of working computers, and particularly miniature ones. By the time it encountered Saturn, another doubling period had come and gone, and the personal computer was arriving. Voyager seemed vaguely archaic. When it encountered Neptune ten years later, another five doubling periods had so pumped up computing power that Voyager stood to an average cell phone as a caravel to a Centaur rocket.

Onboard computers evolved out of sequencers, or preprogrammed counters that triggered a series of operations. Once started, a sequencer performed a series of activities that allowed a spacecraft to disengage from rockets, unfurl, and fly past a planet while recording data and images and transmitting them back. Earth controllers initiated the sequence and uploaded new ones for each phase of the mission. A central computer could oversee several such sequencers. For longer flights, such redundancy was mandatory, and as computing power swelled, programmable computers could replace simple sequencers. Voyager had three such computer systems, each with a backup. During its cruise phases, a "sequence load" was uplinked to the spacecraft every four weeks. The uplinked sequences quickened during encounters.[68]

The Attitude and Articulation Control Subsystem computers assisted with navigation. No ground controller could hope to maintain the constant monitoring that told Voyager where it was in the solar system and how to stabilize its three axes. The AACS computer could, and it kept the spacecraft's antenna directed to Earth. The Flight Data Subsystem computers performed analogous tasks for the scientific instruments and the transmission of data. Overseeing both subsystems, along with the programmable sequencers that monitored the spacecraft and piloted its encounters, was the Computer Command Subsystem. The CCS was virtually identical to the computer used for the Viking mission. The FDS had to be constructed uniquely. To conserve power, the FDS and AACS computers could go into sleep mode.[69]

Command was distributed, redundant, and capable of hibernation. While it ultimately rested with NASA (through JPL), it routinely resided within Voyager. If Voyager went awry, its Earth-based human controllers would have to reclaim and recharter the spacecraft. If within Voyager one computer failed, its backup would step forward, and if one system failed, another might be able to compensate. The ability to upload new software meant that fresh orders could be rewritten as the mission evolved, particularly if Voyager 1's success meant a Grand Tour was possible. The capacity to power down when not in use meant the RTG power packs could sustain the long journey.

Yet autonomy, while necessary, could introduce instabilities of its own. Voyager's understanding was literal and did not always coincide with that of its human controllers. There were occasions when the nominal controllers were confused and onboard computers had to intervene to correct them. (This happened, for example, during Voyager 1's launch.) There were also events in which the people forgot to perform tasks or did them incorrectly, and the spacecraft went into backup or safe modes. And there were incidents in which "the computers on board displayed certain traits that seemed almost humanly perverse—and perhaps a little psychotic," as one staffer observed. The programming was too tight, the instructions too sensitive, the tolerance for the quotidian of flight, even cruising, too fine to accommodate the realities of interplanetary travel. The computers overreacted, a kind of automaton hysteria. Some software therapy in the form of new programming sedated the machine and allowed it to regain equilibrium.[70]

Even mutiny was possible. There could be internal arguments among the onboard computers, perhaps temporarily paralyzing decisions. But most spectacularly it could happen as it did to Voyager while en route to Saturn, in which new commands were uploaded, only to be refused on the grounds that they contradicted more fundamental instructions and would result in catastrophe. Such insubordination could not be flogged away or strung up on an instrument boom yardarm. The source of the confusion had to be found, and corrected. In this case Voyager—which made the proper decision—was able to defend itself without a court-martial conducted through the Deep Space Network.[71]

How much autonomy was technically possible? Some, with increasing degrees of self-control possible as physics improved computing. How much was desirable? That question edged into metaphysics. No explorer has been fully autonomous. All have had restraints imposed by their mission or their sponsor, and all have had orders that would allow for surprise, adaptation, and redirection.

What the robots did not have was a human consciousness and conscience. They knew nothing of the traditional horrors, like ship rats, that had ever accompanied long-voyaging expeditions. They would experience nothing akin to the miseries Magellan's crew faced as they sailed for "three months and twenty days without taking on board provisions or any other refreshments," during which they ate "only old biscuit turned to powder, all full of worms and stinking of the urine which the rats had made on it, having eaten the good." The crew "drank water impure and yellow. We ate also ox hides which were very hard because of the sun, rain, and wind. And we left them four or five days in the sea, then laid them for a short time on embers, and so we ate them. And of the rats, which were sold for half an ecu apiece, some of us could not get enough." The worst affliction was scurvy, which killed thirty-one of the crew. Other "maladies" crippled another twenty-five or thirty.[72]

Yet spacecraft had their own peculiar ailments. Parts broke, mechanical gears became arthritic, instruments lost sight, computers suffered hearing loss. There were leaks. Filters clogged. The critical onboard provisions for Voyager were the hydrazine required by its thrusters and the electrical power derived from the RTGs. Without hydrazine the spacecraft could not maneuver, and without power, it could execute none of its mission or even communicate. It would succumb to robotic scurvy. It would become a ghost ship, at the mercy of solar winds and gravitational currents. Earth, Mars, Venus, and the Moon were littered with wrecked vehicles. The Mars Observer had vanished in space.

Voyager knew none of this—could not fret over what might cripple it, could not leap into the breach or suddenly freeze in fear, could not imagine failure or triumph. Its designers could, and they sought to program the spacecraft to handle both the expected and the unexpected. But the only way to cope with the myriad possible hazards was to grant Voyager some capacity to respond on its own. To find its own way it needed something like its own will.

## VOYAGER AS GRAND GESTURE

Still, if Voyager was less than human, it seemed more than a machine. It had symbolic value. In particular, it will likely stand as the grand gesture of the Third Age.

Such expressions are not the first bold announcements of discovery, which come early, with the shock of first discovery or first expression. Such was the case, for example, with Columbus and Dias, with the campaign to measure the transits of Venus, and with IGY, all of which proclaimed a new age. While that annunciatory first revelation can galvanize the imagination and inspire successors, it comes too soon to capture the still-inchoate features of the times, or it is too tightly bound to a particular people or project. So although the Carreira da India first launched under Vasco da Gama and became the basis for Portugal's national epic, *The Lusíads*, Spain could equally point to Columbus's four voyages; and England to John Cabot's daring sail to the New Found Lands at the same time; and France, always willing to consider history pliable in the interest of *gloire*, to Jacques Cartier. The essence of the First Age was, as J. H. Parry put it, the discovery of the sea, not simply the Ocean Sea that lay beyond the classical world or maybe encircled the lesser seas as the Styx encircled Hades, but the astonishing way in which they all connect and flow one into another. The enduring symbol of the age would highlight that fact.[73]

Grand gestures seem to follow twenty to thirty years later, after exploration and its sustaining culture have worked through the full terms of engagement, as first discovery becomes full discovery. They are the moments of exploring that more than any other capture the general imagination, that fuse place, time, discovery, and yearning in ways that seem to speak to an era's sense of itself. If they do not inform an age, they do display the vital attributes of an age as nothing else can.

Thus, for the First Age, the grand gesture belongs to the circumnavigating voyage of the *Victoria*. Ferdinand Magellan was a paragon of his time, an explorer tenacious, religious, tough beyond reckoning and ambitious beyond yearning. God, gold, and glory—all filled the mold of his soul. When the Armada de Molucca, reduced from five daunting vessels to one straggler, the Victoria, completed its ambit of the world ocean, it defined the limits of what might be done (and had, in fact, consumed Magellan, dead in the Philippines). So bold was a circumnavigation that few expeditions dared to follow. But when the Second Age geared up in the 1760s, it did so with a new wave of Earth-encircling voyages, joining the two ages as one might lay a plank between ships.

Yet however much those new circumnavigations might bedazzle the late eighteenth and early nineteenth centuries, they could not speak to the dramatic discovery or rediscovery of Earth's continents or those lush fragments of it chipped off as islands. The floating excursions of James Cook and Louis-Antoine de Bougainville had to come ashore and penetrate inland. The man who achieved that task with cultural panache was Alexander von Humboldt. Convincing Carlos IV to let him enter New Spain, he became, as his own age appreciated, a "second Columbus," in this case, the *scientific* discoverer of South America. For five years (1799–1804), accompanied by Aimé Bonpland, he explored widely, paddling up the Orinoco, traversing the Andes, climbing Mount Chimborazo, scrutinizing the archives and natural history of Mexico and Cuba. He elaborated Linnaeus's natural-history excursion into a cross section of continents. He carried the Old World's grand tour to the New World, where it could also gaze on monuments and relics of the past, not least those of nature. He gave empirical heft to the misty musings of *Naturphilosophie*. He empowered geographic science with a global reach. In the words of Ralph Waldo Emerson, he was one of "those Universal men, like Aristotle." He personified the explorer as Romantic hero.

While he was not the first European to boat the Orinoco or climb in the Andes, Humboldt was the first of a new kind of European, such that even when explorers of the Second Age revisited sites known to the First, they did so with original eyes and to novel ends. Symbolically, during a layover on his way back to Europe, he dined with Thomas Jefferson a month after Lewis and Clark's Corps of Discovery departed St. Louis. For Americans the latter was the defining moment of the new age, but for Western civilization overall, that honor belonged to Humboldt, not only because he was first but because he followed the journey with decades of exhaustive research and publication, which meant his experiences could transcend mere adventure and politics and become an exemplar for others. His trek through South America defined the grand gesture of the Second Age as the cross-continental traverse.

The task was roughly completed in the 1870s. For the Arctic and Antarctic, a journey to the pole and back became an acceptable substitute, although it never had quite the cachet, and Ernest Shackleton's celebrated and ill-fated *Endurance* expedition sought to complete the circuit by traversing Antarctica. That particular achievement had to wait until IGY; by then the hype could no longer match the mood. It had a sense of cleaning up odd jobs, not of launching bold new adventures, a bit of geographic housecleaning outfitted with Caterpillar tractors instead of sled dogs. Something else would have to define the grand gesture of the Third Age.

An era of exploration characterized by the inventory of whole planets needed something both vaster and newer than Antarctica and a deed more startling than a descent to the abysses of the world ocean. A journey to the Moon might be dramatic, but it belongs with the Norse voyaging to Greenland, or Polynesians to Easter Island, something astonishing but that leads nowhere else. The same might be said of the marvelous robotic missions to Venus and Mars, or even to Jupiter and Saturn, a brilliant expedition to a single place and theme, but not a moment that, like a parabolic mirror, concentrates all the energy around it to a focus. A grand gesture has to be something that encompasses it all, as the *Victoria's* circumnavigation did, and that like its formidable captain, embodies an age in its character; that by speaking with special force to its own time, it speaks to all times; that can travel across history as the expedition does over geography.

Voyager's Grand Tour does this. In its design, in its scope, in its reprogrammable computers, in the ability of its mission to retain the past even as it speeds beyond it, Voyager has embodied the ideas, ambitions, hopes, paradoxes, aesthetics, flaws, and energies of its time. It has made manifest the power and peculiarities of a Third Great Age of Discovery.

# 5. Launch

Launch is a working synthesis—the possible made plausible. It reconciles motive with machinery. It welds the propellant force of rockets to ideas and instruments. It codes vision into operating plans. It is enough for prophets to find words or images to move audiences, but engineers and managers require a more tangible expression, the contact not of ink on paper or electrons on a TV screen but of metal, wires, hydrazine, lenses, and decaying plutonium. Without such labors the Grand Tour belonged in the realm of science fiction.

Expeditions are made, they are scheduled, they occur at designated places and at specified times. They require ports of departure. They must harmonize social timing with environmental opportunities. They are time capsules of a cultural moment. The Voyagers were all this.

## TITAN IIIE/CENTAUR D-1T

The Voyagers could not carry more instruments than what the allotted thrust and dimensions of the Titan/Centaur rocket and Centaur casing allowed. In their liftoff the Voyagers thus continued a long heritage of mixed-technology vessels, and in the staged sequencing of their flight they recapitulated an old history of vessels redesigned, vessels refitted, vessels shed, and vessels lost. And with unintended inspiration their ascent even emulated the multistage evolution of exploration itself.

The Titan was a two-stage rocket originally developed as an ICBM, which then evolved as an all-purpose vehicle for heavier payloads such as surveillance satellites. The first stage had two engines, and the second stage, one, all of which burned liquid fuels that fired spontaneously when mixed, and rendered "ignition" a figure of speech. To the sides were bolted two solid-fuel boosters to add initial thrust. Atop rose an aptly named Centaur rocket powered by two engines that burned by kindling liquid hydrogen and oxygen and gave the hybrid a third stage. Within Centaur nested a Voyager spacecraft attached to yet another engine for a final propulsive push. Each stage fired, and then physically disengaged and dropped away, leaving the next stage to position itself and burn unimpeded without the dead weight of the husk.

Liftoff commenced with the blast of the first stage's engines and double boosters. As the solid rockets burned out, a suite of explosive bolts and small rockets peeled them from the core. When the first stage burned out, it, too, separated, and the second stage kicked in. When that stage exhausted its fuels, it also fell away, and the third-stage Centaur shed its shroud and its own engine caught fire. Here the Centaur demonstrated its value for such missions. It had high specific impulse, it could coast and reignite repeatedly, and it contained its own avionics for navigation and control. It was, in brief, the Centaur's rocket that mediated between the brute blast of the Titans and the onboard solid-propulsion engine of the Voyagers.

For each Voyager mission Centaur fired and coasted twice, once to achieve a low Earth orbit, and then again, after cruising to the ideal spot, in order to reach escape velocity and embark on its journey beyond Earth. The first episode came as Centaur jettisoned its shroud 203 seconds after liftoff and, 4 seconds later, burned for 101 seconds. Almost 43 minutes later, Centaur repositioned itself for the next burn, which lasted 339 seconds, followed by a short coasting for 89 seconds. That, at least, was the scenario for Voyager 2. The last-stage Titan faltered during Voyager 1, the result of a leak, and threatened to abandon the spacecraft short of orbit, which would have left it to finish its combustion by burning up in the atmosphere. Fortunately, the computers aboard Centaur recognized the problem and adjusted by extending their burn. Even so, they barely had enough fuel to compensate; only 3.4 seconds of fuel remained when Centaur shut down. Had this sequence played out during Voyager 2, the mission would have failed, for it had a less forgiving trajectory; Centaur's boost of burning could not have compensated. The Voyager mission thus displayed, from liftoff, that most ineffable and most essential quality of great expeditions: good fortune. If "they had swapped the Titan booster rockets," Bruce Murray observed, "there would have been no Voyager to Neptune. Flip of the coin." "We were lucky," he concluded, "and that's important."[74]

For both Voyagers, the particular trajectories required for Jupiter demanded yet another thrust. With Centaur spent, the action moved to an augmented Voyager itself. The spacecraft had a supplementary solid-fuel propulsion module under its direction. Voyager segregated from Centaur, ignited its rocket for forty-five seconds, and then shed the booster. The mission module—the spacecraft proper—was now at sea, subject to minor corrections (or "nudge" factors) to compensate for its inherent quirks and the unseen winds and currents of gravity.

Hybrid vessels, staged launches, the practice of shedding dead weight or separating close-exploring craft from long-haul vessels—all these had ample precedents dating back to the Great Voyages themselves.

The ships of discovery were a mixed lot, the result of centuries of evolution in the Mediterranean and a recent fusion with Baltic traditions. A critical innovation was the transfer, probably from the Indian Ocean, of the lateen sail. A triangular sheet with a long, stiff edge, the lateen demanded only simple rigging, could work amid a variety of winds, and sharpened maneuverability. It became the norm for modest-size ships and those working along coasts. But it required proportionally large crews, and could not be put about easily; nor could it run with heavy loads before the wind. The solution was to look north, and adapt the bulkier, square-sailed cog typical of the Baltic and North seas an introduction that came during the Crusades. The synthesis of these two styles occurred with exquisite geographic logic in Iberia, where the Atlantic and Mediterranean converged.

Each tradition was further modified to circumstances over several centuries, and then suddenly joined one to the other during roughly two decades in the mid-fifteenth century. Shipbuilders borrowed and rearranged hull design, keels, spars, rudders, numbers of masts, and various sails, all tweaked into an ancestral barque from which subsequently descended miscellaneous and mongrel progeny. Some, such as the carrack, were large and designed for haulage. Some were built mostly for fighting. But the most celebrated was the caravel (*caravella*), a specialty of Portugal and southern Spain. Mostly Mediterranean in its genes, it morphed to accommodate a square sail and a rear rudder, and became the core ship of discovery. Small, versatile, capable of riding over waves, maneuverable around coasts and shoals, the caravel was the vessel with which Portugal mapped the coast of Africa, and along with Castile, colonized the Atlantic isles and discovered the Ocean Sea.[75]

Still, the caravel was small, and long voyages required vessels of modest crews and greater capacity, so most armadas sailed with mixed fleets, of which the caravel was a part. Da Gama had one caravel, as did Cabral, and Columbus took two on his first voyage, the *Niña* and *Pinta.* (Interestingly, it was the larger *Santa Maria,* a *nao,* that foundered on a shoal off Hispaniola.) The caravels did the close work of coasting, carried messages, tested new seas, and got emissaries and explorers to previously unknown land. Few voyages lacked at least one in their company, although the most extraordinary of all, the Armada de Molucca under Ferdinand Magellan, had none. As they acquired experience, explorers turned to smaller-size carracks, which proved more capacious for crews and goods and better suited for long voyages.

Heavy-lift rockets enjoyed a comparable evolution. They were hybrids. As with ships during the Renaissance, there were geographic traditions—in this case, with centers in Russia, Germany, and

America—and some exchange between them, catalyzed by war. The Russians favored a basic model to which they added boosters. The Americans looked to a more mixed fleet, but one largely evolved from the German V-2, a process catalyzed by the capture of German rocket scientists as the war ended (the celebrated Operation Paperclip). From the V-2 evolved the Redstone, and from the Redstone, the first American intermediate-range ballistic missile (IRBM), the Jupiter; from Jupiter came the intercontinental ballistic missiles (ICBM), Atlas and Titan, separately developed for different purposes. Outfitted with a Centaur upper stage, Atlas launched seventy-two payloads over its history, and Titan, eight, of which Voyager was the last.[76]

If such hybridization and tinkering has a long pedigree, so does the freelance, freebooter movement of those who had expertise. German scientists carried knowledge of the V-2 from Peenemünde to White Sands. During the era of the Great Voyages, Italian capital and merchants, mostly Genoese and Venetian, bonded with Iberian geopolitical enthusiasms to drive overseas expansion. Portuguese pilots, like Italian *condottieri*, were the mercenaries of discovery, serving the highest bidder. During the era of continental colonies, Germans in particular were everywhere, furnishing the technical skills in cartography and botany that locals lacked. Charles Preuss did much of the mapping for John Charles Fremont's first, second, and fourth expeditions; Heinrich Möllhausen collected natural-history specimens and illustrated for Whipple's reconnaissance through northern Arizona, and later Lt. Joseph Ives's abortive voyage up the Colorado River, where F. W. von Egloffstein joined and did the critical cartography. Exploration invited the footloose and the obsessed to join whatever party would sponsor them. The Third Age is no exception.

For exploration history add the practice of adapting vehicles developed for one purpose to another. Frijthof Nansen reworked Inuit sledges into sturdier forms for traversing rough pack ice and wrestling over pressure ridges. John Wesley Powell put modified dories into the Colorado River to descend through its succession of gorges. Captain Cook's fabled *Endeavour* was a modified collier. Charles Wilkes's flagship, the *Vincennes,* was a reconditioned sloop of war; Dumont d'Urville's *Astrolabe* was a converted corvette; James Ross's *Terror* and *Erebus* were refitted naval mortar ships. (They later went to the Arctic, where they sank off King William Island with Sir John Franklin.) The HMS *Beagle* that carried Darwin around the world began as a ten-gun brig remade into a barque and modified to slough off rougher seas. Ernest Shackleton's *Nimrod* was a forty-year-old wooden sealer; Robert Scott's *Terra Nova,* a barque formerly used for sealing and whaling.

Only Nansen's *Fram* was designed from scratch for polar exploration. Its heavy oak beams gave it a stiff pliability and its curved hull was designed to have the ship ride up as pack ice squeezed. Several explorers, however, devised boats that could be disassembled into more portable parts, and then reassembled as needed. Nathaniel Wyeth experimented with a design in the American West, and Samuel Baker and Henry Stanley did likewise in central Africa. The Ives Expedition to the Colorado River transported a disassembled steamboat. So it was that rockets for planetary flybys were adapted from ICBMs, and replaced MIRV warheads with multiple-instrumented exploring spacecraft.

Those vessels, by their launch gantries, were as mongrel as the caravel-piloted armadas of the Renaissance. The heavy vehicles had to do their work first and then fall aside, their task completed in minutes and seconds upon reaching escape velocity, while Pioneers, Mariners, or Voyagers cruised through interplanetary space for months or years. In the sixteenth century the first stage could take months, tacking across winds and currents and around continents, while actual contact might exhaust itself in a few days. For this end the vessels might include a pinnace (prefabricated and carried in the hold until needed), or tow a shallop or longboat, ships that could support geographic foraging parties, or "away teams," at new coasts, akin to surface landers dropped from orbiters. The danger was that they might be swamped during the crossing, as they often were. Except for the longest voyages, they could expect to provision themselves en route. Not least, for those expeditions that intended to finance themselves by importing spices or bullion, heavy vessels would be needed for the return, for unlike planetary spacecraft, those exploring parties had to report in person and deliver their discoveries in tangible form; they could not transmit digital data across space. Still, early-stage supply ships could be shed, discarded like empty rocket boosters, until all that remained was the pure mission module.[77]

## AROUND THE CAPE

The Voyager twins departed Earth from launchpads at a site first selected only eighteen years previous.

In the past, exploration had not devised its own point of departure. Expeditions left from places that had evolved organically from sites of enterprise and lines of transport that, from time to time, outfitted parties of explorers as needed. John Cabot sailed from Bristol, long a scene for trade and fishing; Jacques Cartier, from La Rochelle, likewise a well-established natural port; Captain James Cook launched from Plymouth, and Captain FitzRoy's *Beagle,* with Charles Darwin aboard, from Devonport, both naval depots. The Lewis and Clark expedition left St. Louis, the major entrepôt of the Mississippi River. Those countries, such as Russia, that lacked ready access to the sea lagged, or turned overland, until ports were available; for the Bering expeditions to Alaska temporary ports had to be constructed. Once in the New World, Spain promptly erected ports at Isabella, Vera Cruz, Lima, and Acapulco from which new voyages could set forth.

Cape Canaveral, Florida, followed the latter pattern as a place developed specifically to fire rockets. Rocketry's hazards, and to some extent its secrecy under military sponsorship, argued for remote sites, typically deserts such as White Sands or Edwards Air Force Base or Baikonur, Kazakhstan, until their heft demanded greater domains of emptiness. By 1949 the range of the V-2's successors had lengthened sufficiently that President Harry S. Truman established a Joint Long Range Proving Ground on what was then a largely uninhabited belt of Florida shoreline.

The site had plenty of attractions. The surrounding countryside was lightly developed; missiles could be readied, and even fail in violent fireballs, without threatening communities. Launches would arc over the Atlantic, and if missiles wobbled off course, blew up, or had to be destroyed, they again posed no danger to people. The weather allowed for continual operations, save for fleeting thunderstorms and the occasional hurricane. Islands trending southeast, from the Grand Bahamas to Ascension, themselves famous as sites for past discovery, could service tracking stations. There was ample land for future development.

In 1950 the army moved its facilities from White Sands to Cape Canaveral, while the air force acquired a nearby naval air station (Banana River) to convert into its Air Force Missile Test Center. In 1955 the navy relocated its Vanguard missile program to Canaveral. In December 1959 the army transferred its facilities to NASA. When the expanding program demanded still more land, nearby Merritt Island was acquired for development into a major "spaceport," the John F. Kennedy Space Center and Eastern Space and Missile Center.[78]

The site had its quirks, as all ports do, which is why historically pilots were first and foremost versed in local lore. Canaveral's shoals and tidal bores were its weather—its thunderstorms, its frosts. Having an electrostatic match striking randomly around a tower of distilled combustibles could kindle catastrophe, and a cold snap could turn pliable O rings into brittle plastic. These were local lessons, learned over time, and they would take many decades to codify into pilotage for the Cape.

Yet in one respect it did resemble so many of its predecessor ports from which explorers had embarked: Cape Canaveral was an island, or effectively so, cut off from mainland Florida by estuaries and the Banana River. Its development stands to space voyaging as the colonization of Atlantic isles does for the Great Age of Discovery. Those isles permitted a final fitting and provisioning, and then positioned a vessel within the geography of wind and currents that would carry it to the New World or around Africa, which is what also happened at the Cape. Today's equivalent to the Canaries, astride the trade winds, is a locale closer to the equator, which can contribute more rotational velocity from the Earth to a rocket moving east (Canaveral adds 1,666 kilometers an hour). For deep-space ventures, the ideal would be a port of call beyond Earth's gravitational shackles altogether, something like an orbiting space station or a lunar base. Since neither exists, the compromise is a pause in orbit before the final stage ignites.

---

The Voyagers' launches were the climax to a dozen years of meticulous planning, and liftoff was almost ritualistic in its tight choreography. The larger factors were many, and rarely aligned, among them the immutable geometry of the planets, the rocket's capacity for thrust, the size of the payload, the ability to navigate, and the ambitions of the mission. Every consideration varied, and each changed with changes in the others. There could be no singular solution.

Every planet had its unique launch window based on the minimum energy trajectories first elaborated by Walter Hohmann in 1925. For Venus these came every nineteen months; for Mars, every twenty-six; for Jupiter, every thirteen. The thrust available determined velocity, which then set launch opportunities, but as payload varied, so did velocity, and as velocity changed, so did potential routes and hence launch dates. Where one launch was to lead to another through gravity assistance, the options multiplied, and a planning calendar became maddeningly complex. For the Grand Tour, a suitable collective geometry would reappear only every 176 years. Yet at each planet along the way there were sites to visit, so the prospects for routes were almost unbounded. Planners began with 10,000 potential trajectories, which they then whittled down to 100 possibilities.

Voyagers' window was tight. To preserve the Grand Tour option, launch had to occur between 1976 and 1978; to capture Jupiter's gravity, it had to obey a thirteen-month cycle; to reconcile Jupiter's and Earth's orbits, it had to happen within a month of consecutive dates within that recurring near-annual cycle, and for each day within that month, within an hour-long window. There was, as well, a cultural window, for five years earlier a Grand Tour would not have been technically possible, and five years later not politically feasible.

As in the past, ceremonies of departure had emerged. There were bleachers for dignitaries and associates, cameras to record and broadcast to the public, rites of political blessing. It had ever been so. Traditionally, ship launches were a bedlam of last-minute provisioning, the testing of anchors and ropes, the bustle of late-arriving crewmen. There were observers, whether family, dockworkers, the curious, or the emissaries of sponsors. There would be a Mass said, a service recited, a political proclamation, a poem. The expedition's orders and rules of conduct might be read.

There would also be a search heavenward for signs and portents. When Sir John Franklin departed to plot out the Northwest Passage, a white dove settled on a mast, and became for many well-wishers (including Franklin's daughter, Eleanor) "an omen of peace and harmony." As Voyager 2 rested at its gantry the day before launch, black thunderheads and lightning approached threateningly within a mile of the combustible rocket before veering off to sea. In both these cases, the portents misread subsequent events entirely.[79]

Since they had different trajectories, the Voyager twins launched separately. Their names reflected the sequence of arrival at Jupiter, not the sequencing of launch. Voyager 2 went first. Voyager 1 followed sixteen days later. They had met their launch dates and sailed with the cosmic tides.

For expeditions, time can be more powerful than place. It is always possible to reposition ships (or keel boats or entradas) for secondary departures, and this was normal; but it is tricky and often fatal to ignore the seasonal geographies of wind and wave. The need for timing could further argue for staggered embarkations. Always exploration has had its launch sequences.

These could span from hours to years. Spanish fleets from the Caribbean had to sail outside hurricane season; Russian promyshlenniki relied on frozen rivers in winter and flowing streams in summer, and shunned the mires of spring breakup. Voyages to the Spice Isles had to obey the ebb and flow of the monsoons, first from Africa to India, and then to the East Indies. Vasco da Gama caught those winds for 23 days from Mombasa to Calicut, and when he tried to defy them on his return, he nearly lost his crews to scurvy in fighting their contrary flow for 132 days. Expeditions up the Missouri River left early in the spring to reach the mountains before the snows, or else they found a valley to winter over until the following spring. In Australia, the Burke and Wills expedition managed to undertake its trek across the interior with exquisite (and fatal) mistiming, fighting brutal summer heat to Cooper's Creek, and then breasting the rainy season to the Gulf of Carpentaria. In extreme cases it might take a year or more to relocate to a site from which further exploration could proceed. The Franklin expedition to the Northwest Passage expected to winter over in the frozen Arctic before summer melt would allow further progress. Ship-borne Antarctic exploration would likewise break through the enveloping pack ice late in the summer, only to winter over in preparation for late-spring treks. Russian forays to the Pacific might demand a year's travel just to reach the coast, and another year to return, with exploration limited to fleeting opportunities in between.

But apart from seasonal or secular rhythms, there were always local considerations of tide and wind, over which the explorers had often little knowledge and less control. Many a bold expedition toasted the tide, and then sat while contrary winds kept them in port, or took to sea and had to beat against the winds for a week or more until they could properly launch. (The equivalent for rockets are technical glitches that pause or close countdowns.) Tide and winds could make for cruel embarkations, and even crueler returns. These were the launch delays of the past, when the countdown stopped, the expedition slid from élan to ennui, and the drama stalled while the stagehands replaced the faulty curtains and broken limelights. This mattered less when communications were not instantaneous and when the vagaries of earthly weather offered more flexibility than the foreordained orbits of celestial mechanics.

## LIFTOFF

On August 20, 1977, at 10:29:45 a.m. EDT, Voyager 2 lifted off from Cape Canaveral. On September 5, at 8:56:01 a.m. EDT, Voyager 1 followed.

They rose atop a boiling column—a controlled explosion—of flame. If their ultimate thrust derived from a cultural combustion that mixed ambition, curiosity, greed, political posturing, and utopianism, the proximate cause was a violent collision of liquid hydrogen and liquid oxygen into a distilled, full-throated burn. It was a magnificently Promethean gesture, in its audacity—the first satellite had broken into earthly orbit only a scant twenty years before, yet here was a launch intended to traverse the solar system. But it was Promethean, too, in its choice of means, for fire is humanity's signature technology and a chemistry that has no natural presence outside Earth. Once beyond Earth's gravitational tug, the Voyagers would be beyond the realm of life, which is also the realm of fire. That reach beyond was the essence of their mission. But it seemed altogether right that they should begin those journeys to lifeless worlds and a near-eternity of vacuum upon the pillars of the fire its makers had long ago seized from the heavens.

# part 2

## BEYOND THE SUNSET: JOURNEY ACROSS THE SOLAR SYSTEM

Push off, and sitting well in order smite
The sounding furrows; for my purpose holds
To sail beyond the sunset, and the baths
Of all the western stars, until I die.

**—Alfred Tennyson, "Ulysses"**

But the way is dangerous, the passage doubtful,
the voyage not thoroughly known.

**—Richard Hakluyt,**
***The Principal Navigations***

# Beyond Earth

DAY 1–28

# 6. New Moon

Two escape velocities frame the Voyager mission. The first, the escape from the gravitational pull of Earth, allowed the twin spacecraft to begin their quest through the solar system by getting them to Jupiter. The second, the escape from the gravitational pull of the Sun, happened at Jupiter and allowed them to complete the Grand Tour. That second momentum would also define the end of their quest by propelling them beyond the solar system altogether.

Those dual moments of propulsion help to bookend the mission's narrative as well, by segregating the geography of places visited from that of places shunned. Simply departing Earth's gravity was not enough, for there were two other bodies from whose psychological orbits the Voyagers also had to escape. The first was the Moon, whose tidal pull had turned minds heavenward since ancient times and whose political phases had filled the night sky of the space race. The second was Mars, the most Earth-like of planets and the obsession of a culture of colonization. Both the Moon and Mars were the scenes, either actual or imagined, for human space travel, and thus for all the rest of what the segue from exploring to colonizing implied, not least some bond to a national military. A critical fact is that Voyager bypassed both bodies with hardly a wobble. In less than a day the

Voyagers blew past the sublunary realms and began the first of their long cruises between encounters.

## OUT OF ORBIT

Passage from Earth to the Moon was, for both Voyagers, little more than a heartbeat in a long marathon. When they completed their final firings, Voyager 1 and 2 had velocities in excess of 143,000 kilometers per hour and 140,000 kilometers per hour, respectively. They passed the average orbit of the Moon, roughly 384,403 kilometers from Earth, some ten hours after launch.

There were a few launch glitches, some serious, some merely annoying. The most critical occurred when Voyager 1 nearly failed to attain its mandatory velocity, which would have still sent the spacecraft to Jupiter and Saturn but would have scrubbed the Grand Tour. By contrast, Voyager 2's launch crisis involved more of an anxiety attack. This began when onboard computers detected accelerations greater than expected and overly "self-corrected"; then dust particles set in motion by vibrations sparkled and misoriented the spacecraft's tracking system. Together these events left the antenna misdirected, and hence oblivious to ground commands. For a heart-stopping moment, it seemed that the spacecraft might be lost altogether. In fact, it was working as programmed, but its programs had been written for a different set of expectations; the shaking incurred at launch exceeded those forecast conditions, and had jarred overly sensitive fault-protection algorithms into sending the spacecraft into a fetal crouch. The scare passed. When the etiology of the affliction had been diagnosed and cured, one analyst described the episode as an "anxiety attack"; another, as a kind of "vertigo," or perhaps seasickness, part of Voyager getting its sea legs. Bruce Murray observed that the "new, superautonomous robot" lacked the "repertoire of human judgment and experience," and it had "mistrusted itself" before finally being "righted by its own logic." Voyager had already begun to surprise its creators, and it had begun to learn.[1]

The anxiety attack continued in other manifestations, felt more intensely by its human controllers than by Voyaget 2. Signals indicated that the science boom had failed to deploy fully; but this was deemed a failure of the sensor, not the boom. Weirdly, a hydrazine thruster pointed wrongly at the spacecraft itself, which caused the vessel to drift off trajectory, which required more precious hydrazine, until the problem was recognized and repaired and course corrections made. Sensors began to fail (some seventy-two out of several hundred throughout the mission), all pertinent but none fatal. Then the infrared sensors deteriorated when bonding materials crystallized and distorted their mirrors after the protective sheath was removed, until a flash heater corrected the problem. There were odd pitches and yaws, never explained. As a result, several days of testing commenced before Voyager 2 could acquire its navigational guide star, Canopus. For their first two weeks both spacecraft underwent shakedown cruises in which they extended booms, tested instruments, and learned to communicate with a fast-receding Earth.[2]

Similar breakdowns continued throughout the mission and not all were mechanical; some of the worse stemmed from the failures of human controllers. The Grand Tour meant a constant negotiation between robot and human, with each learning from the other and at times compensating for the other's lapses. Voyager 2 overcame its makers' errors in failing to program for a jarring launch, and its makers overlooked the malfunction of its boom sensor and fixed its faulty thruster. Over the years that exchange deepened, though their mutual reliance, begun at launch, never disappeared.

## WAXING MOON

The Moon mattered. After IGY had field-tested the technologies of the Third Age, Earth and its Moon were the place where it took its ambitions. The Earth-Moon system was to the new geography of discovery what the Atlantic isles were to the First Age, and what a resurvey of Europe's interior by naturalists was to the Second.

Proximity was an obvious explanation. The Moon was the closest as well as the dominant object in the night sky, and space travelers advanced upon it step by step. The first Earth-launched satellites were identified as "new moons." The earliest spacecraft targeted near-Earth realms and the Moon. The primary rival to (and disturber of) planetary exploration, the Apollo program, aimed at the Moon. Colonizers imagined the Moon as a port of call for the solar system akin to that enjoyed earlier by Madeira and the Canaries to the Indies and Americas. Planetary spacecraft evolved from lunar Ranger to inner-planets Mariner to outer-planets Voyager. In 1969, in the months prior to the Apollo 11 lunar landing, Mariner 6 and 7 completed magnificent flybys of Mars, and while for true believers Mars mattered as much as the Moon, public attention and politics remained riveted on a lunar landing.

So it was a waxing Moon, not Earth's planetary siblings, that dominated the political sky. For planetary discovery to thrive, spacecraft, and mission planners, had to see the satellites of far planets, and couldn't do so if their eyes were flooded with the reflected light of Earth's. Then the Apollo program, its cold war mission completed, sank into near-parody as astronaut Alan Shepard hit two golf balls down the gray-powder fairways of Fra Mauro. The glittering vision of humans leaping beyond their home planet imploded into Skylab and Soyuz, the shuttle, and a hypothetical space station. Yet even as the shuttle began to drain NASA in a slow-death hemorrhage of money and talent, the planetary program blasted into its glory years.

As the Voyagers quickly rushed beyond the orbit of the Moon, they passed by the first planetary marker on their grand traverse. When the Third Age began, the Earth-Moon nexus was its Pillars of Hercules, the limits of humanity's practical powers and its blinkered imagination.

That setting served the Third Age as Europe's bounding seas had the First Age, and as the scientific rediscovery of its internal geography did for the Second when the grand tour shifted from art and classical literature to natural history. The botanical excursions of Linnaeus replaced a pilgrimage to the ruins of Rome, and the geological collecting of Christian Leopold von Buch amid the Alps superseded oils and watercolors of Mount Etna in eruption. In effect, the eyes of discovery turned to Europe, not to witness new scenes but to see old ones with new insight. The emergence of modern biology, geology, and ethnology revived the tired geography of Ptolemy into something fresh and vibrant. That experience was the essence of the Second Age, much of which had been known in some fashion to formal learning, and only a tiny fraction of which did not have indigenous inhabitants.

Each era of exploration mingled new discovery with rediscovery. The First could celebrate Madeira, previously unknown, and the Canary Islands, known to the Ancients as the Happy Isles, and even included in Ptolemy's *Geographia* and Pliny's *Natural History*. But before becoming a critical platform for launching the Great Voyages, the isles had to reestablish contact. Both Madeira and the Canaries were later rediscovered yet again by the voyaging naturalists of the Second Age, as the Enlightenment redefined what was interesting and important about them and how a rational nature might be understood. Joseph Hooker and Alfred Wallace saw the isles with very different eyes than those of Amerigo Vespucci and Pedro Cabral.

That was how it went: the reexploration that blossomed into the Second Age had learned its craft in Europe, and then propagated throughout Europe's emerging imperium, and beyond. But it had a special catalyst, and it may be well worth pausing to examine it. Much as the Voyagers coasted in orbit in order to reposition themselves properly for the final thrust of the Centaurs' RL-10 engines, so historical narrative needs to resituate itself from time to time. The final propulsion for the Third Age came from the International Geophysical Year, which blasted the era into a trajectory the Voyagers followed across the solar system.

Meanwhile, the two spacecraft sped past the nominal lunar orbit and so broke the bonds of imagination and political bureaucracy that had shackled the early years of space travel and that subsequently tethered NASA to an orbiting shuttle that, like a parodic Moon, only went in circles around Earth. One of the three pairs of hands that had held high the embryonic space program in that canonical photo of Explorer 1, von Braun's, was dropping out, and with him the vision of the Voyagers as robotic scouts blazing a Martian trail for migrating pioneers.

## THE GREAT GAME

Exploration, like war, was politics by other means. Especially when grafted to science, exploring ventures provided a cover for the more devious endeavors of the twentieth century's version of the Great Game. Yet that had been no less true for Kipling's Kim, who had found himself dodging Russian "geologists" in the Himalayas, as each imperial contestant sought less an understanding of mountains than the high ground of geopolitical advantage.

The American space program began under military sponsorship, and then was laundered through participation in the overtly civilian IGY and the creation of NASA; but by the time Voyager commenced its long trek, the program was seemingly returning to something like military oversight. In 1982, while Voyager 2 flew between Saturn and Uranus, military spending on space outstripped civilian by a widening margin. If the planetary program escaped this trend directly, as the space shuttle did not, it was a segregation more apparent than real, because so many indirect costs and breakthrough technologies built upon an infrastructure erected with military monies. The Soviet space program remained squarely under military command.

Exploration has long and often hybridized civilian and military personnel, vessels, and purposes. Most expeditions went armed, and most of those that enjoyed government sponsorship—which was the great majority of them—relied on naval vessels or army escorts and had an implicit military liaison if not an overt military purpose; overwhelmingly, expedition leaders had some military background. Not least among expeditionary goals was national security, and not least among the reasons that the peoples encountered by explorers proved so often hostile was the indigenes' suspicion that exploration proclaimed in the service of untrammeled curiosity could segue seamlessly into spying, which might announce a future invasion of soldiers, prospectors, settlers, and officials—and often did. While colonization remained the ambition of many enthusiasts, it had lost momentum in an age of earthly *de*colonization. Even *Star Trek* has imagined its exploring adventures as a benign arm of the United Federation of Planets' defense force, Star Fleet, while discarding imperialism and keeping its military in check with a "prime directive" that makes noninterference a standing order.

The uneasy alliance of military and civilian has taken many forms. One is simply that exploration can be a reconnaissance in force. Since exploration in the early days had to pay for itself, exploring parties had to trade or fight, or both. The "voyages of discovery" down the coast of Africa orchestrated by Henry the Navigator, the titular inspiration for the opening age of discovery, were, as Malyn Newitt observes, "openly and explicitly a series of raids designed to obtain slaves for sale or important 'Moors' who might be ransomed." When Vasco da Gama met resistance to his plans, he seized ships, bombarded cities, hanged hostages, and otherwise intimidated both potential enemies and trading partners. With sure martial instincts, Afonso de Albuquerque directed his voyages to precisely those pressure points that regulated traffic through the Indian Ocean–Aden, Hormuz, Molucca–all patrolled out of fortified naval bases such as Goa. Portugal remained in open war with Morocco and a cold war with Castile, and it is no surprise that members of military orders commanded its expeditions; its coastal fortresses were the military-industrial complex of their day. That sentiment extended even into the more scientifically ambitious surveys of the Second Age. The wide-traveling Russian explorer of central Asia Nikolai Przhevalsky openly declared himself a conquistador. "Here," he exulted, "you can penetrate anywhere, only not with the Gospels under your arm, but with money in your pocket, a carbine in one hand and a whip in the other." Here, "the exploits of Cortez can be repeated."[3]

Of course not everyone went with rifle and whip. Missionary priests fanned out along ancient routes of travel, established themselves even in venerable centers of learning such as China, became fluent in local languages, and reported on terrain, customs, and exotic curiosities, all done with the hand on a cross instead of a carbine. The Jesuits were founded as a missionary order, roughly the same year Coronado returned from his entrada across the American Southwest; and they targeted the new lands unveiled in the Indies. Francis Xavier preached in India and Japan and was preparing to visit China when he died; appropriately he was buried in Goa. When Charles-Marie de

La Condamine readied for his descent along the Amazon, recording exact coordinates by latitude and longitude, he was handed a map of the region by Father Samuel Fritz, a Czech Jesuit who had successfully traced out its major hydrography and settlements. Jesuit fathers, at immense personal cost, recorded the contours of the Great Lakes and Mississippi River.

The Second Age secularized these tendencies, substituting the proselytizing naturalist for the missionary friar, and found more private sponsors as old-money aristocrats and new-money industrialists occasionally funded expeditions as they might museums or universities. Some, such as John Ledyard or Thomas Nuttall, simply traveled, untethered to anything beyond their own wanderlust or obsession for plants or birds or beetles. Some accompanied private expeditions or had the wealth to sponsor themselves, such as Prince Maximilian of Wied into Brazil and later the upper Missouri. A few mounted major expeditions, of which Alexander von Humboldt's remains the exemplar and Ernest Shackleton's Imperial Trans-Antarctic Expedition a fitting valedictory. But expeditions that could not claim an attachment to national security or quick plunder were not likely to attract much support from national treasuries or commercial sponsors. Most serious expeditions had somewhere in their chain of causation at least a nudge, if not something more substantial, of military assistance.

The United States was no exception. It was a nation whose origins coincided almost precisely with those of the Second Age and whose national epic, an expansion westward, occurred through explorers outfitted with guns as well as sextants. Meriwether Lewis and William Clark were of course army officers, and their Corps of Discovery operated under military discipline. As the country grew, especially by war, the nation turned to its military to help wrestle those unknown lands into understanding if not control. The Army Corps of Topographical Engineers, organized in 1836, oversaw much of the grand reconnaissance that mapped out the bulk of America west of the Mississippi. The Corps of Engineers geared up for more in the aftermath of the Civil War, sponsoring two of the four so-called great surveys (with one under nominal civilian leadership). Moreover, in antebellum America, the United States launched fifteen major oceanic expeditions, which, however many "scientifics" they carried, were fleets under the command of naval officers.[4]

But equally significant is the peace that follows conflict. The immediate effect is a surplus of equipment and officers. The vessels will be mothballed, sold, or junked, and the officers cashiered, confined to dismal barracks duties, or placed on impossibly long lists for promotion. An obvious solution is to put them all to the service of exploration. The British Admiralty after the Napoleonic Wars sponsored globe-circling surveys for several decades. As second secretary to the Admiralty John Barrows put it: "To what purpose could a portion of our naval force be, at any time, but more especially in time of profound peace, more honourably or more usefully employed than in completing those details of geographical and hydrographical science of which the grand outlines have been boldly and broadly sketched by Cook, Vancouver, and Flinders, and others of our countrymen?" The resulting outpouring scouted from the Great Lakes of Africa to the Ross Sea of Antarctica.[5]

The United States had surpluses following the Civil War, some of which it redirected to western exploration. After World War II it donated a goodly portion of its otherwise dry-docked fleet to the service of earth and oceanographic sciences. Perhaps the most astonishing expression was the U.S. Navy Antarctic Developments Program, Task Force 68, otherwise known as Operations Highjump and Windmill, in 1946–48, in which the navy dispatched thirteen ships, including an aircraft carrier and a submarine, ostensibly to survey an unmapped Antarctica, but equally to test its fleet against polar conditions that replicated those they might face against the country's Arctic rival, the USSR.[6]

The fact is, exploration by itself cannot long command the political will and cultural commitment it needs unless it can rally deeper justifications than curiosity, appeals to a genetic imperative, hyperspace rhetoric, or very expensive (and often arcane) science. The relationship between civilian exploration and military reconnaissance, that is, has been symbiotic. Exploration has repeatedly been a means of beating swords into plowshares, and ICBMs into vessels of discovery. But a perceived need for national security, and a military infrastructure, has often proved a useful and necessary incentive to encourage exploration. It is a strategy the military has repeatedly recognized and even sought. Those plowshares can serve as sheathed swords.

There was no direct military involvement with Voyager: no funds, no seconded personnel, no secret experiments. Its singularly civilian control places it within a small fraternity of major expeditions that have been wholly autonomous from the military. But there was plenty of covert context, beginning with JPL itself.

The Jet Propulsion Laboratory that designed and oversaw Voyager could not have thrived only as a dedicated laboratory for civilian spacecraft. Its origins lay in amateur and academic enthusiasms, but it became permanent when the army endowed it as a facility for experimenting with jet propulsion and then with rockets, and its future depends on a partial rerooting in that same soil. When it transferred to NASA, funding became more episodic, which left staffing unstable and technological know-how far from cutting-edge. Repeatedly, the lab has sought outside support, almost always governmental and eventually military as the only large and relatively reliable pool of money. While defense funding never dominated the budget, it was a thumb on the scale that made the final weightings balance. JPL could assemble Voyager without military money; but beyond the Grand Tour, it could not likely survive, particularly with changes in national administrations hostile to aerospace outside Pentagon control. Military funding frames the JPL story before and after Voyager, but the plucky Voyagers slipped through unscathed, as they did the Van Allen radiation belts.[7]

As it showed with IGY and NASA, and as it had done with the U.S. Geological Survey a century before, America preferred where possible to segregate its civilian from its military institutions. When scientific inventories had greater claims than absorbing new conquests, the nation gladly turned to civilian agencies. It sublimated the cold war into an International Geophysical Year and tweaked military rocketry into NASA. By such maneuvers the Third Age relocated the Great Game from the Pacific Ocean to the Sea of Tranquility, and from the Khyber Pass and the Himalayas to the Valles Marineris and Olympus Mons. Almost always there was some military money in the chain, if only covertly.

The cold war concentrated its rivalries on Earth and its Moon, and secondarily on Venus and Mars; and by avoiding all of those bodies, Voyager seemingly escaped the implicit militarization of space. Yet it was there, its presence a perturbation in the political field, like those massive if invisible bodies that ripple through and bend gravitational fields. Voyager felt the tugs of them all. But it could bypass the ideological fixations of the inner planets, and it could escape the Sun-like attraction of the cold war. It had the pull of the Grand Tour and the push of momentum from a golden age that with an assist from Jupiter, would give it an escape velocity to take it beyond its own times.

## LOOKING BACK

On September 18, two weeks after launch, some 11.65 million kilometers from its home planet, Voyager 1 turned its cameras back to its origins and took a photo of Earth and its Moon together. It was the first of the Voyagers' stunning images, and it reminded observers that however far the Voyagers traveled, whatever new worlds they might discover, they always looked homeward as well and spoke ultimately to Earth. By so doing they challenged not only the bounds of our understanding and ambition, but also our sense of what exploration itself might be.

That haunting image recalled how much Voyager shared with other exploits of its era, and how much it differed. The earliest images of Earth from space were photographic maps from weather and surveillance satellites and lunar missions. A breakthrough moment occurred with Surveyor's first lunar orbiter. While it circled the Moon to photograph potential sites for Apollo, controllers adjusted the camera to take two shots that had the Moon as foreground and Earth behind. The images were not part of mission orders, and required tinkering with the spacecraft's attitude to reorient the lens. The maneuver was a "calculated risk," and when it worked, it impressed officials powerfully. This was something the public could understand. We could see ourselves from the Moon.[8]

Apollo 8 repeated the Surveyor shots in color and created one of the genuinely iconic images of the space age. But a required change in attitude was not restricted to the spacecraft: the Apollo program to the Moon went nowhere, withdrawing to the virtual solipsism of the space shuttle and a near-Earth space station. The great images of Apollo had been scenes of self-reference. That majestic earthrise was one, the blue and white gem of a living planet looming over the horizon of a dead Moon and capturing precisely the closed-circuit character of the enterprise. From the beginning the point of discovering new worlds had been to improve the old one.

Yet the Voyagers did something different, even as it remains a truism that their messages were for us and that even their gold-plated records with a greeting to some hypothetical Other beyond the solar system were in reality a dispatch to ourselves about how we wanted to be seen. Voyager widened the perspectival field. It put Earth into a planetary context, pushed beyond the range of rediscovery, and promised a reference outside our self-image, of ourselves seeking to portray ourselves as exploring. Voyager could not take pictures of itself, nor bring back images others had made of it. When it was represented for public display, the constructed field of vision typically came from behind the spacecraft as it looked toward its discoveries, not from the discovered scene looking at a spacecraft posturing as an explorer.

That first look back was a gamble, if a benign one. It might cause an instrument failure or jam the scan platform (as a test a few months later did) in ways that could compromise the mission before Voyager even reached the edge of the Grand Tour. But perhaps the greater hazard was the potential for narcissism, that the mission might look too far inward instead of outward. This peril also passed. What spared Voyager from preciousness and self-absorption was the sheer immensity of its geographic sweep—a trek too vast to remain within the frame of Earth and its Moon—and its existence as a robot, which shifted the focal plane of its cameras from itself.

Voyager compelled us to move beyond ourselves—that is what great exploration has always done. If it occasionally turned around, the intent was to measure how far it had come and to shift our perceptual frames. Eventually it would move so far away that Earth could no longer be seen at all.

DAY 29–94

# 7. Cruise

Their journeys were chronicles of spasms and calms. The flurry of launch and the frenetic bustle of planetary encounters were mere moments amid long lulls in the effective void of space, ruffled only by gusts of solar wind, the stray meteorite, and the subtle nudges of gravity. The interplanetary seas probably consumed 98 percent of the voyage, and beyond Neptune, all of it. Magellan's crew had found the wearying calm of the ocean they accordingly named "the pacific" to be maddening; but the weeks of Pacific tedium were nothing compared to the emptiness of interplanetary space.

Like far-sailing ships, spacecraft had their routines, and they attended to onboard maintenance. Each Voyager took about two weeks after launch to complete its circuit of instrument tests and navigational orientations. From time to time there were midcourse corrections to realign trajectories. Voyager 1 had two during its shakedown cruise; both spacecraft continued with minor corrective burns almost monthly.

The long cruise continued. On December 15, 1977, Voyager 1 passed Voyager 2. Some 445 days remained before it would make its closest encounter with Jupiter.

## CAPTAIN VOYAGER

In 1632, having crossed the South Atlantic six times and the North Atlantic twenty, sometimes as a crewman and some as a captain, Samuel de Champlain set forth his conceptions of the ideal expeditionary commander, or what he termed the good seaman. He should be "above all" a "good man, fearing God." He would not blaspheme, would attend to liturgical duties, and if possible retain a "churchman" to keep his crews "always in the fear of God, and likewise to help them and confess them when they are sick, or in other ways to comfort them during the dangers which are encountered in the hazards of the ocean."[9]

The captain should be robust, alert, inured to hardships and toil, "so that whatever betide he may manage to keep the deck, and in a strong voice command everybody what to do" and should do so as "the only one to speak, lest contradictory orders, especially in doubtful situations, cause one maneuver to be mistaken for another." He should share dangers, duties, and rewards. He should be able to eat whatever the circumstances permit. He should avoid drunkenness. He should not delegate by default. He should know everything that concerns the ship, for "great care and constant practice" are the means of safe passage. He should have practical experience "in encounters and their consequences." He should be "pleasant and affable in his conversation, absolute in his orders, not communicating too readily with his shipmates, unless with those who share the command. Otherwise, not doing so in time might engender a feeling of contempt for him." He will punish evildoers, reward achievers, and avoid occasions for envy, "which is often the source of bad feeling, like a gangrene which little by little corrupts and destroys the body," and can spark outright conspiracy or worse. He should, in brief, lead by both skill and example.[10]

The nature of geographic exploration left little slack for fumbling or fussiness; and to the extent that the perennial attraction with exploration resides with the character of the explorers, not simply with destinations reached and collections gathered, how they achieve their goals has as much interest as whether they reached them. What exploration could not tolerate was irresolution in purpose, and what leaders could not overlook was insurrection within the corps. Yet uncertainty was the norm and unrest a commonplace, the one often inspiring the other. Unable to locate the Strait where forecast, Magellan nearly fell to mutiny at Puerto San Julián in Patagonia. Henry Hudson, endlessly promising successes he could not deliver, was set adrift by crews in his eponymous Bay. Inadequate leadership was as lethal as rebellious crews; expeditions with weak captains could expect to suffer and fail like Bering's second voyage to Alaska, the crew shipwrecked and wasting away on Bering Isle, practically within sight of their destination port on Kamchatka. Expeditions led by indomitable captains were almost certain to succeed, no matter the cost, as Henry Stanley's extraordinary traverses through Africa demonstrate.

People judge exploration by the explorer as much as the things explored. The perennial fascination is not simply with destinations reached and collections gathered, but with how explorers achieve their goals. By the self-conscious end of the Second Age, even explorers openly accepted this standard. Apsley Cherry-Garrard described his intrepid band as "artistic Christians," as keen to test themselves with winter journeys around Cape Crozier as to unveil the mysteries of evolution as expressed in penguin eggs. The greatest expeditions had both great personalities and fabulous quests. Which is why the Third Age stands so awkwardly.

What is the personality of Voyager? What is the meaning of its leadership? Where is the commanding presence when a captain is called upon not to stare down a smoldering mutiny but to upload a software patch? Where is the moral drama of exploration? Where is that second, latent act of discovery, the unveiling of character? Who can be named as leader when a single Voyager of discovery will continue across a score of bureaucratic and careerist lifetimes? What might be the defining traits of great explorers—their invincible will, their curiosity, their passion for fame—was here hardwired into computers. Yet Voyager has variants of all these features.

What an act of transfiguration requires, however, is what many engineers and most philosophers disdain: anthropomorphizing a robot, even if only indirectly as a device by which to refract beliefs, passions, and intentions. It means having Voyager stand as a cipher for the hundreds of people and dozens of leaders who assembled, remotely navigated, and advised the spacecraft, and who shared its encounters with new worlds—who could supply the perception that pixilated images could merely copy. It means investing in a machine such virtues as fortitude, resolution, and discipline. There would be no impressive, fallible, undaunted explorer to stand on the quarterdeck or the mountain pass and issue commands, record first contacts, and meditate on the panorama before him. The explorer was being replaced by the public he traditionally reported back to.

Its small rebellions, blown circuits, and cranky scan platforms remind us that the Voyager spacecraft did not make itself, nor was it birthed from cyborgs. It was a very human creation, and as much as a painting or a novel or a political constitution, it embodies the flawed character of its creators. To give expression to such a machine was an act of imagination and passion, and to guide it was leadership of a high order. Ultimately it was the project managers who commanded: Harris "Bud" Schurmeier (1970–76), John Casani (1976–78), Robert J. Parks (1978–79), Raymond Heacock (1979–81), Esker Davis (1981–82), Richard Laeser (1982–87), Norman Haynes (1987–89). They got Voyager to Neptune and pointed it beyond, where another succession of project managers is today overseeing a vastly reduced crew staffing a vastly diminished vessel as it plunges into the new Sea of Darkness that lies beyond the solar system.

What was Voyager? To those who built and guided it, *they* were Voyager. They made it happen; they told it what to do and received its responses; they interpreted its discoveries. And yet Voyager was something more. It existed with or without them; they had to grant it a degree of autonomy; and they sent it where no one truly knew what it might encounter. It was capable of reprogramming, which is to say, of learning and discovering. It has lasted far beyond its life expectancy, which was guaranteed only to Saturn, but which proponents hoped would extend to Uranus, and might, just might, last to Neptune, but which is persisting through the heliosphere.

Voyager did things no one predicted, found scenes no one expected, and promises to outlive its inventors. Its autonomy did not reside solely in its primitive software and magnetic tape memory, but in its capacity to inform, and inspire, and to force us outside the ordinary. Like a great painting or an abiding institution, it has acquired an existence of its own, a destiny beyond the grasp of its handlers. If it had no consciousness of itself, it could provoke consciousness in others. The mission, with its implausibly grandiloquent Grand Tour completed, continues; the spacecraft, so obviously a piece of engineering, endures as art; a project sold as science persists as saga.

## JPL AS EXPLORING INSTITUTION

The deeper threat on long voyages, however, was a loss of discipline. The problem was not so much insubordination by the onboard computers as inattention or slacking on Earth. It was one thing for a spacecraft to coast; another for its commander, in this case, mission control at JPL, to go on autopilot.

At Pasadena the sense congealed that the crisis had passed. The Voyagers were launched, they were working, they would cruise through a veritable void for eighteen to twenty-two months before beginning their first encounters with Jupiter. The tension lifted. Attention turned to the next project. Operations relaxed, planning fell behind, anomalies passed by without prompt action. In December 1977 a tricky maneuver was aborted when it demanded a management decision that no one on the flight-control quarterdeck could give. In April 1978, with mission attention directed to malfunctions in Voyager 1, the officer of the watch forgot to send the mandatory weekly message, which prompted Voyager to default to safe mode. There were good reasons why classic expeditions adopted military discipline.[11]

Such breakdowns forced a NASA-mandated reorganization upon JPL. If it was to function as an exploring institution, it had to recognize that operations were continuous, not limited to launches and encounters, and that even quasi-autonomous spacecraft required a vigilant chain of command. Orders had to be restated, adjusted to circumstances, and enforced, even if they took the form of software patches and uploads. While Voyager could not put into port for an overhaul, JPL's mission crew could. Even as the spacecraft sailed on, the administration was in effect careened and scraped.

The fact is, while the Voyagers might travel in a vacuum, they were not created in one; nor could ideas, however heroic, express themselves of their own volition into scan platforms and antennas. They were the work of people organized into institutions. With particular force, Voyager was an enterprise of the Jet Propulsion Laboratory. It may well constitute JPL's finest hour. But when JPL failed to meet routine obligations, the episode not only threatened the mission but exposed the peculiar status of JPL within NASA.

The facility began in 1936 when a small group of enthusiasts organized themselves around Hungarian émigré Theodore von Kármán, a founder of aerodynamics, then a professor at the California Institute of Technology. The Guggenheim Aeronautical Laboratory, as it became known, evolved into a major center for rocketry and, as Clayton Koppes observes, "could plausibly claim that they–not Robert Goddard or the German V-2 experimenters–laid the foundation for the development of American rocket and missile technology." In 1944 Caltech operated the embryonic facility for the U.S. Army Ordnance Corps, one of a growing archipelago of labs run by universities under military contract. The facility changed its name to the Jet Propulsion Laboratory.[12]

But while JPL developed the nation's first tactical nuclear missiles, Corporal and Sergeant, a postwar status as a military shop became less attractive, particularly granted JPL's association with Caltech. The lab's real passion was for space exploration. As early as 1945, its WAC Corporal succeeded in escaping Earth's atmosphere, a first. It supplemented missiles with payloads and, together with von Braun's group, launched Explorer 1, America's first successful satellite. Once it was amalgamated into NASA, the way to the Moon and the planets opened, and JPL sent by itself or assisted in sending Ranger, Surveyor, Pioneer 4, Mariner, Viking, and Voyager.[13]

Its partisans early appreciated that propulsion was only as good as guidance: that was the difference between a rocket and a bomb. When it was reorganized in 1944 into JPL, the lab began acquiring expertise in tracking, telemetry, and the kind of communications and electronics that made missile guidance possible. The program flourished under William Pickering, a New Zealander with a physics PhD from Caltech, then a professor in its electrical engineering faculty involved with IGY's Upper Atmosphere Research Panel and early plans for an orbiting satellite, and finally director of JPL from 1954 to 1976. When military (and later NASA) support for rockets went mostly to von Braun's group, JPL reoriented itself to emphasize satellites and guidance systems. Its research into solid-fuel rockets may have given the lab liftoff among space institutions, but it thrived by designing spacecraft and the electronics that made it possible to communicate with, guide, track, and instruct the payloads once off Earth. And travel to other worlds was what JPL's staff wanted.

As Oran Nicks recalled, JPL was from the outset "aching to begin planetary missions," and it started with imagined missions rather than with existing capabilities. It would craft technology to suit missions, not limit missions to what was presently possible. JPL soon strengthened its communication capabilities with an eye beyond the sublunary realm. (Revealingly, it was the lab's chief of guidance research, Eberhard Rechtin, who proposed a "visionary program of lunar and planetary missions" and, unauthorized, commenced improvements in antennas.) In July 1958 JPL submitted the formal request to back that vision with a "Proposal for an Interplanetary Tracking Network." Five months later JPL proposed to a still-groggy NASA that it become the agency's "major" facility for space flight, or as Director Pickering expressed it, "the national space laboratory." Shortly afterward it dispatched for NASA's approval Project VEGA, an upper-stage rocket and spacecraft, along with a bold package of probes to the Moon, Mars, and Venus; a meteorological satellite for Earth; and a small clutch of unspecified interplanetary missions. In January 1959 NASA accepted JPL's concept for a deep-space communications network, and in March it approved Project VEGA, partially substantiating JPL's bid as the prime center for extra-lunar space flights.[14]

The euphoria couldn't last. VEGA got canceled, JPL sat awkwardly within the NASA administrative matrix, and after the elation over Explorer passed, projects stumbled. The lab faced rivals both inside and outside NASA. Director Pickering might proclaim that "it is the U.S. against Russia, and its most important campaign is being fought far out in the empty reaches of space," and JPL might declare itself the nation's primary institution for pursuing that contest throughout the solar system, but the reality was that the lab faced competition within the United States, and especially within NASA, as intense as any from its Soviet counterparts.[15]

The difficulties were several, both institutional and intellectual, and they began at conception. The first lay in competing visions of "space"; but the most serious—the bureaucratic version of original sin—dated from the organic act of July 29, 1958, which created the National Aeronautics and Space Administration and forced a merger of the erstwhile National Advisory Committee for Aeronautics labs with upstart facilities, mostly military. JPL now found itself competing with NACA's old Ames Research Center, NASA's new Goddard Space Flight Center, and even Langley Research Center, all of which had charges and ambitions with regard to satellites.

Unlike the others, JPL also had a costly and confusing affiliation with a university. From NASA's perspective Caltech extracted maximum fees for minimal supervision. JPL had a managerial style that placed it outside the bureaucratic norm: its university connections, which were vital for many of its participants, conveyed a sense of academic freedom, intellectual élan, and institutional insouciance that budget-conscious and politically harassed bureaucrats at NASA headquarters found often annoying and occasionally dysfunctional. JPL's sense of itself and its mission did not always agree with NASA's. It all might be tolerated so long as JPL performed well. When Ranger spacecraft after Ranger failed, NASA stepped in. It would do so again and again, even during the Voyager mission, when it perceived that JPL's demands for autonomy and Caltech's nominal supervision interfered with programmatic needs.[16]

There was an alchemy at work, but whether of white magic or black depended on perspective. NASA wanted more programmatic discipline, JPL sought standing as an intellectual institution not simply a government job shop, and Caltech coveted the funds but worried about how to reconcile a contract lab with university purposes, particularly when military research continued at JPL, as it did, and brought with it requirements for secrecy. Defense spending would in fact increase as NASA's budget plunged, a situation that worsened when the Reagan administration sought either to shift space onto the military or to privatize it. The three parties found themselves in a state of more or less constant turmoil. Major contract negotiations occurred in 1964 and 1966 as the planetary program began to hit its stride. They continued throughout the Voyagers' traverse across the solar system.

Such confusions and contests have been common over the centuries and reflect the culture's changing understanding of what exploration means and why it is done.

The Great Voyages had commercial and geopolitical purposes. There was, on one hand, an urge to open up and unleash discovery on the theory that the first comer would grab the most spoils, while, on the other, a concern that exploration serve its society, which argued for a controlling agency. Portugal placed oversight with the Concilho da Fazenda (Treasury), into whose vaults the secret discoveries were entrusted. Spain gave primary responsibility to the Casa de Contracción in Seville. Both emphasized the commercial significance of voyaging and the implacable interest of the state. Competitors might unload duties onto chartered bodies such as East India companies, a Company of Adventurers, or outright privateers; and in the Second Age, scientific associations could become promoters and occasionally sponsors. Still, expeditions rarely diverged from government interests, nor did state-sponsored exploration stray far from military support, if not leadership. The major exception is the heroic age of Antarctic exploration that concluded the Second Age. But whether within or between countries, there was competition aplenty.

There was precedent for NASA's dilemma during the Second Age, when exploring institutions proliferated and began to trespass on one another's turf. Instead of interplanetary space, the contestants sparred over the near-nation Great Plains and intermountain space of America's unexplored West. In the nineteenth century the United States had turned over most of the responsibility for exploring its continental acquisitions to the U.S. Army, and for oceanic probes, to the U.S. Navy. From 1836 to 1863 the army created a Corps of Topographical Engineers, which organized expeditions, surveyed in the company of naturalists and artists, and prepared composite maps, until the Civil War directed its talents to other landscapes. After the war, the army was keen to renew its role, even as civilians were claiming priority for more science-based institutions. What emerged were four separate programs that became known as the Great Surveys, each with a different sponsor. They soon began to crowd into one another.[17]

King, Wheeler, Hayden, Powell—each leader's name became shorthand for his survey, and each survey conveyed a style as much as a geographic locale. One emphasized high-caliber geology, one cartography, another natural resources, and another land reform. One advertised the Yellowstone and galvanized Congress into creating America's first national park; another, Grand Canyon; another, Death Valley. They appealed to commercial lust, national security, and cultural pride; they enlisted art to promote where science couldn't. Their rivalry was ferocious, not only between military and civilian, but among civilians, and often among scientists, who disputed what constituted real geology and which group made proper maps, which institution of higher learning deserved the government's patronage and which did not, and who was a genuine scientist and who a mere celebrity. So long as the West was open and money plentiful, each survey could find popular partisans and political champions in Congress, and take to the field year after year.

But after they began to stumble over one another, after the sleaziness of the Grant administration was replaced by an emphasis on sobriety, reform, and retrenchment, and after a second, more scientific grand reconnaissance had completed its survey of the West, the exuberance for unbridled exploration waned and some consolidation was mandatory. A political brouhaha ensued, with scientists fighting as ferociously against one another as against military dominance. Each survey denounced the others as a disgrace or an outright fraud. The outcome: the establishment of the civilian U.S. Geological Survey in 1879, with Clarence King as director.

The King Survey had, in truth, proposed the better hybrid of interests, and Clarence King himself, "as Yale man, adventurer and clubman, litterateur, scientist, and exposer of the Diamond Hoax," was the epitome of "that peculiar alliance between very big business, the socially acceptable intellectuals, and the advocates of limited reform" who finally rallied behind the chosen program. That scroll should sound familiar; such were the syntheses of personality and programs that allowed a comparable compromise to proceed a century later.[18]

The similarities then and now are striking—and misleading. The greatest is timing: the U.S. Geological Survey helped close out an era of frontier exploration, while NASA helped to open one.

Consolidation in 1879 purged the resulting institution of much of the conflicts that had riven the Great Surveys; it moved the enterprise from geographical survey to formal science; its charter closed a debate that had flourished for a decade. In 1958 consolidation drove external rivalries into NASA, an agency kindled into being by impulsive reactions to Sputnik, without close argument and decades of consensus. When funding shriveled, those absorbed tensions could become unbearable. In 1879 the country was able to continue demobilizing from the Civil War. In 1958 the cold war retarded a full transfer from military to civilian operations. By the 1979 centennial of the USGS—in fact, while Voyager 1 was approaching Jupiter—the system was poised to reverse that trend, and under the Reagan administration to remilitarize America's space program. Voyager threaded through that gap as though it were a historical Strait of Magellan.

JPL had made Voyager happen. Its staff had discovered the potential for the Grand Tour, had recognized in gravity assist the propulsion to yield the necessary trajectories, had designed a durable and redundant spacecraft, had equipped the robot with the semi-autonomy required to operate at the edge of the solar system, and had granted Voyager a persona—had stamped it with a JPL style. More than anything else, JPL had simply made Voyager possible. It had argued for it, fought for it, brought it back from the dead, and ultimately willed it into being. Voyager has traveled in a JPL manner, and its trek speaks in a JPL idiom. The quirks and powers and personality of the one are those of the other.

Yet that observation can be turned around with equal force. Whatever becomes of JPL, whatever stresses tear at its peculiar institutional arrangements, whatever future missions take shape or dissolve, the Voyager saga will immortalize JPL in the history of exploration. VGR-77 is a kind of Pygmalion story in reverse: the creation, begun in very mortal flesh and then preserved in imperishable form, that will outlive its creator.

DAY 95–124

# 8. Missing Mars

The cruise might have ended, as so many planetary expeditions have, at Mars. More than anywhere else, Mars had fused the culture of exploration with the culture of space, distilling and distorting each. Here the boosters sang loudest, and the scorners scoffed longest. In the Viking mission to the Red Planet that preceded it by a year, Voyager had its greatest competitor and a rival vision. If the goal of the Third Age was Mars, then Viking could well stand as the era's grand gesture. But exploration had greater unknowns to plumb than the Red Planet.

In November and December 1977, the Voyagers slid past the orbit of Mars without a twitch. Their meeting with Mars was a nonencounter; but it was rich with significance nonetheless, for what Voyager did not do—was never intended to attempt—could be as revealing as its announced goals. In seeking newer worlds, it bypassed the world that has most mesmerized the imagination of space partisans, that best expresses their effort to control the direction of Third Age exploration, and that best boils down the motives of those who have most fervently wished to project exploration into space and those who have most doubted its value.

## MOTIVES

The Voyagers were machines on a mission. In their choice of instruments and in their design, in the passions and intentions of their creators as coded in trajectories, they carried a legacy of Western exploration. In their motivations, too, there were continuities as well as disconnects.

The Great Voyages had expressed their times. They were as much a part of the Renaissance as its commercial bustle, flamboyant arts, endless warring, and renewed learning. Over and again, explorers and their sponsors repeated the same trilogy of reasons to justify expeditions. They went for gold, for God, and for glory. They went to get rich, and thus acquire power. They went to spread the Gospel, weaken religious rivals, and generally ennoble the spirit and enhance the intangibles that endowed life with meaning. They went for fame and status, to rise in their societies and to become known and to have the future look upon them as they did the demigods and heroes of the past. Geographic exploration was an amalgam of quest, crusade, and commerce, with discovery and new knowledge as a means to those ends. With remarkable tenacity these reasons, or their reembodiments, have persisted.

The sponsors and captains of the First Age made no effort to disguise their purposes. Henry the Navigator sought new Madeiras, the gold and slaves of Africa, and a further means to wage Portugal's perpetual conflict with Morocco. Columbus's Enterprise of the Indies was founded on a premise of gold, which is "most excellent; of gold there is formed treasure and with it whoever has it may do what he wishes in this world and come to bring souls into Paradise." Gold was the font of all other motives: Columbus promised he would find as much of it as his monarchs would "require," though the forecast lands of bottomless gold were always a bit "further west." When Vasco da Gama arrived in India, he declared he had come for co-religionists and wealth. The expedition had sailed to establish trade, by force if necessary.[19]

In his epic retelling of that voyage, Luiz Vaz de Camões includes long passages on both fame and greed. The "giant goddess Fame" was

*Hot-blooded, boasting, lying, truthful,*
*Who sees, as she goes, with a hundred eyes,*
*Bringing a thousand mouths to propagandize.*

But it was gold that "conquers the strongest citadels," that turned friends into "traitors and liars," debauched nobles and maidens, and that could buy "even scholarship." The blinding, mesmerizing, corrupting, all-commanding power of gold could pervert; it could also drive men to astonishing deeds. Marching with Cortés to Tenochtitlán, Bernal Diaz noted simply that they came "to serve God, and also to get rich," for "all men alike covet gold, and the more we have the more we want," and it was the dazzling gold of the Aztecs that kept them fighting.[20]

Expeditions had to pay for themselves, if not by trade, then by looting. The prospect for plunder drove Iberians across the New World in search of further Mexicos and Perus, and that example inspired repeated forays by Portugal into Africa and prodded other nations to emulate them. Richard Hakluyt, for example, recounts how "certain grave citizens of London, and men careful for the good of their country," upon "seeing that the wealth of the Spaniards and Portuguese, by the discovery and search of new trades and countries was marvelously increased, supposing the same to be a course and means for them also to obtain the like, they thereupon resolved upon a new and strange navigation," in this case three ships to discover a northern route to Asia—which ended at Muscovy, but which flung greedy would-be discoverers around the littoral of the Ocean Sea.[21]

In truth, the New World would require a couple of centuries to become commercially self-sustaining; until then, its preponderance of trade goods were luxury items, plunder, and drugs, with the occasional tourist junket. Colonization was a distraction. Fantasies dissolved upon actual contact. After the glowing visions conjured by Columbus's first voyage, for example, everyone wanted to go to Hispaniola, "a land to be desired and, seen . . . never to be left," and after his second, no one did. Trade produced wealth; colonies siphoned it off. What Europe wanted was riches, and what it sought from its explorers were routes to get them.[22]

Yet the intangibles mattered, too. They all professed to serve the Cross, or the prestige of their monarch, and their own fame. The most successful had a sense of personal destiny, of themselves as ciphers for a cosmic purpose. Gomes Eannes de Zurara, chronicler of Henry the Navigator, explained that "the reason from which all others flowed" was Henry's faith in his horoscope, which had declared he would make "great and noble conquests" and "uncover secrets previously hidden from men." Columbus gave such sentiments a Christian baptism by which he believed himself to be God's instrument, and the discovery of a new route to the Indies, his destiny. He had no choice but to proceed, for by doing well by himself he would do good for all Christendom. (A remarkably similar conviction kept a fever-stricken David Livingstone in his traces.) Their personal purposes were sanctioned by their assertion of a larger consummation that both compelled and justified their implacable ambition.[23]

Still, the sum of greed, pride, and devotion does not seem enough to kindle the fuels that amply lay about. Some other spark had to set those piles ablaze. Despite endless attempts, historians have fared no better at snaring it than Renaissance mariners did in plying the Northwest Passage. In reviewing the Great Voyages to the New World, Samuel Eliot Morison asked, "What made them do it?" Was it, he continued rhetorically, "mere adventure and glory, or lust for gold or (as they all declared) a zeal to enlarge the Kingdom of the Cross?" He could not say. "I wish I knew." But he did identify as a common theme "restlessness," a physical and perhaps spiritual disquiet that led the restive to wander afield and, in the right age and with the right backing, become explorers. Felipe Fernández-Armesto proposed a more formal explanation: that "European culture" of the era was peculiarly "steeped in the idealization of adventure," that exploration was another manifestation of a "code of chivalry." The role models for such questers were "the footloose princes who won themselves kingdoms by deeds of derring-do in popular romances of chivalry—the pulp fiction of the time—which often had a seaborne setting. The hero, down on his luck, who risks seaborne adventures to become ruler of an island realm or fief, is the central character of the Spanish versions of stories of Apollonius, Brutus of Troy, Tristram,

Amadis, King Canamor, and Prince Turian, among others, all part of the array of popular fiction accessible to readers at every level of literacy in the fourteenth and fifteenth centuries." For some, then, glory might substitute for gold. Antonio Pigafetta ingenuously opened his account of Magellan's voyage by listing as his reason for travel, "that it might be told that I made the voyage and saw with my eyes the things hereafter written, and that I might win a famous name with posterity." He did.[24]

What happened with literature happened also with geography. Fabled places of legend—prodigious with lore, wealthy beyond avarice—kept appearing just over the horizon, and when not found there, reappeared at another horizon. Moreover, "at the margins, chivalric and hagiographical texts merged"; there was a "divinization" of old legends; the romance went to sea with sword and cross in an expectation of earthly and heavenly rewards. Such romance proved impervious to facts and deathless to actual discovery. It simply reincarnated and relocated, and where adventurers went, historians have followed.[25]

The Second Age secularized those motives and laundered them through the Enlightenment. A more aggressive commerce replaced simple plundering, a generalized Civilization substituted for Christendom, and glory softened into national prestige and professional reputation. The Enlightenment allowed science standing as a justification, such that new discoveries of nature could serve as the font from which all else might flow.

Consider Captain James Cook's orders for his 1768 voyage. He is first to perform his observations for the transit of Venus, this from Tahiti, an island barely discovered before his departure. He is then to proceed southward to "make discovery" of a new "Continent" (Terra Australis), for whose existence there is "reason to imagine." If that voyage falters, he is to search for the continent westward until he either discovers it or encounters "the Eastern side of the Land discover'd by Tasman and now called New Zeland." Whatever he found he was to render into detailed surveys, complete with collections of flora and fauna, and records of its inhabitants, if any. If uninhabited, he should "take Possession for His Majesty by setting up Proper Marks and Inscriptions, as first discoverers." If inhabited, he should "cultivate a Friendship and Alliance" with the indigenes and establish "Traffick." Columbus's Golden Chersonese has morphed into Terra Australis, and gold bullion into astronomical measurements, naturalist collections, and commerce. But the ambition for prestige has endured; and science and geopolitics have converged, as religion and geopolitics had earlier.[26]

The instructions written by Thomas Jefferson for Meriwether Lewis and William Clark tacked closer to Cook than to Cortés. More and more, Enlightenment learning became the public declaration of the motives behind expeditions and the means for understanding what they discovered, even if commerce followed the tracks of the explorers closely, as the Rocky Mountain fur trade did the two captains. Still, the trend was to deepen such exploits with science, and to replace individual senses of destiny with national ones. Thus America's century of discovery ends with a sprawl of Great Surveys that managed to combine adventure, science, and geopolitics in classic fashion, as the instructions to Clarence King demonstrate: "examine all rock formations, mountain ranges, detrital plains, coal deposits, soils, minerals, ores, saline and alkaline deposits . . . collect . . . material for a topographical map of the regions traversed, . . . conduct . . . barometric and thermetric observations [and] make collections in botany and zoology with a view to a memoir on these subjects, illustrating the occurrence and distribution of plants and animals." All this is a long way from looting gold and searching out souls to save; but it is a recognizable descendant, as Przhevalsky's horse is from *Eohippus*.[27]

Where wealth of any kind was unlikely, fame could substitute, as before. The Antarctic offered only frost and struggle, and perhaps death, but there was also glory promised to explorer and sponsor. Ernest Shackleton's celebrated advertisement in the London *Times* says it all: "Men wanted for hazardous journey. Low wages, bitter cold, long hours of complete darkness. Safe return doubtful. Honour and recognition in event of success." That was the voice of pure adventuring tied to a geographical goal, a traverse across the continent through the pole; this was General William H. Ashley's corps of mountain men heading to the ice sheets.

But the Second Age had come to expect more. Apsley Cherry-Garrard stated it blandly and abstractly by calling exploration "the physical expression of the Intellectual Passion." He urged those who have "the desire for knowledge and the power to give it physical expression, go out and explore," although not everyone would accept such endeavors as worthy. They would deny glory and disdain anything without an economic payoff. "Some will tell you that you are mad," Cherry-Garrard wrote, "and nearly all will say, 'What is the use?' for we are a nation of shopkeepers, and no shopkeeper will look at research which does not promise him a financial return within a year." Yet even here, even if one had to sledge "nearly alone" across the proverbial frozen wastes, there could be expectations of social recognition and a contribution to the civilization of science.[28]

Exploration had become an entity in itself: the explorer could claim honor apart from where he practiced his craft. One of the greatest explorers of the day, and Scott's rival to the South Pole, Roald Amundsen, did no science, but he did reveal geography and personified an ascendant Norwegian nationalism. For the larger culture exploring was, as Cherry-Garrard said of the "sledging life," the "hardest test." In the end it was enough that its practitioners explored their own character, and the larger civilization, its.[29]

The Third Age was different—and the same.

With astonishing tenacity, the classic rationales have persisted, not least because exploration has become a valued enterprise the culture is unwilling to discard. Gold takes the form not of wealth discovered but of wealth created through government-sponsored jobs, near-Earth satellites, and the scientific information that serves as the precious spices and bullion of an information economy. Glory has softened into national prestige, a hazy pride, and a devalued currency of the cold war. And God has permutated into vague yearnings for Something More—explanations for the fundamentals of existence, an atonement for lost virtues, and even, for some, contact with an intelligence beyond the solar system. Across five centuries, while the vocabulary of exploration has changed, its syntax has remained intact.

The biggest change is economic. Explorers do not head to known markets or seek to cultivate new ones—they do not probe new routes to great entrepôts or unveil natural resources along the routes they travel. No one can even pretend to imagine an Aztec or Incan empire to sack, or a bonanza of gold waiting to pan out of the lost streams of Mars, or a "traffick" in useful goods with the far Indies of the solar system, perhaps a triangular trade between Earth, Titan, and Ganymede. Private commerce is restricted to its offshore environments, the near-Earth of satellites and the occasional tourist, and the near-ocean of continental shelves rather than the abyssal plains.

A modern Muscovy Company would not load its holds with trinkets for trade to discovered peoples and draw up rosters of what they have that we want, and what we have that they want, but would determine what instruments to send for reconnaissance. What can the planets and their moons tell us that we want to know? What traffic can exist between our instruments and their data? How much bullion must we expend to get the cinnamon and rubies of knowledge? Instead of red ocher and black coney skins, vessels sail with ultraviolet spectrometers and plasma wave detectors. But no more than the merchants of the Great Voyages do scientists know exactly what will trade best. They bring what they have.

## BELIEVERS

Where there were motives, there were also motivators. Exploration has overflowed with publicists, seers, prophets, soothsayers, boosters, propagandists, the putatively all-knowing and farseeing, the promoters of fame and fortune, and the augurers of collective destiny. Through them the crassest lust for money and power could alloy with the most exalted aspirations. In some cases boosters kindled the necessary enthusiasms; but mostly they captured the sentiments of the zeitgeist and placed them into the holds of ships, on sledges, and into the software of spacecraft, for, in addition to food, water, navigational instruments, and maps, explorers needed a vision of where they were going and why.

The roster of publicists begins with the explorers themselves. Almost all were experts in self-promotion. No one believed in their mission more than Christopher Columbus and John Cabot, Henry Stanley and David Livingstone, and it was often the very intensity of their conviction that proved decisive in their convincing others. The greatest welded an unyielding faith to an iron will, the self-certainty that if the actual scheme proved wrong, their sense of destiny was right, and if destiny proved muddled, they could succeed by strength of character alone. Even when unexpected landmasses blocked passages, when rivers did not run or seas failed to appear where promoters had insisted they would, when the Isle of Brasil and the Seven Cities of Cibola migrated from one imaginary locale to another, they completed their expedition; they found what did exist and temporarily banished what did not, and returned with that knowledge. The poorest foundered on incompetence, ill luck, and the violent collision between a lofty rhetoric of inspiring visions and the hidden shoals of an indifferent geography.

Delusions could prove deadly. For every Ferdinand Magellan and Henry Stanley who simply would not be deflected from his self-assigned destiny, there was a Martin Frobisher and a William Baffin who could not force a Northwest Passage whatever their determination, and a Walter Raleigh and a William Paterson whose delusional bombast led to geographic fantasies and, for their followers, death. Even the greatest of explorers could not do what was impossible, and the impossible was as much a matter of historical timing as of principle. In prophecy, as in medicine, the toxicity resides not in the substance but in the dosage. Besides, explorers too often died; some other, more enduring mechanism had to keep the enthusiasm alive. Prophecy, too, needed its institutions.

That task fell to promoters. Someone had to persuade the monarchy or a Company of Adventurers to sponsor expeditions; someone had to rally both motives and money; someone had to speak in the media of the time to influence and inspire; someone had to record and interpret what discovery had unveiled and forecast what it might yet find. As exploration became institutionalized, it found its propagandists and publicists. During the First Age, they relied on the printing press, to which in the Second, they added newspapers and popular magazines, and in the Third, film and television. Each age, too, found distinctive ways to bond commerce and culture, to fuse and transmute mixed motivations from outright greed and vainglory to national prestige, and scholarship into a popular genre that could rouse a more general ardor without which voyages were no more than fireflies in the night. Each, that is, found a suitable Romance.

Over time the English became particularly adept at promotion. The history is worth tracing, since it leads directly to those who, apart from the Russians, have most boosted interplanetary exploration. The condensed version is this: England needed trade, which meant it needed explorers to identify places, goods, and routes.

In late medieval and Renaissance Europe, news of the earliest voyages of discovery was mixed. Some reports were quickly promulgated, such as Columbus's *Letters*; others were hoarded as state secrets. But if details were locked away, the general outlines of discovery became known, if only to ensure claims to lands, seas, peoples, and trade. At the Spanish court, Pietro Martire d'Anghiera, better known as Peter Martyr, became the official chronicler of Iberian discovery, his register updated as a series of *Decas* (*Decades of the New World*). It was Peter Martyr who appreciated, as Columbus never did, that the latter's four voyages had not made landfall on Golden Chersonese but on what Martyr termed an *otro* (or *novo) mundo*, a New World. The printing press helped to propagate the news of discovery, and something of its excitement, throughout the elite of Europe.

England, however, lagged. Though the first press arrived with William Caxton in 1477, it specialized in what the public wanted to read, and works of geographical discovery were not among them. Those sponsoring merchants who needed to know the latest voyages did so through other means; the reading public still preferred pilgrim guides and updated encyclopedias. Despite Bristol merchants, England's interests pointed to Europe rather than New Found Lands.

Not until 1554, with the crisis of Mary Tudor's marriage to Philip of Spain, did the *Decas* appear in English. It did so amid a political crisis, economic opportunities, and a spiritual call to arms for what would evolve into a prolonged struggle with Spain.

The critical voice was Rycharde Eden, England's "first literary imperialist." Eden began by publishing *A Treatyse of the Newe India*, arranged to have the *Decas* translated along with Martin Cortes's *The Arte of Navigation*, and urged his countrymen to learn from the major colonial powers and then expand for themselves, searching out new markets for the "commoditie of our countrie," which was wool. In dedicating the *Treatyse* to the Duke of Northumberland, Eden listed his reasons for urging travels of discovery. They could improve the national economy, they served God, and they could inspire both an entrepreneurial and an adventuring spirit essential to a robust people. Those who demurred from overseas enterprises showed, he felt, a contemptible lack of courage. The earliest endeavors focused mostly on the Northeast Passage, culminating in the Muscovy Company and trade with Russia, and then renewed interest in a Northwest Passage. As Eden's first book suggested, the Indies were the prize.[30]

As England quickened the tempo of its overseas probes, the basic rhetoric laid down by Eden endured: that only by joining the scramble overseas could England secure its future at home. Yet the governing classes and reading public remained largely apathetic. Not until 1580 did the tide begin to turn. In that year, England determined two routes to the wealth of the East: a trade treaty with Turkey and the astounding return of Sir Francis Drake from "the world encompassed," not least with the *Golden Hind* stuffed full of the plunder from unsuspecting Spanish ships and towns in the Pacific. More important, England found its great publicist of voyaging. That year Richard Hakluyt, then thirty years old, commenced his pleas for an English overseas empire of commerce and colonies.[31]

An older Hakluyt recalled that, as a much younger Hakluyt, he had chanced upon "certain books of Cosmography, with an universal Map" in his cousin's study, and that he then and there resolved that he would "by God's assistance prosecute that knowledge and kind of literature." His 1580 appeal on the occasion of Humphrey Gilbert's first try at an American colony, was followed by a 1582 compendium, *Divers Voyages Touching upon the Discovery of America*, in which he exploited historical events to argue the cause for English expansionism. *Discourse of Western Planting* followed, this time aligned with schemes by Walter Raleigh. More than a simple shill, Hakluyt was a Renaissance scholar, versed in many languages and intent on amassing accurate accounts from any source, and thus eager to talk with pilots, merchants, and captains. His masterpiece followed the year after Elizabethan England defeated the Spanish Armada.[32]

*Principal Navigations, Voyages and Discoveries of the English Nation* was an immense encyclopedia of travel that celebrated the new age of adventuring while seeking to establish continuities with an ancient English past. It argued that the English were, by heritage, a voyaging and trading people, that an expansion of enterprises was both possible and essential, and that accurate knowledge was the font of such inspiration. The *Principal Navigations* was in equal measure patriotic and practical. A second, expanded edition published in 1598–1600 weighed in at a million and a half words distributed over three volumes. Here was history in the service of commerce and politics.

The outcome could not help but awe and, through awe, inspire both admiration over what had been achieved and ardor to further it. Yet Hakluyt was candid about his purpose: "our chief desire is to find out ample vent for our woollen cloth, the natural commodity of this our realm, the fittest place I find for that purpose are the manifold islands of Japan and the northern parts of China, and the regions of the Tartars next adjoining." To this goal were gradually added colonies in America, where trade might be supplemented by the natural wealth of those lands. The second edition made the point even more explicit by adding "Traffiques" to the title.[33]

Others, such as Samuel Purchas (*Purchas His Pilgrimage*), were less fastidious about authenticating sources and less willing to subject readers to vast catalogues of prospective trade goods. They abridged the long chronicle into a brisker narrative, pumped up the emotional receipts relative to empirical expenditures, and bequeathed a saga that made the voyage of discovery and the planting of colonies a distinctive and necessary feature of the English identity. Complicated accounts of commercial traffic became tales of derring-do and patriotic glory. The exorbitant costs of exploring were only apparent: the enterprise would pay for itself many times over.

Yet for some journeys, mirages work as well as landmarks. Samuel Eliot Morison observed that "although the French seaports had every possible advantage for Atlantic exploration that the English had; although the French crown gave its merchant marine more support and encouragement than the Tudors ever did, there is one English asset which the French lacked—a Hakluyt." France lacked that enormous compendium, the comprehensive vision, that sense of urgent inspiration. Later French enthusiasts for a Greater France lamented the similar lack of Robinsonades—French versions of Robinson Crusoe, popular stories of overseas adventuring.[34]

In truth, Richard Hakluyt and his counterparts did more than chronicle an age of discovery. They placed it into a national (and civilizational) narrative, they created valences with other vigorous elements of the culture, they implanted it into the minds of the educated and governing classes. They helped institutionalize exploration. They ensured that the Great Age of Discovery could lead to others. Exploration became complex, and because of that cultural complexity, it could survive. It endured in part because Western civilization could no longer imagine itself not exploring.

During the Second Age, exploration found new prophets, new romances, and a more popular media. Not only had books become far more common, but also newspapers and popular magazines abounded to spread stories of discovery widely throughout a vastly more literate public. Particularly among former colonial societies such as a newly independent America, now bent on its own imperial project, explorers became iconic figures, founders of a national creation epic. They were American Aeneases, the guides and seers of national destiny. The frontiersman became the new knight-errant of a romance later called the Western. The explorer into unknown lands became a Romantic hero.

Their accounts were enormously popular. Let William Goetzmann tell how the reclusive Henry David Thoreau spent days "in his quiet study at Concord, at his cabin at Walden Pond, or stretched out under a poplar tree in his backyard behind the family pencil factory" reading exploration literature. A master of synecdoche, Thoreau could see the whole world in Walden Pond, and he could observe also in his journal that "the whole world is an America, a New World." So, too, we might have his personal absorption stand for that of his culture. He read Humboldt's *Personal Narrative*, Cook's *Voyages*, and Darwin's *A Naturalist's Voyage Round the World*. He read Alexander Henry's adventures in Canada, Thomas Atkinson's travels across Tartary, John Dunn Hunter's narrative of Indian captivity. He trekked across Africa with John Barth, Hugh Clapperton, and David Livingstone; sledged to the Arctic with Isaac Hayes and Elisha Kent Kane; traveled over China with Evariste Régis Huc; searched for Mount Ararat with Frederick Parrot; tracked down the source of the Mississippi with Giacomo Beltrami and Henry Schoolcraft; sailed down the Amazon and Orinoco with Lts. William Herndon and Lardner Gibbon; slogged along the Mosquito Coast with Ephraim George Squier; and accompanied Ida Pfeiffer on *A Lady's Voyage Round the World*. He read the five-volume narrative of the United States Exploring Expedition in its entirety, steering around Lt. Charles Wilkes's cloying prose as the USS *Vincennes* did Antarctic floes. He was, in William Goetzmann's words, "another of Humboldt's children."[35]

The alliance with Enlightenment science had created powerful incentives. What was good for science was good for empire, and what was good for empire was good for science, and both looked to exploration to advance their interests. The frontiers of science were, as its promoters never ceased to proclaim, endless. Institutions of science became boosters of geographic exploration and forums for announcing their findings, which quickly moved from the pages of obscure academic proceedings to the front pages of penny newspapers.

The prophets of exploration experienced another transfiguration for the Third Age. With renewed tenacity the old justifications were refurbished, given a high-tech gloss, and sent to explore ice, abyss, and space. A new genre of romance—science fiction, or more properly technological romance—reoutfitted the knight-errant with the armor of a spacesuit, a grail quest to find intelligence or at least life elsewhere, and adventures across galaxies. The HMS *Beagle* acquired warp drive; astronauts unveiled new Botany Bays on the moons of Saturn; a New Jerusalem arose on Mars. New media that substituted light and electrons for ink added to popular enthusiasms through film and television. The pioneering rocketeers must be numbered among the emergent seers, for so they saw themselves, and it was a vision of extraterrestrial exploration (and colonization) that drove their engineering imaginations. A Great Migration beyond Earth had inflamed both Tsiolkovsky and Goddard.

But the godfather of the Third Age in space is surely Arthur C. Clarke, who knotted together as no one else could the technical, the institutional, the literary, and the visionary. It is not just that Clarke was a type or a precursor, but that he was himself the dominant oracle for an age of space voyaging. Almost every defining feature of the space age finds echoes in his voice. As seer and revelator, his immense, and immensely successful, literary output became a kind of Testament of the Space Age. Born in Britain in 1917, he outlived all his contemporaries and even the first generation of successors. He was in Britain when V-2 rockets rained down on London, and he commented by teleconference on the Cassini mission to Saturn.

He could not afford college; he learned his craft by working and by writing science fiction in his spare time. During the Battle of Britain he served in the radar units of the RAF; after the war he earned a first-class degree in mathematics and physics at King's College London, and applied his literary talents to predictions of telecommunication satellites and space travel generally. In 1949 he committed himself to a full-time career as an author of short stories, novels, and semi-technical studies. Of necessity for one who lived by his wits, he evolved a graceful, popular style. In 1950 he wrote a technically informed but accessible book on astronautics, *Interplanetary Flight*. By then he was chair of the British Interplanetary Society, an institution devoted to promoting space travel, a successor to such Second Age organs as the African Association and Royal Geographical Society. A year later he expanded from astronautics to a survey of what might be possible not only technically but socially, and perhaps spiritually, with a foundational book, *The Exploration of Space*. In 1956 he moved to Sri Lanka, only recently unyoked from British colonial rule. A year later, Sputnik made Clarke's prophecies seem clairvoyant; within three years the first communication satellite was in orbit.

Interestingly, from his tropical island Clarke enthused over the *two* emerging realms of Third Age discovery. He became, as he put it, "amphibious." The sea beckoned. He took to underwater scuba, toured reefs, and wrote a string of sea novels that paralleled his space fiction. His motivations? He liked the experience, and after he suffered paralysis in 1962, he relished the sensation of weightlessness. He regarded the two environments, sea and space, as similar, so much so that he believed the seas could train future astronauts and serve as a nursery for space treks, since "submarine exploration is so much cheaper than space flight." It could also prepare travelers for exotic life forms and, with whales and porpoises, for encounters with alien intelligences.[36]

During the era of IGY preparation, and before the space race became dominant, Clarke wrote sea and space novels in a kind of thematic fugue, which culminated in two parallel collections of essays: *The Challenge of the Spaceship* (1959) and *The Challenge of the Sea* (1960). But the deep oceans became decoupled from colonization and popular imagination, leaving space as the more untrammeled realm for the imagination, and that is where fiction, and Clarke, went.

Clarke brought to an apex the alloy of exotic technology, latent spiritualism, and futurist settings so common to the genre. From his once-colonial island he imagined brave new worlds, bold encounters with wonders and monsters, and humanity's assisted evolution into godlike stature. With the 1968 release of the movie *2001: A Space Odyssey*, based on a Clarke story and then a Clarke–co-authored screenplay, the hard engineering of rockets met utopianism, and with his *Rama* trilogy (1973–91) he created a Romantic epic, a *Lusíads* for space imperialism. Throughout there is the specter of a humanity advancing beyond the limits of its genetics and ancestral geography, with perhaps links to alien civilizations in its distant past and unbounded spiritual promise, even immortality, in its future. This vast literary output and commentaries established Clarke in the public mind as the Delphic oracle of the emerging space age. In 2000 he was made a Knight of the British Empire. He outlived, outwrote, and outprophesied every potential colleague and rival.

What did Clarke foresee? That "the conquest of space is possible must now be regarded as a matter beyond all serious doubt" and that "the conquest of the planets" was now both necessary and inevitable to renew the human mind and spirit.[37]

It was necessary because geographic expansion enlarged "mental horizons" and stasis contracted them. Exploration and colonization were essential because "when Man loses his curiosity one feels he will have lost most of the other things that make him human," a legacy of wanderlust that "the long literary tradition of the space-travel story" showed was rooted "in Man's nature." Like the Great Voyages, space exploration would spark a New Renaissance, "an expansion of scientific knowledge perhaps unparalleled in history"; would forge a common Earth identity and force Earthlings to perceive the "true" place of their "single small globe" amid the cosmos; and most thrillingly, would hold the "prospect of meeting other forms of intelligence," for it would address "one of the supreme questions of philosophy," whether or not "Man is alone in the Universe," and "at last learn what purpose, if any, life plays in the Universe of matter." The Earth is not world enough. The Intelligent Visitors would of course be benevolent, and they would force Earth into a Reformation before which those of the past would seem laughably puny.[38]

This entire Romantic edifice was built on the mixed sand and stone of history and literary precedent. Clarke commenced his serious space romances as the British Empire began its dissolution; he resided at one of its former colonies; and his vision that Earth's future necessarily lay in expansion—in exploration and empire, though of benign forms—was a lineal descendant of Rycharde Eden and Richard Hakluyt urging that England's future lay with far-voyaging and overseas colonies. What arguments he did not invent, he revived, absorbed, and endowed with literary expression.

In striking ways few contemporaries have added significantly to the oeuvre. Virtually the entire corpus of Carl Sagan's passionate pleas for a future in space, for example, are not only foreshadowed in Clarke but present in almost identical language: The potential of technology. The power of scientific curiosity. The significance of expansion to forestall decadence. The search for intelligence beyond Earth and the belief that it is human destiny to make contact with such intelligence. The assumption that such contact would revolutionize human civilization.

What Sagan added was a stronger argument from biology. In Clarke's day rockets were the critical motive force: they made interplanetary travel possible. In Sagan's time, rockets and gravity assist were adequate to explore the solar system, but the motives to use them needed bolstering. The search for life could serve as an intermediary to the search for intelligence. If Clarke implicitly sent a benevolent British Empire into the heavens, Sagan drafted a less chauvinistic master narrative that made the space age another link in a great chain of purposed human wandering. Significantly, the two writers' passion for planetary exploration converged on Mars.

## DOUBTERS

The rhetoric for space was vibrant, loud, and, for many, compelling. But where there are arguments, there are also counterexamples, and where prophets flourish, apostates abound. For every Arthur C. Clarke and Carl Sagan, there were Cassandras and chroniclers to count the cost.

In *The Sirens of Titan*, published the year after IGY ended, as satellite after satellite lofted upward, Kurt Vonnegut mordantly observed that "the state of mind on Earth with regard to space exploration was much like the state of mind in Europe with regard to exploration of the Atlantic before Christopher Columbus set out." There were differences, of course, almost all tending to make the contemporary scene even more formidable. "The monsters between space explorers and their goals were not imaginary, but numerous, hideous, various, and uniformly cataclysmic; the cost of even a small expedition was enough to ruin most nations; and it was a virtual certainty that no expedition could increase the wealth of its sponsors." In the face of such facts only one conclusion was possible: "on the basis of horse sense and the best scientific information, there was nothing good to be said for the exploration of space." His satire places Vonnegut in a long tradition that has turned the rhetoric of prophets, in this case the siren call of space wrapped in technological romance, against itself.[39]

From the beginning the skeptics and the scoffers were at the docks, in the committee rooms, and all over the press to rebut dazzling thesis with grubby fact, and to scorn forecast bright visions of the future with recalled dark visitations from the past.

In the greatest of exploration literature they come together, as in Luiz Vaz de Camões's *The Lusíads*, which casts the founding voyage of Vasco da Gama to India within the elevated style of the classical epic. The Olympians themselves are astonished by the audacity of the Portuguese. Jupiter exclaims

*Now you watch them, risking all*
*In frail timbers on treacherous seas,*
*By routes never charted, and only*
*Emboldened by opposing winds;*
*Having explored so much of the earth*
*From the equator to the midnight sun,*
*They recharge their purpose and are drawn*
*To touch the very portals of the dawn.*

He tells a worried Venus that they will exceed even the exploits of Ulysses and Aeneas, that "Your greater navigators will unfold / New worlds to the amazement of the old."[40]

This exchange occurs while the mariners are leaving the piers at Belém, ready for their passage to the Indies, but as demanded of an epic, the tale had begun in medias res. When the text returns to narrate the fleet's departure, Camões inserts an astonishing scene, one of *The Lusíads'* most moving, in which the crew try to avert their eyes from the sight of loved ones left behind and close their ears to those petitioning them to cease. But they cannot avoid an old man, his eyes "disapproving," who harangues the departing ships from the shore, hurling prophecies and mockery from a wisdom plucked out of a "much-tried heart."[41]

One by one, the Old Man of Belém demolishes the presumptions that fill the sails of the unmoored ships. Honor is no more than "popular cant"; fame, but vainglory; "visions of kingdoms and gold-mines," delusions; bold discoveries, mere folly; idealism, a disguise for greed. Crusading zeal is better satisfied closer at hand. Adventuring will only lead to "new catastrophes" and wreck "all peace of soul and body." The glitter of gold is a seductress's call that will deplete rather than enrich. Subtly, yet "manifestly," the Carreira da India will "consume the wealth of kingdoms and empires!" The voyage is but the latest example of an interminable, tragic restlessness:

*In what great or infamous undertaking,*
*Through fire, sword, water, heat, or cold,*
*Was Man's ambition not the driving feature?*
*Wretched circumstance! Outlandish creature!*[42]

There lies the meaning of the Old Man's lament: the unalterably flawed character of humanity. The founding epic is a tragedy. And so it proved. Six years after *The Lusíads* was published, with its admonition to renew the grand adventure, King Sebastian led a crusade to Morocco that destroyed his army, cost him his life, and shortly after Camões's death, drove Portugal into a reluctant union with Spain. Without the quick riches of the India trade, such foreign adventures would not have been possible, and without the elevated rhetoric of redirected romance, they could likely not have rallied the obligatory enthusiasms.

But as the Old Man knew, the ships were already unmoored and riding the tides of the Tagus out to sea. His voice has to call to them over the waters. His warning comes too late.

---

As the Great Voyages proliferated, scoffers found ample experiences to criticize. Only the incorporation of exploration into national creation stories has allowed the chronicle to tilt toward the visionaries.

Columbus proved almost delusional in his systematic inaccuracies and willful refusal to acknowledge what he had actually discovered rather than what he wished to have discovered, much less what he had promised would result from his discoveries; and this stubbornness resulted in a loss of interest in the New World, certainly among prospective colonists. Until global warming melted back the pack ice, the Northwest Passage was a fata morgana that called challengers to their death. The Castilian sagas of conquest in Mexico and Peru sent ravenous conquistadors forlornly over much of the New World in a vain, often disastrous, pursuit to find another Tenochtitlán or Cuzco, and inspired a flagging Portugal to dispatch exploring warriors to unveil fabulously rich Mexicos rumored to be hidden in the interior of southern Africa, all with ruinous outcomes. Still, the call to oars was stronger, or at least louder, than the injunction to mind one's store.

But as the First Age became moribund, as exploration was tamed into trade, the doomsayers outwrote the soothsayers. The prevailing perception among elites was that exploration was a loser's game, the geopolitical equivalent of buying lottery tickets. Many of the dominant figures of the age, if they commented at all, turned against what they regarded as irrational, wasteful, and destructive voyaging. The classics of travel literature that span the early Enlightenment and continue through Britain's Augustan age are almost all inoculated against the riotously extravagant rhetoric that accompanied the Great Voyages and ridicule the voyaging visionaries. The era begins with a cautionary tale from Daniel Defoe, passes through brutal satires by Jonathan Swift and Voltaire, and ends with a moral epistle wrapped in a story by Samuel Johnson.

Defoe's *Robinson Crusoe* is not what abridged juvenile literature today often makes it: the celebration of a bold adventurer who finds and then fashions a new world in his own image. Rather, it is a paean to destructive willfulness and wanderlust that leads to a loneliness that leaves Crusoe only a stone's throw from all-consuming despair in what he openly calls his "captivity." Appropriately, the story opens with Crusoe's father in the role of the Old Man of Belém. Sensing his son's intentions, he asks "what reasons more than a mere wandering inclination" Robinson might have to throw away his prospects. "He told me it was for men of desperate fortunes on one hand, or of aspiring, superior fortunes on the other, who went abroad upon adventures, to rise by enterprise, and make themselves famous in undertakings of a nature out of the common road; that these things were all either too far above me, or too far below me; that mine was the middle state," which he had found by long experience was "the best state in the world, the most suited to human happiness" and the one "which all other people envied." Robinson père urges his son to reconsider, to be satisfied with his station in life, and cease his footloose folly, or else he would have ample opportunity to reflect upon his heedlessness. In that, Crusoe notes ruefully, his father proved "truly prophetic."[43]

The great satirists of the Enlightenment went beyond a chronicle of misplaced ambition to ridicule outright appeals to free-floating discovery and utopian adventuring. The godfather of the genre was Jonathan Swift's classic, nominally written by a Lemuel Gulliver, *Travels into Several Remote Nations of the World* (1726; revised in 1735). In it Gulliver, a surgeon forced to turn to the sea for his livelihood, finds himself "condemned by Nature and Fortune to an active and restless life" that results in four voyages of discovery. The first takes him to Lilliput, a land of tiny people northwest of Van Diemen's Land; the second, to Brobdingnag, a land of giants east of the Moluccas; the third, to Laputa, an island magnetically levitated, and to other islands leading to Japan, all full of "projectors" and other visionaries; and the fourth, as captain of the *Adventure*, after being overthrown by mutineers, marooned in the country of the Houyhnhnms.

*Gulliver's Travels* turned the emerging genre of the travelogue, or more properly, the journal of a voyage, back onto itself, for what Gulliver finds are commentaries upon his own society, and a critique upon discovery as it was practiced. He was particularly disgusted with the prevailing logic by which exploration must lead to claims, and claims to conquest, since he did not believe the lands he had visited had "a desire of being conquered and enslaved, murdered or driven out by colonies, nor abound either in gold, silver, sugar, or tobacco," he did "humbly conceive they were by no means proper objects of our zeal, our valour, or our interest." He scorned travel writing as "fables" and thought it "better, perhaps, that the adventure not have taken place, or if occurring, that its discoveries remain unknown."[44]

To Voltaire outrage invited a bitter satire, the short story "Micromégas" (1752), likely inspired by the 1735 journey to Lapland under Pierre Maupertuis. Voltaire reverses the usual relationship between discoverer and discovered by imagining a giant from Sirius and a dwarf from Saturn who visit Earth. As with Swift, the obsession of the age with mathematics and travel combine to produce a work of measurements and proportions, though in the end these are not so much mathematical as moral. They become an argument for a golden mean. The real need is not to look outward with idealistic awe but to look inward with greater realism. A missionary turned entrepreneur turned visionary of colonization such as Pierre Poivre might want more stories of travelers who, like Crusoe, must create gardens on deserted isles, but most preferred, after Voltaire, to tend their own plots outside the kitchen.

The same year that those paired expeditions set out to Lapland and Ecuador was the year Samuel Johnson translated into English *A Voyage to Abyssinia* by the Portuguese missionary Father Jerome Lobo. In 1759 that project birthed *Rasselas; or, the Prince of Abissinia.* Rasselas grows up in Happy Valley, a hidden Eden in a remote valley of Amhara, in which every desire is sated. Yet the prince is restless. "That I want nothing," or "that I know not what I want," he concludes, is "the cause of my complaint." The only solution, it appears, is for him to escape and see for himself "the miseries of the world, since the sight of them is necessary to happiness." The journey begins.[45]

This of course is a moral trek, a Pilgrim's Progress, brought into alignment with the geographic travels that the past age had thrown up. The journey reverses the expected route by flowing down the Nile from its source to its delta; it thus begins with a putative paradise and ends with a knowledge of misery. The conclusion is that the travel was pointless, except to confirm what the sages had said originally.

Marooned during the Nile's flood, the party abandons the ambitions that launched its travels, the search for a scheme of happiness, and resolves to return home.[46]

But that had always been the true bullion brought back by travelers. They saw themselves and their society differently. They had to engage with Others in what could ultimately be described only as an encounter between moral worlds. They did not merely climb mountains or search out the fountains of great rivers: they had to talk with other peoples. Gulliver talks, Micromégas talks, Rasselas talks, and only after Friday arrives on his island can Crusoe cease to talk to himself and begin the real task of discovery. But for such knowledge, one did not have to sail to undiscovered worlds. Equivalent lessons existed in history, for the past, too, was "a foreign country," and the neoclassicists could argue that Pliny and Plutarch might inform as fully as the Houyhnhnms and the Laputans. This, however, was an argument for learning and languages, a scholar's plea, not a call for geographic adventuring.

Yet even as *Rasselas* the text saw print, a Second Great Age of Discovery was aborning. An inextinguishable restlessness would be channeled into a renewed era of far-voyaging. On March 31, 1776, James Boswell informed Samuel Johnson that he had found a copy of Johnson's translation of Lobo's travels, which Johnson dismissed and was content to have forgotten. Two days later Boswell described a meeting he had just had with Capt. James Cook, recently returned from the South Seas, and found Johnson "much pleased with the conscientious accuracy of that celebrated circumnavigator." Boswell then burst out that "while I was with the Captain, I catched the enthusiasm of curiosity and adventure, and felt a strong inclination to go with him on his next voyage." It was easy, he continued, to be "carried away with the general grand and indistinct notion of a Voyage Round the World." The august doctor agreed but dismissed the sentiment when he considered "how very little" one "can learn from such voyages."[47]

But the culture at large sided with Boswell, and with Joseph Banks, the aristocratic naturalist who declared that *his* Grand Tour would be a voyage around the world. The romance of exploring science would sweep the moral realism of skeptics aside and would pull literature in its train.

Once Earth had been again encompassed by the Second Age, some sought to look heavenward. The means to go wasn't there, however, and the culture meanwhile had wandered into the labyrinths of modernism. When, after World War II, technology made such travel possible, the issue rekindled, and a new generation of boosters arose. Untethered to Earth, prophets could claim a high ground of hope and idealism, new creation stories unmarred by tragedy, though to critics they might seem like the Laputans on their Floating Island. Still, the rhetoric of the future rode upward on ever-more-powerful rockets. Satire had flourished as exploration sagged; now boosters could express their fantasies in steel and rocket fuel, not merely words; critics could muster nothing equivalent. It was a great age of science fiction and, by the time Voyager sailed, of Hollywood space Westerns and costume epics.

Yet the critics persisted. The scoffers noted that boosters confused geographic exploration with the virtues with which it alloyed: that curiosity was not limited to travel but could be found in libraries and laboratories; that adventuring and hardihood did not demand untrodden geographies but could be found in rock climbing, NASCAR, and extreme-site science; that wanderlust could be satisfied by internal migrations from country to city and back, by tourism, by vagabond communities of retirees in RVs, by walkabouts in virtual worlds; that idealism did not demand a ridgeline below a setting sun but could be found by trying to make a better world through political reform or social activism; that synthetic cyberworlds might satisfy precisely those human urges that in the past could come only from hazardous travel, that they might sate curiosity as a candy bar could the craving for carbohydrates. The boost to economies, science, and technology that accompanied a government-sponsored space race could follow from *any* massive spending program, any stimulant to scientific research, and any technological investments. America as a "people of plenty" required an open society, not open lands; the economic frontiers lay with building not the space equivalent of railroads across the Far West but information highways spanning the world wide web. Skeptics such as Amitai Etzioni dismissed the hype of the Apollo program as a "moon-doggle" and systematically demonstrated how it was a "monumental misdecision" that acted as a "drag" on the larger civilization.[48]

The doubting doctors remain. Space historian Roger Launius notes that, after Apollo, the NASA budget has steadied at 1 percent of the national budget, and that popular interest waxes and wanes not with launches but with Hollywood space movies. The "inescapable lesson" of Felipe Fernández-Armesto's brisk survey of humanity's exploration of Earth was "that exploration has been a march of folly, in which almost every step forward has been the failed outcome of an attempted leap ahead." Explorers were often the "oddballs or eccentrics or visionaries or romancers or social climbers or social outcasts, or escapees from the restrictive and the routine, with enough distortion of vision to be able to reimagine reality." The exploration of space he regarded as a "gigantic folly."[49]

All true, and all more or less irrelevant. The Old Man of Belém may stalk the launchpads of Canaveral and Tyuratam, but the fleets still sail.

Yet the Third Age *is* different. The reason lies not in the ultimate rightness of the boosters and prophets, not because uncontaminated idealism and techno-utopianism can at last prevail, but because of the age's peculiar geography. Amid ice, abyss, and space it is possible to shear away the moral ugliness and ultimately tragic core of exploration because there is no Other to confront, and without an Other, there is no need for a human self. Robots are not only the most practical explorers but the only ones that make sense of the age's distinctive terrains.

That is the tradeoff: exploration by robots of places fit only for robots. If exploration can proceed without the ethical burdens of the past, it also comes without the past's inspiration, moral drama, and narrative tension. Voyager would undergo no mutinies. It could harm no one. It reduces criticism to that of comparative costs or relative returns on cultural investment. It can find new worlds without the horrors that have marred past discoveries. All this, however, comes at the cost of a sanitized encounter cleansed of the messiness and ambiguities and tragedies that had defined the founding discourse and its supporting story.

No place had shown the ambition to revive and project the old age into the new more than Mars, and nowhere did the Voyagers demonstrate better both their mission's novelty and its purity than when they bypassed Mars without a whisper. In late November 1977, Voyager 2 crossed the Martian orbit; a month later Voyager 1 repeated the transit. But both avoided Mars itself—did not even pay respects before moving on. In shunning Mars, Voyager bypassed all the agendas, and much of the inherited apparatus of boosting and scoffing, that had waxed and waned with the Great Ages of the past.

## ALWAYS-KNOWN MARS

For the space program, Mars held many attractions. After the Moon it was the obvious common goal. For geopolitics, it was a potential sphere of influence, analogous to the hinterlands of Second Age imperialism. For space science, it offered the nearest best target to extend the package of IGY instruments beyond Earth, and it might, just might, have some form of life. For spacecraft designers, it promised to extend the range of technology and ambition incrementally. And for colonizers, it was the ideal arena for migration and the test planet for experiments in terraforming. After the Moon, more spacecraft went to Mars than anywhere else. By the time Voyager launched, America had sent eight vessels to Mars (six successfully), and the USSR thirteen (none really successful). Between them they had made nineteen attempts to Venus, one to Mercury, two to Jupiter (Pioneers), and one to Saturn.

Moreover, Mars had far more cultural associations than any other planet. That was its glory—and its burden. What made it attractive to popular culture also made it potentially an exorbitant distraction, for it proved impossible to shear the fantasies from the facts, each of which was renewed after every encounter. So while Mars was actually smaller than Earth, its gravitational attraction, as measured by cultural interest, was far greater than the giant planets visited by Voyager. In fact, Voyager's rival as a grand gesture was its immediate predecessor, the 1976 Viking mission to search for life on the Red Planet. For Mars partisans, Viking proposed an alternative narrative and interpretive prism. It is worth pausing, as Voyager did not, to examine how that happened and what it meant.

For the visionaries who promoted space as the next arena of colonization, Mars had been the realm of the technological romance, the target for colonizing rockets, the founding planet. When Edgar Rice Burroughs reached beyond the landscapes of the Second Age for exotic settings, he sent Tarzan, in the person of Virginia gentleman John Carter, to Mars (which he named "Barsoom")—morphing the classic Western *The Virginian* into the science fiction pulp novel. When H. G. Wells sent imperialism beyond the sublunary realm, he projected an invasion from Mars. Over and again, Mars loomed as the specter before the kindled imagination of those who looked to space for inspiration, be they rocketeers from Robert Goddard to Wernher von Braun, or visionaries from Arthur C. Clarke to Carl Sagan, or the romancers of contemporary literature from Robert Heinlein to Ray Bradbury. Mars was the bidding siren of space-voyaging history, the challenge to technology, the test case for life beyond Earth, the first port of call for the dissemination of humanity throughout the universe. Earth-orbiting space stations, lunar bases, robotic reconnaissance—all had meaning to the extent that they contributed to the settlement of Mars.

The drumbeat began early in the postwar era. Robert Heinlein published *Red Planet* in 1949. Ray Bradbury wrote *The Martian Chronicles* in 1950. The science that such fiction demanded, however, came with Wernher von Braun's *The Mars Project*, an "algorithm of spaceflight" to establish colonies, published in 1952 (English edition in 1953). Soon afterward, *Collier's* magazine ran a series of eight features on space activities. Von Braun and journalist Cornelius Ryan argued in 1954 that such a mission was technologically possible. With television emerging as the popular medium of the time, Disney Studios made the transition from print to screen by creating three animated features for TV, aired between 1955 and 1957, the last appearing two months after Sputnik. Sandwiched between the shows, von Braun and Willy Ley assembled the *Collier's* series' arguments into a summary book, *The Exploration of Mars*. By 1962, Marshall Space Flight Center, under the directorship of von Braun, was busy projecting a future beyond the Moon. Early Manned Planetary Roundtrip Expeditions (EMPIRE) would first send a major exploratory party to Mars. Immediately after Apollo 11, von Braun submitted detailed plans that would repeat the triumph of the Moon landing on Mars by 1982. And Arthur C. Clarke wrote a short story in which he neatly framed the exploratory impulse that had been announced with the eighteenth century's expeditions to measure transits of Venus by imagining the sole survivor of an inaugural expedition to Mars who watches the transit of Earth across the Sun in 1984.[50]

Missions followed. JPL had early established a Mars group, believing that Mars was where vision and practice would converge. Mariner 4 made the first planetary flyby of Mars in 1965. By then JPL was immersed in the elaborate Martian program centered on a new spacecraft called Voyager. When that program got scrapped, the name floated free until it was reclaimed for the Grand Tour. Then came Mariner 9, in November 1971. As the spacecraft entered Martian orbit, JPL convened a panel of tribal elders and young oracles to discourse on the subject of "Mars and the Mind of Man." Bruce Murray, Walter Sullivan, Ray Bradbury, Arthur C. Clarke, and Carl Sagan—all save Murray were literary figures, journalists or novelists, and science popularizers. Mars could attract that kind of cultural event.[51]

Unsurprisingly, it was Mars that merited for planetary exploration the first experiment in big science, as the abandoned Mars Voyager mission metamorphosed into Viking. When the two spacecraft landed in July 1976, NASA Langley convened another soiree, again at JPL, expanding the theme to encompass "Why Man Explores." The panel's luminaries included James Michener, Norman Cousins, Philip Morrison, Jacques Cousteau, and Ray Bradbury. Michener tidily summarized the cultural tethers to Mars: "All my life I have followed the explorations of Mars intellectually, philosophically, imaginatively. It is a planet which has special connotations. I cannot recall anyone ever having been as interested as we are in Jupiter or Saturn or Pluto. Mars has played a special role in our lives because of the literary and philosophical speculations that have centered upon it." He concluded: "I have always known Mars."[52]

The panel veered into metaphysics, meandering poetry, and loose allusions that sought to equate exploration with curiosity. But Michener was right. Attempts to replicate the event with "Jupiter and the Mind of Man" and "Saturn and the Mind of Man" never caught fire. The literary imagination returned to where there were people, or the prospects of people, or at least the works of previous literary people to contemplate. For space philosophy, Mars was less a planet than a strange attractor, perturbing all the intellectual fields around it.

At the time, the smart money would have bet that Viking would become the grand gesture of the age. It seemed to unite perfectly the classic motives behind exploration with a modern knot. With its dual orbiters and two landers, it was Apollo come to Mars, and instead of collecting Moon rocks, its robotic astronauts sampled soil for evidence of life. Both settled into the Martian maria (or *planitia)*. The first was named Chryse, and the second, Utopia. "Chryse" is Greek for gold, and "Utopia" is a word invented during the Renaissance to describe an ideal if imaginary place in the distant sea. In an appropriate if eerie way the spacecraft had carried God, gold, and glory to Mars.

The Viking mission thus fused into one expedition all the inherited parts—science, with the exotic search for life; politics and national prestige, with its landings staged to coincide with the American bicentennial; engineering and journeying encoded within its complex choreography of guiding, landing, and robotic sampling, all at the edges of technology; cultural contact, that sense of always having known Mars. Here, it seemed, was the ideal narrative for the Third Age.

Yet it never quite took, even with Sagan shilling its story, and not solely because it failed to find life or because the country at large was in a dark, sour mood—exploration had often in the past countered such gloom. A likely explanation is that what Mars gave it also took away. If the density of cultural overlays added meaning, it equally burdened the expedition. In looking forward to the human explorers who it assumed would surely and shortly follow, Viking looked back to a past that the Third Age had to selectively discard. Though Mars could join space to those futures the past had imagined, it could not fling that past into the future by itself. Besides, the mission had a target: a suite of experiments to test for life, and these were at best inconclusive. When the experiments ended, so did the Vikings' narrative.

The mantle of the Third Age would require a different kind of journey in which the trek would not end—in which the journey itself contained its drama. This demanded another kind of story, one bonded to exploration and quests, not to colonization. It is the narrative of Voyager.

Voyager could not avoid the Mars mystique altogether. In presenting the Mariner Jupiter/Saturn 1977 mission to the readership of *Astronautics and Aeronautics*, JPL authors framed their account with lengthy quotes from Arthur C. Clarke's *2001: A Space Odyssey*, which had bypassed Mars for the Jovian moon Iapetus. Yet it is doubtful that even the most addled techno-romancer believed that Voyager's journey was a prelude to settlement or the end of earthly childhood.[53]

Voyager was, instead, a modernist machine loosed onto the cosmos. The Voyagers would not be blinded by gold or the mirage of fame. They would not abandon wife or child, or enslave unwary indigenes. They could not despair, could not be crippled by loneliness, could not fight for the cross or suffer for science, would not know epiphanies or endure tropical fevers. They would lay no claims, issue no proclamations of sovereignty, raise no toasts to king or republic, sign no treaties of trade or military alliance, nor send out reconnaissance parties to lay out routes for folk migration. They conducted no conversions, collected no soil samples or ore assays, erected no missions and outposts. The Voyagers confronted no Other, or even life. Instead the Voyagers would carry a tradition of exploration—one bonded to an older and brasher tradition of vision quest—into new times and to new worlds.

For this the Voyagers had to shun Mars, as they shunned the dominant passions of Martian prophets. By the end of 1977 both spacecraft had rushed by Mars's orbit without a blink and in so doing slid past all the fantasies and utopianism that Mars has provoked. They had left one of the founding trilogy of narratives, colonization, behind.

# Beyond the Inner Planets

DAY 125–413

# 9. Cruise

Between the orbit of Mars and the asteroid belt, the Voyagers did what they did mostly through their long trek. They cruised, and cruised, and cruised. For Voyager 1 the first full-fledged encounter, at Jupiter, would not occur until nineteen months after launch, and for Voyager 2, almost twenty-three months. Between them lay the shoals of the asteroids, the first and most visible of the hazards the Voyagers had to face. Before then, however, they had to navigate through the immensity of space.

The real hazards lay on Earth. The Voyagers did not know boredom and could not be distracted; their minders could. As programmed, the Voyagers routinely sent back information gathered about interplanetary space, and they expected only in reply JPL's weekly radio acknowledgment. If they did not receive it, they assumed their primary receiver had failed, and switched to a backup. In early April 1978 a week passed with nary an electronic whisper from JPL, and Voyager 2, as preprogrammed, switched to its secondary receiver. Controllers, now alert to the glitch, instructed the spacecraft to return to the primary, a command it ignored.

Now hardware and software began a toxic scenario in which each minor failure cascaded into another potentially more fatal one. The attempt to reconnect caused a tracking loop capacitor to fail, which meant the backup receiver could not adjust to a wide range of frequencies out of which it could automatically pluck commands, but could only tune to a precisely defined one, a frequency that Voyager 2's earthbound minders would have to search for. Identifying that required frequency is complex, because it varies with all the motions of Earth and spacecraft, along with temperature and other idiosyncrasies. A week of frantic scrambling eventually reestablished the primary receiver, but only temporarily, before a short developed that blew both the receivers' redundant fuses. After another week of scrambling, communication reverted permanently to the backup.[54]

## CRUISE CONTROL

To the uninitiated, space travel might seem no more complex than sending a billiard ball across a Euclidean void. It was only necessary to avoid collisions with large hard objects, which were few and obvious. The reality was, travel beyond Earth was jarring, quirky, and often rough, with slight margins for error.

That process began with launch—a vast shaking by powerful rockets that propelled awkward payloads through a scruffy atmosphere, and for which even seconds of timing and meters of positioning could amplify across many years and millions of miles of interplanetary space. Then the spacecraft introduced imbalances as they unfolded and performed. Their ongoing navigations were no better than their sensors, thrusters, and gyros. Success depended on an exact trajectory, for which it was necessary to know as accurately as possible just where the spacecraft were and what speed they traveled at, and hence what course corrections to make.

Nor was space itself truly neutral. That interplanetary realm, which seemed like a void, actually consisted of hard and soft fields, each full of irregularities and neither fully mapped. Not all the hard geography of planetary objects was known. There were gravitational tugs from bodies unrecognized from Earth, not least undiscovered moons; the planets themselves had masses not previously measured with the accuracy required for precision navigation, and were surrounded by debris-laden rings, some visible, some only hypothesized; there were hits from dust that did little direct damage but that could nudge trajectories microscopically. The soft geography of magnetic currents and solar winds, perhaps even cosmic radiation, exerted its own subtle pressure and was occasionally aroused into tidal surges and storms. Even the minuscule movements of the spacecrafts' magnetic tapes deflected attitudes and velocity. The Voyagers required ceaseless corrections by their internal guidance systems and earthbound steersmen and pilots.[55]

Voyager's targets—planetary encounters—had relatively unforgiving windows. If the spacecraft passed too far away, its instruments might not record what observers wanted; if too close, the immensity of the giant planets would blot out readings in a blur. The spacecraft had an exact route to travel, as intricate as threading the coral-infested Torres Strait, and they could make the pass only once. There would be no second chance. There was no opportunity to pause, to put to shore or lay at anchor while plans were reconsidered or advice sought or local pilots acquired. The reliance on gravity assistance, too, demanded precise steering as to not only speed but also direction, or the Voyagers might be flung wildly into space. The Voyagers had a "delivery error" of one hundred kilometers. The Grand Tour was premised on an ability, as the prevailing metaphor put it, to tee up a golf ball in New York and sink a hole-in-one in California.

An ancient distinction exists between navigation and pilotage. Pilots guided ships through local waters they knew from long apprenticeship, perhaps assisted by logbooks or rudders that spoke to the details of maneuvering through the hazards of particular harbors. The geographic oddity that rendered Europe a peninsula broken fractally into smaller peninsulas that in turn dissolved into isles meant that pilots steered along coasts and over small seas. They crossed the Mediterranean not all at once but through a series of lesser crossings, strait by strait, sea by sea. Not until mariners undertook long traverses over blue-water oceans did navigation, or the means to determine location when out of sight of land, become essential, and pilots and navigators begin to merge. And then they again searched sea by sea and strait by strait.

The means for navigating were primitive. A compass; an astrolabe, a Davis quadrant (or backstaff), or sextant; a wooden log dragged on a knotted rope—by such means one could estimate location relative to the Earth's magnetic pole, or its cosmological setting, and perhaps something of its speed. But all were faulty, little better than calculated magic. Experience counted more than instruments. The distribution of planetary magnetism—specifically the declination of the magnetic pole from true north—was unknown. Sightings of the Sun and polestar could determine with relative ease one's latitude, which argued for sailing along fixed latitudes wherever winds and currents made such passages possible.[56]

At the onset of the Great Voyages the Portuguese assembled scholars to formalize, and if possible improve upon, known practices. To the rule of the North Star, they added the rule of the Sun, created a table for the rule for "raising the pole," and made analogous calculations for the Southern Cross, all of which were gathered into a manual of navigation and nautical almanac, the *Regimento do astrolabio e do quadrante*, taught in formal schools for navigation established at Lisbon and, later, for Spain, at Seville. The exercise did for latitude what later state sponsorship did for longitude. It was thus no accident that the rediscoveries of the New World came from voyages that traveled from east to west, with the Norse sailing from isle to isle along a rude line of latitude (and, barring storms, never more than a couple of days out of sight of land) and with Columbus following favorable trade winds that blew roughly east and west.[57]

Far trickier was to determine longitude, for which there was no practical solution until the nineteenth century. The state of learning was not merely wanting but often dangerously inept. The fact is, the Great Voyages achieved their goals without adequate navigational techniques, and the Second Age accomplished most of its task before chronometers were both accurate and abundant enough to make a calculation of longitude generally sufficient. Instead explorers relied on an artful, if not quasi-superstitious, appeal to eclectic methods and a personal alloy of experience, hunch, whim, prayer, and luck, or the unfortunately named "dead reckoning." Samuel de Champlain, one of the few explorers equally successful on both land and sea, shrugged his shoulders. He had never been able to learn from any mariners with whom he talked how they did it, "except that it be done by fanciful rules, all different, some better than others." He trusted those who had actually voyaged more than those "others who often pretend to know more than they do."[58]

The common practice was to recycle pilots, as da Gama did to double the Cape, or to seize them, as he then did at Malinda, and as Albuquerque did at Java before sailing to the Spice Isles.

By the time Voyager launched, pilotage referred to the minutely choreographed acts associated with planetary encounter; and navigation, to black-space sailing. For the latter, the Voyagers commanded a navigation team (NAV) of twelve. For the former, the spacecraft relied on an elaborate procedure for identifying the hundreds of tasks that an encounter required and then worked out a second-by-second sequence, coded it for the onboard computers, and uploaded the package.

The duties of the NAV team were three. The first was trajectory, or mission design. It tested options for routes by balancing size of payload and launch capacity with where program scientists wanted to go and what they wished to do when they arrived. There were tradeoffs, an infinity of tradeoffs. Unlike Cabot or Cabral, Voyager could not put to port to refit or replan or decide the season was too advanced to proceed or elect to revisit a site of special interest. Each planetary encounter had to be exquisitely choreographed down to seconds, which meant that the mission had to determine core trajectories well before launch, since prospective launch dates varied according to the desired routes past the planets.[59]

As chief navigator, Charles Kohlhase had overseen some 10,000 prospective trajectories for the Grand Tour, a change in any one of which could ripple through all the rest. Eventually mission planners winnowed that unruly swarm into a handful. The first charge was of course to survey Jupiter and Saturn, but behind that was the vision of the Grand Tour, such that the prospects for Voyager 2 depended on

Voyager 1. Besides the anticipated hazards of asteroids and the Jovian radiation field, the mission had to make a critical assessment of Titan and avoid the Saturnian rings. Even after launch, various possibilities abounded through midcourse corrections, but only within a single-minded, multitasking furious passage at 39,000 kilometers per hour that resembled a descent through a cataract. Still, trajectory corrections were both possible and necessary.

These—the "orbit determinations" that specified where Voyager was and the "nudges" that refined or redirected its path—were the largest of the NAV group's assignments. Orbit determination relied on both Earth-based and spacecraft-based methods. Earth-based navigation tracked the spacecraft through its telemetry, the messages it sent back to the Deep Space Network's dishes. Because a regular Doppler shift occurred, it was possible to calculate distance, and this could be done over and over across months, if desired, to know exactly where the spacecraft was and how fast it was moving. In a sense, the triangulations took the place of historic methods for determining latitude, and the Doppler shift assumed the role of onboard chronometers for determining the equivalent of longitude.

But among the unknowns that made precision trajectories so daunting was the range of uncertainty about the target planets themselves. If navigators were to steer the spacecraft within a one-hundred-kilometer window, they required more precise measurements of diameter and mass—better than those obtainable from Earth. A mistake of a thousand kilometers in the diameter of a giant planet was entirely within the range of instrumental error, yet could prove ruinous for Voyager. For such measurements, navigators needed an optical navigation apparatus housed on the spacecraft itself. Images of the planet against the background of fixed stars refined its dimensions, much as small variations in acceleration (by which the spacecraft felt the pull of the planet) honed its mass.

In this way the guidance team juggled with two numbers. One forecast distance to target, and the other, time of encounter. Both were inevitably flawed, but the magnitude of error could be trivial. Distance was the more critical, since the positioning of flyby decided what the instruments and images would record. As the spacecraft approached closer, both numbers sharpened, and argued (or not) for a final tweaking of Voyager's trajectory. Such corrections were programmed into the formalized sequencing of encounter.

The third navigational duty was to decide how to make those course corrections. The sooner the adjustments, the lesser the variance as encounter approached. Some deviations resulted from the sum of minor perturbations. Others came about as the exact specifications of time and place for encounter made for a more precise if frenetic scenario of maneuvers. But either way, the exercise was harrowing, for it demanded that the Voyagers temporarily abandon their typical mode of navigation.

In normal flight, the Voyagers, like the Mariners from which they descended, stabilized themselves around three axes. To hold the craft's position, controllers had to triangulate from two fixed points in space. One was simple: the Sun. For near-voyaging craft, the second point could be Earth itself. For far-voyaging craft, however, Earth could be confused with its Moon, and both lay too close to the Sun, so another mark was needed. The star of choice was Canopus, in the southern constellation Carina, the second brightest light in the sky. Between those two sensors, one on the Sun and one on Canopus, the Voyagers constantly recalibrated their location. But when they underwent a burn to accelerate and reposition, they had to surrender that cosmodetic baseline. They needed another means to stabilize.[60]

The procedure began by repositioning the spacecraft so its engine would propel it in the proper direction, then turning off the celestial guidance system and yielding stability to a set of three gyroscopes, one for each axis. At such times, along with a temporary loss of its navigational sensors, Voyager would no longer point its high-gain antenna to Earth; it had to surrender the thermal balance that its formal stability had allowed, and then, after the burn, rely on a small omnidirectional antenna to reacquire Earth's location while its star sensors recaptured the Sun and Canopus. Until then the spacecraft was on its own, and dependent on gyros, devices long and well understood but still machines and therefore subject to their own electrical and mechanical gremlins. The maneuver was fraught with hazards, and given the difficulties of telemetry and tracking, it might be weeks before the exact outcome could be assessed.

The Voyagers' long cruises were the ideal times to correct trajectories, for the coasting phase lacked the frenzy of encounter, and without knowing the correct velocity, both position and speed, well in advance, the encounters would fail. On his second voyage Columbus missed the westerly trades and nearly foundered. Da Gama mistimed the monsoon winds to Africa and narrowly escaped disaster.

## STAR STEERAGE

Across the ages navigation had relied on mixed technologies, the search for a new celestial referent, the power of judgment, and simple trust to luck. What Voyager did was accelerate the level of technical knowledge and transfer more of the burden to the spacecraft machinery: the mission was itself a kind of midcourse correction in the trajectory of exploration history. Pilot, helmsman, rudder and log, sextant and compass metamorphosed into a complex machine over which human controllers exercised ever-shrinking capacity for tactile guidance even as the demands for precision maneuvering swelled. As with everything about Voyager, its mission fused the hoary with a high-tech modernity.

The explorer still looked to the stars for guidance. At the onset of the Great Voyages, this meant the Sun and Polaris, the polestar. Yet the latter's value lay in its constancy about the North Pole, and as the Portuguese probed southward, it fell lower toward the horizon; and beyond the equator, it disappeared. Still, one could coast along Africa, though only at the cost of dreary daily tackings that made the Indies seem more remote rather than less so. If they wished to find those distant lands of their fevered imaginations, exploring *marinheiros* would have to sail from the mundane shorelines and into the Sea of Darkness, which was now all the murkier because the travelers had left behind the lights of both familiar constellations and the polestar itself.

Yet an astounding sight greeted them. Looming up was a

striking constellation that resembled a kite or, to the eyes of the Portuguese, a cross. Instead of terrifying them with its novelties and its perhaps unknowable heavens, the southern sky seemed to beckon, as though they were crusaders. The Southern Cross summoned them to new worlds, and new possibilities, of navigation. As swaths of strange stars appeared, so would novel methods emerge to pilot mariners across those untracked seas and even an ocean an order of magnitude broader than Europe itself. Eventually Galileo's telescope unveiled the inner moons of Jupiter and made a calculation of longitude feasible; surveys of geomagnetism plotted maps that traced the deviations that made a recalibration of the compass possible. Exploration and experiment would continue, each one provoking the other to new exertions. The Voyagers now raced toward the Jovian—the Galilean—moons that had caused a revolution in cosmography and navigation and were poised to announce another.

As the Portuguese sailed south, they spied also an extraordinarily bright star that rose in prominence as they pressed on, more luminous than anything in the night sky save Sirius. With providential calculation, it seemed to circle about the southern pole. They named it Canopus.

DAY 383–444

# 10. Encounter: Asteroid Belt

In late 1977 Voyager 2 entered the asteroid belt; Voyager 1 soon followed. In mid-December, as planned, some 124 million kilometers from Earth, Voyager 1, on a faster, tighter trajectory, passed its twin.

The swirl of asteroids was the first of the known hazards awaiting beyond the Earth-like planets. They were, collectively, a kind of failed planet, a vast composite of planetary particles, some, such as Ceres, as large as Texas, most smaller and more worrisome. Mission planners envisioned the cluster as a diffuse shoreline of sandlike debris that could strike with the speed of bullets. Moreover, the belt could stand, in surrogate, for the rings, both known and yet to be discovered, that encircled the giant planets. It had to be threaded or rounded for the Voyagers to find the open seas beyond. Within the barrier, the biggest blocks were mapped, and could be avoided; the others were a matter of blind reckoning and, for the frozen sandstorm of meteorites, a question of luck. The only way to know was to do it.

Fortunately, the Voyagers did not have to sail blind. A plucky pair of spacecraft, Pioneers 10 and 11, had already blazed a route to Jupiter and Saturn. Thanks to them, the passage to the Indies of the outer planets was open.

## O, PIONEERS

Their names were appropriate. Pioneer was the hardy frontier scout, the first to Jupiter and Saturn, the tough traveler who discovered the pitfalls and fords, the pathfinder who made the Grand Tour possible. Voyager was the scientific sojourner, the one who rediscovered and elaborated the Jovian and Saturnian systems with rigor, and then plunged into the far unknown before overtaking the trailblazers. Pioneer was the indomitable Jedediah Smith, hunkered against the wind, crossing over South Pass; Voyager, the Pacific Railroad Surveys' grand reconnaissance of the West.[61]

The Pioneer spacecraft were designed to be simple, durable, cheap, and, if necessary, expendable. The largest difference between their design and that of the Mariner series was that they maintained stability by spinning, rather than using a three-axis arrangement of thrusters. The earliest prototypes struggled through multiple launch failures until 1960, when Pioneer 5 roamed the interplanetary domains around Venus. Pioneers 6–9 were arrayed throughout the inner solar system as a "space weather network." But the famous missions were Pioneers 10 and 11 to Jupiter and Saturn. In what they found and how they did it they not only foreshadowed the Voyagers but often led.

They were, as NASA put it, "precursor missions." They were the first spacecraft to the two giant planets, they discovered the first of the new moons that the far-traveling robots would find, they helped sharpen the communications apparatus for deep-space communication, they confirmed the capacities of gravity-assist trajectories, they demonstrated how a lead spacecraft could inform and redirect the purposes of a second, and they blazed a trail through the asteroid belt, Jupiter's lethal radiation, and Saturn's rings. They were the first spacecraft to travel beyond the orbit of Pluto, and the first to carry engraved messages to any other intelligence that might lie beyond the solar wind. Dispatched the year after Voyager, Pioneer Venus 1 and 2—the last of their breed, turning inward to spiral around Venus—concluded the golden age of American planetary exploration.[62]

---

Pioneer 10 launched in March 1972, and Pioneer 11 a year later, in April 1973. It was Pioneer 10 that first encountered Jupiter, warned of its lethal radiation, and sent back real-time images of the planet as it approached periapsis. Its instruments sketched a new cartography of the solar system, a geography of magnetism and radiation. But the reports that captivated the public were the pictures that took shape, scan line by scan line, from Pioneer's Pulsed Image Converter System (PICS). Its spin-stabilization design meant Pioneer did not have a camera, but it could broadcast data from the imaging photopolarimeter's narrow-angle telescope in strips as it rotated. These were assembled into color images by computer, which added green to the reds and blues that Pioneer sent back, and then projected the result on television.

By December 2, 1973, the images exceeded those from Earth-based observatories, and as the spacecraft approached periapsis the next day it proclaimed a new perspective on Jupiter, the greatest advance since Galileo had trained his telescope on the planet. Even more, Pioneer changed the process as well as the pictures. While there were no human eyewitnesses, as with the Apollo lunar landings, the public could see what Pioneer could as it swept around Jupiter. The images were grainy—clunky by later standards, and far from the bewitching scenes recorded by Voyager. But they came in real time, and they thrilled, and they helped announce an era of virtual exploration. For its imaging triumph, the Pioneer program even received an Emmy award.[63]

The near-death, and ultimate success, of Pioneer 10 warned JPL that Voyager had to harden its electronics and tweak its trajectories. But the Grand Tour required that the mission have similar information regarding Saturn. This task fell to Pioneer 11, which followed a trajectory to Jupiter that would also allow it the option of going to Saturn.

Pioneer 11 made its closest encounter on December 2, 1974, a third as near to Jupiter as its predecessor, and on a trajectory that gave the first views of Jupiter's poles (invisible to Earth-based observation), and then slung it to Saturn. By passing through the Jovian radiation ring vertically in this way, it would experience only a short burst of irradiation, which would allow it a nearer encounter. Then it flung around on a trajectory to Saturn in which it trekked across the solar system by passing over the plane of the ecliptic. As it approached Saturn, its controllers and scientists debated between an inside or an outside passage—whether to cross the rings sharply or swing around them altogether. Either route would reveal new knowledge, but NASA determined that the heavy scientific lifting for the planet would come from Voyager, and that Pioneer would test Saturn's shoals for the explorers to follow. As Pioneer 10 had allowed Pioneer 11 to proceed to Saturn, so Pioneer 11 would be used to allow Voyager 2, if all went well with Voyager 1, to proceed to Uranus. What Voyager required was more precise knowledge about the outside passage.[64]

On September 1, 1979, Pioneer successfully navigated through the outer rings, and then passed through them again on the other side on September 3. Along the way it dispatched five images of Titan, another target of Voyager; and through occultation, it discovered a new Saturnian moon, later validated and named Epimetheus.

That achievement was perhaps not all that Pioneer 11's most visionary proponents had wished for, but it was all the spacecraft had to do. Among routes over America's Continental Divide, South Pass is virtually inconspicuous, a rise of land in a broad plateau, rather than a rift through towering mountains like a Khyber Pass; yet it was the way west. So it was with the outside passage at Saturn. The route lay open for the Grand Tour.

Both Pioneers had yet more to do, this time probing a path that would ultimately take them beyond the solar system. The first, Pioneer 10, crossed the orbit of Uranus on July 1979, and of Pluto in 1990, completing a traverse of the solar system's hard geography. The soft geography, as defined by the solar wind, was much farther away. Moreover, it was asymmetrical, compressed in the direction of the Sun's travel and stretched in the opposite direction, leaving a long ionic tail. Pioneer 10 traveled down that tail. Despite its early launch, its slower speed and trajectory thus put it behind in the race to the frontier. Pioneer 11, meanwhile, coasted toward the heliosphere's bow shock.

Each carried two messages. One was a plaque attached to the chassis of the spacecraft that, overtly, stood as a communiqué to any future finder of the hulk. It was a kind of calling card that identified who sent the machine and from where; in this, again, it tried out what the Voyagers would elaborate. The plaques became the lasting logo of Pioneer. The second message was its radio signal to Earth. Both spacecraft continued to broadcast—Pioneer 11 up to 1995, and Pioneer 10 into 2003; and the Deep Space Network continued to tease out their ever-feebler transmissions. In this, again, Pioneer assisted Voyager by prodding and refining the DSN. As one observer remarked, the total energy from Pioneer's radio signal at the edge of the planets "would have to be collected for several million years to light a single 7.5-watt nightlight for a millionth of a second." To extract that signal from the cosmic background buzz was a daunting achievement. If the two Voyagers were to succeed—Voyager 1 blitzing to the heliopause, and Voyager 2 completing the Grand Tour—they would need extraordinary communications. Those technologies were first devised to listen to Pioneer.[65]

In June 1983, while Voyager 2 was halfway between Saturn and Uranus, Pioneer 10 slid beyond the orbit of Neptune. In 1995 Pioneer 11 suffered a broken Sun sensor, and despite some slight reserves of power, it could no longer direct its wispy transmissions to Earth, and fell silent, still heading toward the constellation Aquila. Pioneer 10 continued to broadcast, but so painfully and weakly that its scientific value was nil. It sank into a virtual coma, and on March 30, 1997, at 11:45 a.m., life support ended, as the last signals were downlinked to the DSN station in Madrid. Still, hope lingered, and Pioneer 10, oblivious to Earth, sent a signal captured by the radio telescope at Arecibo. But like a Newcomen engine in an age of diesels, the equipment required to track and interpret Pioneer was beyond obsolescence. Even heroic measures could barely sense the messages sent across 7.5 billion miles and 20 hours. On March 2, 2002, Pioneer exchanged greetings with Earth on its thirtieth birthday. On February 7, 2003, across 8 billion miles, it finally fell silent, and Pioneer 10 joined its sibling as a machine corpse hurtling through the cosmos.[66]

---

The tragedy of pioneering is to create conditions that eliminate the pioneers—and that was the fate of Pioneers 10 and 11. Compared with the Voyagers, the Pioneers were slow and kludgy, and despite an impressive early lead and first contacts with Jupiter and Saturn, the Voyagers overwhelmed their scientific results and their capacity to evoke wonder, and then overtook them physically in 1998 to become the most distant objects to leave Earth. Unlike the Pioneers, the Voyagers have enough power and working instruments to function until perhaps 2020. They would likely survive to pierce the veil of the heliosphere and to report that event.

The frontier belonged to Voyager.

## BEYOND THE BARRIER

The hard geography proved less formidable than feared, and for mission planners, surprisingly less daunting than the solar system's soft geography.

Pioneer 10 found (and Pioneer 11 confirmed) that most of the interplanetary realm's perilous particles clustered around Earth or at the center of the asteroid belt, where they thickened to almost three times the density apparent in interplanetary space. High-velocity micrometeorites and dust particles were fewer than anticipated. Both spacecraft whisked through with hardly a scratch. Likewise, the planetary rings proved porous or thinned to the point where they could be skirted. There was only one near-collision. At Saturn, Pioneer 11 found, by occultation, a new inner moon on its approach, and then nearly collided with it upon its departure.[67]

The Voyagers threaded through the asteroids, slid around and through the rings, and avoided undetected satellites.

What did threaten Pioneer was interplanetary space's soft geography. An immense solar storm, the largest ever recorded, shook Pioneer 10 on August 2, 1972. Pioneers 6 through 9, already deployed around the inner planets to measure solar activity, had sensed and then tracked a solar wind of unprecedented power and velocity (3.6 million kilometers per hour) that exploded outward. By the time this ionic tsunami struck Pioneer 10 some three days later, it had shed half its velocity, although by transforming that energy into higher temperature, and its wind had become so diffuse that Pioneer 10 sensed its passing only through the traces it left on a battery of delicate instruments.[68]

What did nearly cripple the spunky spacecraft was its scrape with the radiation belts of Jupiter, a blast completely unanticipated by designers. As it approached closest encounter, each half-hourly data dump notched up radiation to levels that would saturate and then overwhelm its onboard instruments—more than a thousand times what theory had predicted. Then, just as unexpectedly as rates had risen, they fell. The radiation belts of Jupiter, it was soon realized, were not only far more potent than those of Earth-like planets but also more unstable, and wobbled around the gaseous giant at an eleven-degree tilt. The gusty ionic wind had veered, as it were, just enough to permit Pioneer to pass through without foundering on the magnetic reefs. Two cosmic ray telescopes became temporarily infirm, and the blast of irradiation gave fatal cataracts to the optics of the Sisyphus instrument.[69]

The Voyagers' designers hurriedly reworked the still-abuilding spacecraft to shield its sensitive instruments. Thanks to Pioneer, the hazards of space had become sufficiently well known to build protection into the Voyagers' apparatus before launch. But the Voyagers had one other enormous advantage over Pioneer: their programmable computers. They could self-analyze, receive new instructions, and operate with a degree of autonomy. Pioneer 10's flyby of Jupiter required 16,000 commands from Earth, and a flaw in any one could prove fatal. On Voyager an error could, within limits, be repaired or circumvented. Just as the Voyager Grand Tour could incorporate the lessons of Pioneer, so new lessons could become encoded into practice as the spacecraft sped across the solar system.

The asteroid belt does not track a simple orbit but those of a slushy cluster of planetoids and debris scattered across a wide swath. It took the spacecraft most of a year to traverse the full field. Voyager 1 and 2 exited the realm unscathed in August and September 1978, respectively.

DAY 413–602

# 11. Cruise

The Voyagers sailed on, and on. Week passed week, month passed month. Between the asteroid belt and Jupiter lay 370 million kilometers of interplanetary space. Mars was half again as far from the Sun as Earth was, the asteroid belt almost three times as distant, and Jupiter was twice that. The journey could seem endless.

The euphoria of launch had long passed, and the frenzied elation of encounter lay far in the distance. Other planetary projects beckoned, and JPL siphoned off engineers to work on those. Attention wandered. The Voyagers passed through a void of concentration, filled, when possible, by anticipation and analogy. Since there was a lot of interplanetary space, and a lot of earthly time to pass before the first, defining encounter, the Voyagers spent a great deal of time measuring and weighing analogies, and were in turn measured and weighed by them.

## SPACE FOR ANALOGY

What did space mean? In practical terms the choices were those represented by von Braun, Van Allen, and Pickering as they hoisted Explorer 1 over their collective heads. It meant colonization, science, and journey as exploration, and the American space program found it hard to hold them all together. But for some partisans, the space program overall meant more. It stood for—promised to catalyze—an immense social reformation destined to transform the American commonwealth on a scale with few precedents. In the words of one techno-prophet, "the thrust into space will change the ideas and lives of people more drastically than the Industrial Revolution."[70]

As early as 1962, NASA commissioned the American Academy of Arts and Sciences to explore the "secondary" impacts of the rapidly metastasizing national investment in space, anticipating that the vaunted spin-offs would extend far beyond simple economic stimulus. Good academics, the group devoted most of its exegetical energies arguing the limitations of historical analogy in principle. In the end they opted to examine the advent of the railroad as a possible model, and again, good scholars, they so hedged their analysis with qualifiers and caveats that the analogy virtually fizzled out on the launchpad. By then, too, political realities were quelling the rush of spending; by the end of Apollo, the NASA claim on the federal budget had reached a very modest steady state. The space program would not become the transcontinental railroad of the twentieth century. Partisans for utopian revolution would have to look elsewhere.[71]

What the exercise did emphasize was the complexity of novelty amid a thriving culture. A new machine, program, or idea derived its power from the richness and intensity of its interactions with the rest of society—and the space program *was* an interaction. Had the AAAS Committee on Space shifted its attention from the startling monies funneled into NASA and the promise of endless spin-offs, a kind of political perpetual-motion machine, and turned to exploration history, it would have probed and poked at very different data sets but come to a similar conclusion. Voyages of discovery derived their power from the fullness of their cultural engagements. A new machine or idea may reform society but only after it first fits in. An enterprise such as Voyager could not inspire or remake what did not already exist.

Similarly, there was no lack of analogies to exploration history. Space travel was imagined as a stage in earthly evolution, a metamorphosis in humanity's maturation, and manifest destiny gone to Mars. The planetary program was Leif Ericson, Columbus, and Magellan. What such allusions all shared was dissociation from particular sustaining societies; they were universal. Yet what boosters and designers needed was a closer look at just how those predecessors interacted with their cultures, and for that, since the Third Age was still aborning, they might profitably have examined the founders of the Great Voyages. They might have looked, especially, at the Portuguese paradigm.

The Portuguese ignited the First Age, traveled the widest, contributed the greatest number of pilots and mariners, and established a framework of trade, conquest, and discovery that the other nations subsequently emulated or seized outright. Portugal's overseas empire was an astonishing commitment that drained perhaps a tenth of the population out of Portugal and compelled the country to defend what it quickly appreciated it could not. The flush of early wealth it acquired it soon destroyed with foreign wars. Others, notably the Dutch, picked off its best Indies holdings. But while the interlopers poached pieces, a remarkably robust imperium endured. Portugal's saga told not of an endless quest that, once launched, became unstoppable, but of a life cycle, one that expanded and then contracted.

The Portuguese experience established the default setting for exploration's software. The degree of interpenetration between geographic discovery and Portuguese society was astonishing, of which the suite of exploring ships was only a down payment. Consider the founding explorers, all of whom combined exploration with some other enterprise: affairs of state, commerce and conquest, proselytizing and poetry. Henry the Navigator, late-medieval prince, blurry-eyed speculator, who began the fusion of discovery with state policy. Vasco da Gama, merchant and administrator, representing the bonding of commerce with exploration. Afonso de Albuquerque, soldier and strategist, seizing at gunpoint the critical nodes of traffic through the Indian and South China seas. Saint Francis Xavier, tempering the sword with the cross, missionizing in India, the East Indies, and especially Japan, with plans to proselytize in China. Luiz Vaz de Camões, adventurer turned litterateur, author of *Os Lusíadas*, which cast contemporary explorers into the mode of classical heroes and which became the national epic. "Had there been more of the world," Camões wrote, his bold mariners "would have discovered it." Revealingly, the founders of the Enterprise of the Indies all died overseas.

Enthusiasm there was for expansion, but the questions of where and how and to what end kindled a furious debate amid competing claims. Nine years passed after Bartholomeu Dias rounded the Cape before da Gama headed to India, and exploring energies in the Indies were soon dissipated in Morocco. The whole enterprise began winding down within sixty years after its founding. That astonishing outrush had lasted less than two generations, exhausted by overreach and especially by heedless military adventures. The colonies took on lives of their own, populated not by fresh émigrés from Portugal but by mixed-blood societies whose ties to the metropole were largely language and faith, and ever-more-tenuous memories. And even as Camões penned his epic amalgam of triumph and tragedy, Fernão Mendes Pinto was recording a parallel, equally fabulous tale of adventure, mishap, and squandered opportunities that makes his *Peregrinicão* read as though it were an early draft of *Gulliver's Travels*. In brief, Portugal did it all, and its Great Voyages transformed it as little else could. They also destroyed it as few national undertakings might. The outcome depended on what else exploration bonded to.

Did the program that launched Voyager resemble the Portuguese paradigm? Not in its particulars; not even by analogy. What they shared was that both Voyager and its Portuguese antecedents rallied and merged many complex enthusiasms within their sustaining societies. Those exploration analogies work best that are older and more diffuse and less subject to empirical qualifications. The more remote the historical sources behind an allusion, the more nebulous are its details and the more susceptible it becomes to multiple meanings. Columbus could mean anything. Lewis and Clark could mean everything.

Where commentators also groped for analogies was to depict the terrains of the Third Age. The "new ocean" of space was one, although its use was encumbered by the presence of that real new ocean of the Third Age, the recently explored abyss. Even more elaborate was the analogy to Antarctica. To both partisans and critics Antarctica was the future. They just saw that future in diametrically opposite ways.

Antarctica was an ideal venue—the geographic and historical transition from the Second Age to the Third. Its contrast with the Arctic expresses that status perfectly. The Arctic is a sea surrounded by land; the Antarctic, a continent surrounded by ocean. The Arctic, for all its hostility, harbors life around and under it, and to some extent on it, not just random organisms but functional if spare ecosystems. For the Antarctic, save trivial oases of exposed rock, life ends at the continent's edge. Even more pronounced are their divergent human histories. The Arctic is encased by, accessible to, and hence bonded to, human history. It has its indigenous peoples, its imperial claims and colonizing epochs, its ancient economies of hunting, fishing, and trade. The Antarctic has none of this. No one has ever truly lived there; no enduring natural assets bind it to the world economy; no colonization or claims to sovereignty have global recognition. Its population is scientists; its trade, information; and only an immense expenditure of will and money has forged even these tenuous links. The Antarctic's isolation is so complete that it seems less an intrinsic feature of the planet than an extraterrestrial presence accidentally slapped onto its surface, as though an icy moon of Uranus had slammed into Earth.

It is a place that is isolated, abiotic, acultural, and profoundly passive. One goes there in defiance of natural impulses. The scene reflects, absorbs, and reduces. It acts as a geophysical and intellectual sink. It takes far more than it gives. With implacable indifference it simplifies everything: that is its essence, the synthesis of the simple with the huge. It reduces an entire continent to a single mineral taller than Mount Whitney and broader than Australia.

The scene acts on people as it does on other earthly features. There is no genuine society. There are no children, no families, no schools, no social order, and no matrix of interlocking institutions. It is more like a mining camp or those shore-based trading outposts typical of the First Age. But worse: at least the residents of earlier outposts could intermarry with indigenous peoples, and the resulting mestizo societies—pioneered, as so much of European expansion was, by Portugal—did the heavy work of exploring and settling the interiors of South America and southern Africa, and of parts of south and southeast Asia. Exploration meant a transfer of knowledge from one group to another; explorers relied on native guides, translators, hunters, collectors; and they typically adopted native clothing, if not native mores. None of this was possible in Antarctica. Yet without such cultural contact and without a true social setting there could be no great literature or art. The ultimate Antarctic saga is Douglas Mawson's solo trek, slogging alone across broken ice fields, with no guide and little direction, nothing to record but conversations with himself, nothing at all but his own will to continue.

To this sketch there are seeming exceptions. Chile maintains a small army base, complete with families, on King George Island. The Southern Ocean swarms with krill, fish, seals, whales, and penguins. There are microlichens on some exposed rocks, and bacteria in sandstone and perhaps under the ice sheet itself. Tourists visit sites on the peninsula annually. But all these activities occur along the continental fringe, or on minor outcrops along the margins or, in the case of the Chilean base, on an island outside the Antarctic Circle (roughly equivalent to the latitude of Helsinki). The social order, often quasi-military, is akin to that of a ship, quite independent of place. America's McMurdo Station is not a Plymouth colony but a Virginia City, and Amundsen-Scott South Pole Station not a St. Louis but an icy St. Helena. Biotic fragments do not make a sustaining ecosystem. The Antarctic analogy begins only when you cross the barrier ice and step onto the ice sheet. Then you enter the Third Age.

The Antarctic analogue suggests that the Third Age will be dominated by near-Earth terrains. Exploration will happen through remote sensing and robots. Outposts may be permanently established but not permanently staffed, their inhabitants coming and going routinely, like migratory flocks or marching penguins. They will traffic in the luxury goods of an information society. The requisite rivalry that must power exploration will probably derive from the competitive character of science itself, conducted as a cultural pursuit with national prestige, not national survival, as its payoff. It will mean, in a curiously postmodernist way, talking with ourselves.

These traits identify Antarctica with the Third Age. Some of them are more intensely manifest on the Ice than elsewhere. Mars is richer in information; the Canary abyss far lusher with life. But Antarctica is accessible to people at relatively little cost compared with space travel, and can be seen without the extraordinary cocoons that shield the senses of human travelers from the environment. You can breathe on the East Antarctic plateau without special oxygen tanks. You can smell, taste, touch, and hear, as well as see (there just isn't much to smell, taste, touch, or hear). In the abyss and in space, only sight is possible, which is why robots and instruments can seamlessly replace human observers. It remains to devise programmable computers to similarly guide mechanical hands and supplement, if not supplant, the respiring brain.

To partisans of space settlement, the Antarctic analogue is a glass half full. It is the beachhead for a more remote and complex colonization. To critics, it is a glass as full as it will ever become. They note that humans have wintered over in Antarctica for more than a century and that the dynamics of such outposts have not changed significantly. For them, a Voyager, not a base at Dome C, is the future of geographic discovery.

## ANALOGY'S END

Analogies fill in what we don't know. They are a form of social anticipation, of prophecy or prediction. Understanding abhors vacuums as much as nature. Certainly that adage has applied to the vacuum of space.

As Voyager moved from idea to machine to functioning spacecraft, analogies buzzed around it, all forecasting what it might do, might become, and might mean. But once it began its tasks, those analogies faded; the mission moved from prediction to history. Voyager was poised to become itself a source rather than a recipient of analogy. It was no longer something to be imagined but rather an event recorded and a source of hard data.

In June 1978 Voyager 1 began transmitting photographs of Jupiter, followed by increasingly dense recorded radio emissions, plasma wave data, and other readings. On December 10, 1978, still eighty million kilometers from Jupiter, Voyager 1 transmitted photos of the giant planet that exceeded in detail and color any ever taken from Earth.

DAY 491–681

# 12. Encounter: Jupiter

Planetary encounter—this is what Voyager was made for. Jupiter was the first, and it established the pattern of sequenced phases: observation, far-encounter, near-encounter culminating in periapsis, post-encounter. That ritual cadence, with modifications, like a slow spiral of activities that quickened with approach, would repeat six times across the outer solar system. Voyager 1 initiated that astonishing series, achieving its closest approach on March 5, 1979.

Preparations began while the Voyagers were still cruising. There were course corrections, as always, and in late August 1978 both spacecraft received new programming intended to sharpen the performance of their imaging systems. In early November, a month after Voyager 2 cleared the asteroid belt, flight controllers commenced training exercises, some four months before nearest encounter. A paradox of planetary exploration was that discovery depended on an intricate scenario of preplanned maneuvers that determined the exact trajectory, decided what instruments operated for how long and in what sequence, and arranged for when data was downloaded—all this calculated to the minute or even to seconds over a 39-hour period that constituted nearest approach as an accelerating spacecraft hurled toward Jupiter at 46,000 kilometers per hour and was flung away at 86,000 kilometers per hour. Such detailed choreography required precision drilling. A rough rehearsal followed on December 12 through 14.[72]

The next day Voyager 1 left its cruise phase for its observatory phase.

## VOYAGER 1

Observation commenced officially on January 4, 1979. There was some flexibility as to when it might actually begin, and once tasks were uploaded, the spacecraft could do its chores with or without close ground supervision. Originally, observation had been scheduled for December 15, but the holidays complicated matters, so rehearsals had been moved up and the official opening slipped back. This left Voyager 1 still some sixty days out from Jupiter.

Beginning on January 6, Voyager took photographs of the planet every two hours; from January 30 to February 3, it began photographing every ninety-six seconds, and did so over a one-hundred-hour period, until Jupiter became too large for its narrow-angle camera. For the next two weeks, photos came in sets of two-by-two pictures, and by February 21, three-by-three images, later assembled into mosaics. Meanwhile, Voyager directed its other instruments—the ultraviolet spectrometer, the infrared spectrometer, the polarimeter, instruments for planetary radio astronomy, and plasma waves—to absorb, scan, and record.[73]

The milestones rushed by. Ground crews waited anxiously for the bow shock, the ionic reef of Jupiter's magnetosphere. On February 10 Voyager entered into the realm of the Jovian system, crossing the orbit of Sinope, the outermost moon. On the seventeeth it photographed Callisto, and on the twenty-fifth, Ganymede. It was now scanning images every forty-eight seconds, and instruments were recording auroras, ion sound waves, and the shifting shoreline of the magnetosphere. The real payoffs, though, would follow periapsis as Voyager made close flybys of the Galilean moons. Scientists and the press began converging on JPL. On February 27, with Voyager some 7.1 million kilometers from Jupiter, public TV in the Los Angeles area began broadcasting a "Jupiter Watch" that allowed the public to view virgin images at the same time as mission scientists. On the twenty-eighth Voyager finally felt bow shock, and time-lapse movies revealed the mechanics of the Jovian atmosphere and its star performer, the Great Red Spot. On March 1, now 4.8 million kilometers away, Voyager crossed the magnetosphere, and photos of the distant satellites were released. The next day heavy rain at the Canberra Deep Space Network site blocked signals for several agonizing minutes, but ever more detailed images continued: the eyes of discovery belonged to anyone with a TV set. On March 4, the eve of near-encounter, Caltech hosted a symposium on "Jupiter and the Mind of Man." There was surprisingly little to engage, for Jupiter had never commanded interest comparable to that of Mars or Venus. Within hours, however, its cultural clout would change.[74]

Closest encounter came at 4:05 a.m. PST on March 5, 1979, some 780,000 kilometers from Jupiter, at a velocity of 100,000 kilometers per hour. The message took thirty-seven minutes to reach JPL. Voyager 1 now flung about the planet for what would be a survey of Jupiter's satellites. It crossed the orbits of the Galilean moons, one after another, and that evening took a full-frame sequence of Io. A failure at the Madrid DSN station caused a gut-wrenching loss of fifty-three minutes' worth of data, and an hour later, another eleven minutes washed away. Meanwhile, as Voyager passed through Jupiter's equatorial plane, it focused its cameras on the space between the gaseous clouds of Jupiter and its tiny rock of a moon, Amalthea. Three days later, analysis confirmed that Voyager 1 had discovered an unknown ring. Now Voyager trained its cameras on Io. At 8:14 a.m., having completed its *gran volta*, the spacecraft disappeared behind the giant planet. For an anxious two hours and six minutes, invisible on the far side of Jupiter, it could only store data, before broadcasting its hoarded information back to Earth when, an hour later, having made its closest approach to Europa, it emerged from the shadow zone altogether. Before the day ended it passed by Ganymede, and the next day, March 6, it flew by Callisto. Two days later encounter officially ended, and as JPL shut down the daily press briefings, project scientist Ed Stone said simply, "I think we have had almost a decade's worth of discovery in this two-week period."[75]

The cavalcade of wonders had stunned nearly everyone. Voyager had photographed a gallery of Jupiter's inner moons, each of them revealed as a "different world." There was Io, pocked but uncratered; Ganymede, cratered but also grooved, apparently through faulting; Amalthea, an asteroid-like body; Europa, smooth with ice and mysteriously dark-streaked; and Callisto, icy, densely cratered yet also smooth and boasting the "largest single contiguous feature" yet discovered, an immense impact crater named Valhalla. Reviewing data, scientists quickly confirmed a Jovian ring; and other, more arcane analyses tumbled out. On March 8, as encounter concluded, Voyager looked back again on Io, now a luminous crescent against the Sun. The intent was mundane, an exercise in navigational backsiting. Within several days, however, the photo became the canonical image of Voyager 1 at Jupiter. In that unexpected outcome the episode might well stand for the entire encounter.[76]

Jupiter is a miniature solar system. It is a gaseous planet with a diameter of 142,800 kilometers, one tenth that of the Sun but more than ten times that of Earth. It features the dual geography of the solar system, the hard field of planets and gravity and the soft field of electromagnetic radiation and ionic gases. It has moons the size of small planets, each distinct. It has a magnetosphere to complement the Sun's. The instruments Voyager carried could measure them all.

Both realms are dynamic, both held surprises, and Jupiter warped each. In its soft geography, researchers found lightning whistlers, mapped the rude dimensions of the Jovian magnetosphere, calculated radio emissions, documented plasmas of various ions, identified the source of exceptional radiation (sulfur from Io), and plotted the magnetic torus around Io. The dominant topic was the magnetosphere, where Jupiter's electromagnetic energies collided with the Sun's to power a kind of invisible weather of strong radiative winds and rough magnetic seas. If rounding Jupiter was akin to doubling the Cape of Good Hope, then it is worth recalling the cape's original name, the Cabo das Tormentas, the Cape of Storms.

The magnetospheric border between the Sun and Jupiter was fluid and blustery; it was not where Pioneer 10 had found it; and when Voyager 1 did cross that frontier, the spacecraft had to pass through its electromagnetic shoreline, buffeting against ionic whitecaps and tides, not once but five times, as violent solar winds pressed and pushed against their strenuous Jovian rivals. The radiation blast, despite hardening against it, was sufficient to damage parts of Voyager 1's payload, notably its clock and the synchronization of the two central computers, with the result that the camera and scan platform operated on a schedule some forty seconds apart. What should have been some of the highest resolution images of Io and Ganymede returned blurred.[77]

Even where fuzzy the images of Jupiter's hard geography were what gripped public imagination. Here was revealed, for all and at the same time equally, amid a collective public gasp, the character of new worlds. While there was no solid surface to survey, Jupiter's fluid, multicolor atmosphere was ready-made for time-lapse movies. Here was the weather of another planet, all recorded in gorgeous Technicolor. Hurricanes the size of Earth's Moon that lasted for centuries; stormy eddies that roiled past like boiling Mississippis; trade winds that would shred and crush sailing ships; cloud depths that could vaporize comets. As Voyager approached, and the images sharpened, the Great Red Spot became a cosmic celebrity. A photo of tiny Io against the backdrop of overweening Jupiter spoke volumes about the majesty of the Jovian system and the awesome scales of planetary discovery. With its Mondrian hues and Pollock swirls, Jupiter became a gallery of abstract, modernist art.

But it was the inner satellites that ignited what evolved into a Voyager mystique. Unlike gaseous Jupiter, they were hard bodies, and could be imagined by analogy to familiar moons and planets. As Voyager 1 trained its cameras on them, everyone could sense that these were, as mission geologist Larry Soderblom observed, "new worlds" ready for discovery. Yet none resembled Earth or its Moon, or even Mars or Venus. None looked like the others. Amalthea, Io, Ganymede, Europa, Callisto—each had its own geological evolution, and operated on principles unlike those of the Sun's inner planets. The images were fresh, startling, spellbinding. The geology they exhibited was, at first sight, inexplicable.

Relief mingled with wonder, like the colored-thread meanders and complex eddies of Jovian clouds. What first overwhelmed observers was a sense of elation that Voyager 1 had journeyed so far and so successfully. They then surrendered to that deep wonder that is what the discovery of new worlds has always promised.[78]

## VOLCANOES

As Voyager whisked beyond Jupiter, it turned and took one parting shot of Io, now 4.5 million kilometers away and a luminous crescent against the deep black of space. It was a long-exposure image intended for use with its celestial navigation system.

In processing the image, Linda Morabito, a member of the optical navigation group, noted an anomaly. A domelike cloud seemed to bubble up from the surface. Yet Io had no atmosphere. Others, meanwhile, had also spotted and begun to ponder the image. An independent discovery by John Pearl of the IRIS team led to a consensus conclusion: Voyager 1 had witnessed a volcanic eruption. An investigation of all the images over the next week identified eight such events. Io in eruption became the defining image of Voyager's encounter with Jupiter.[79]

Suddenly a dozen inexplicable oddities about the Jovian radiation belt made sense. Its exaggerated intensity and the sulfuric emissions recorded by ultraviolet instruments were the result of sulfur ions blasted into it by Io. Geologists now had an explanation for the moon's weird colors, hot spots, and gaseous bubbles. Instead of impact craters, Io had massive calderas. Io was nothing like Earth's Moon, inert and cold. It was far removed from the dead bodies expected to litter the outer solar system, far distant from what theory had predicted. Subject to Jupiter's immense tidal bellows, Io warmed, melted, and blew off eruptions of sulfur that entered into and altered the magnetosphere. At Io the hard and soft geographies of the solar system collided with a display of planetary fireworks.

Such scenes may have been what researchers secretly hoped for, but it was not what they had expected. After Io, they sensed they could only anticipate surprise. They knew the mission had become a true voyage of discovery.

---

That the iconic image should be a volcano seemed particularly apt. Throughout Europe's half-millennium expansion, volcanoes had been a constant feature, both as a practical referent and as a symbolic emblem of exploration. Volcanoes guided and inspired, and became themselves prime objects of inquiry. They can serve as a useful index for the Three Ages.

The shipborne First Age traveled from island to island, and the critical isles were almost all volcanic. The Atlantic isles that served both Portugal and Spain were volcanoes: Madeira, Cape Verde, the Azores, and especially the Canaries. So were their destinations, the East and West Indies. There were some notable exceptions, such as the Bahamas, where Columbus first made landfall, and those critical offshore isles that so often served as protected ports of call. But the Antilles were a chain of larger and smaller volcanoes, as were the fabled Moluccas. The very Spice Isles themselves—Ternate, Ambon, Tidor, Banda—were active volcanoes. Between them the twin peaks, Tenerife and Ternate, port of embarkation and port of destination, defined the passage to the Indies.

After Spain finally conquered the Canaries, Tenerife's Pico de Teide became the tallest peak in Europe, a land's end and a point of departure for far-voyaging mariners. The isles governed the oceanic routes west; they served as both an economic and an intellectual entrepôt for receiving and disbursing the gathered flora and fauna of Europe, Africa, Asia, and the Americas; they were the proving ground for Spanish colonization, and for the naturalists who would, over centuries, replace missionaries and conquistadors; the islands served as a standard reference by which to compare what explorers might discover elsewhere in the world. In particular, Tenerife's Teide stood as a template, a rite of passage for ambitious adventurers and a scientific stele by which all else beyond Europe might be measured and its meaning deciphered. This was where one left the Old World for worlds both newer and older. What popular lore attributed to Prince Henry's mythical college of scholars at Sagres belongs in truth to Tenerife.

Even when physically bypassed, Teide persisted as symbol. The role it enjoyed in the First Age for conquest, the Second Age revived for science. "The port of Santa Cruz is in fact a great caravanserai on the route to America and India," noted Alexander von Humboldt, in this, as in so many matters, the oracle of the Second Age. "Every traveler who writes his adventures begins by describing Madeira and Tenerife, though the natural history of these islands remains quite unknown." He promptly set out to correct that defect, scaling the mountain and recording its rocks and flora, effectively field-testing the new sciences on its slopes. By celebrating Tenerife in his *Personal Narrative*, Humboldt fixed its image in the mind of his endless imitators. Then he went further. When he created an atlas by which to compare Earth's great mountains, he implanted Tenerife at the center, flanked by the Andes and Himalayas, establishing Teide as the standard by which to measure all the others. In 1831 Charles Darwin declared that he "would never rest easy until I see the peak of Teneriffe and the great Dragon Tree." When he had the chance for his own expedition, he regretted that the HMS *Beagle* sailed from Madeira, but as it passed by the Canaries, he glimpsed Teide and recited to himself "Humboldt's sublime descriptions." Then he went on to find new icons for the era at the Galapagos, another cluster of volcanic isles.[80]

When Humboldt subsequently climbed Mount Chimborazo in Ecuador, he recapitulated his ascent of Teide in more Romantic style, and through their comparison he began to formulate his grand synthesis of plant geography. Chimborazo shone as the site where Enlightenment met exploration, the summit of Second Age discovery. It replaced Ternate, the Spice Isles, as the vision quest of journeying. The image appealed powerfully to artists, now making their Grand Tours to classical Europe and discovering in Etna and Vesuvius points where ancient history met natural history. Would-be Humboldts relocated Chimborazo wherever they traveled, using it to assert claims of cultural significance. (Heinrich Möllhausen, for example, inserted a Chimborazo, complete with blowing snow on its summit, alongside the Colorado River in the Mohave Desert.) What the *padraõ* was for the First Age, Chimborazo equivalents were for the Second.

The Third Age, too, has its Möllhausens, keen to establish continuity and parity with previous exploration. A NASA-commissioned painting of a future astronaut exploring party on Mars, for example, featured a Chimborazo looming on the horizon, exactly in the place where it appears so consistently in nineteenth-century renditions. When Arthur C. Clarke imagined his Eden on Mars, in *The Snows of Olympus*, he set it on Olympus Mons, a real if dormant volcano, and the largest then known in the solar system. He thus combined the iconography of the Great Voyages with the mythology of the Ancients.

That was literature. The eruption of Prometheus on Io was fact. So were the volcanoes, hot vents, black smokers, and dormant seamounts that dappled Earth's deep oceans and became a defining discovery for the Third Age on Earth. What these features shared was less shape, or lofty panoramas, than geologic dynamism. When Voyager launched, the expectation was still rife that the satellites of the outer planets would resemble those of the inner planets. Like Earth's Moon and Mars's Deimos and Phobos, they would be inert masses, cosmographic fossils, left from the eon of planetary origins. Active volcanoes and geysers showed otherwise: they testified to a still-vigorous geology. The moons were not dead lumps, like geologic shards left from the creation and preserved in orbit like museum pieces. They warmed, they moved, they belched. They had their own distinctive histories. They spoke to our times, not just to the solar system's antiquity. The volcanoes of Io were for space science the Ternates and Teides by which to triangulate the exploration of the interplanetary seas and new worlds beyond.

## VOYAGER 2

Now came Voyager 2 to confirm and amplify its twin's discoveries, and to demonstrate that it could perform agilely enough to warrant flinging itself not only to Saturn but also to Uranus. If the spacecraft had problems, it also, after Voyager 1's traverse, had acquired some opportunities.

The problems were that its polarimeter was broken, as its twin's had been, and more seriously, its primary radio receiver was still tone-deaf and balky and could not track an Earth-broadcast radio signal through its Doppler shift. The contact frequency drifted. The DSN found tricks to work around the stickiness, but when connections were good, flight control seized the moment and sent new commands and programs, and understood that communications might blink off at critical times.[81]

The opportunities were that its trajectory should help shield it from the kind of radiation damage experienced by Voyager 1. At periapsis it would be twice as distant from Jupiter, and if comparable damage did occur, a program was uploaded that would resynchronize clocks hourly and shrink the potential slurring of programs. Two months before closest approach, a new suite of commands adjusted Voyager 2's sequencing to target topics of supreme interest raised by Voyager 1's reconnaissance. Specifically, the spacecraft would maintain a ten-hour Io volcano watch, better measure the Jovian torus near Io, listen for whistlers, watch for lightning and auroras while on the planet's dark side, and photograph the Jovian ring, which it would cross twice. It would see the inner moons differently, and most of them much closer. Overall, it would reverse the sequence of Voyager 1's trek: it would begin with the moons and then cross the planet on its dark side. The new instructions were uploaded in early April.

On April 24, 1979, seventy-six days before periapsis, Voyager 2 officially entered its observation phase. From May 24 to 27 it took long-range photos of Jupiter, now experiencing a different season of weather. On July 2 Voyager first crossed bow shock. Far-encounter commenced on July 4, some 5.3 million kilometers from Jupiter and 921 million kilometers from Earth. As before, the Jovian magnetosphere was malleable, the result of solar winds and uneven eruptions on Io, and Voyager 2 passed in and out of the billowing cloud. By July 5 it had made eleven crossings and photographed Io so that researchers might see how many of the eight identified volcanoes were still erupting. The press, however, was more interested in a space story with stronger human interest, the imminent death plunge of Skylab. The two encounters—Voyager 2's at Jupiter, Skylab's at Earth—made a media counterpoint until the space junk fell into the Indian Ocean and onto Australia on July 11. On July 6, with ample media attention, Voyager continued its surveillance of Jupiter's weather and Io's vulcanism. The Red Spot alone now demanded a mosaic of six frames.[82]

Then came the main event. Voyager photographed the Jovian ring and Callisto on July 7. Voyager 2 saw that satellite far more closely and from the opposite side than had Voyager 1, so it seemed a relatively new discovery. A last set of commands went to the spacecraft to direct its imminent near-encounter. On July 8 the final sequence commenced: the ring again and Callisto up close; then, on July 9, Ganymede, Europa, and Amalthea, followed by Jupiter's dark side and a long farewell scrutiny of Io. Each sighting offered a closer and fresher perspective than with Voyager 1. At closest approach, 3:29 p.m. PDT, Voyager 2 came within 206,000 kilometers of Jupiter and sped around the giant at 73,000 kilometers per hour, more than ten times the muzzle velocity of a .38 special.

The views were again stunning, and startling. Callisto appeared to be the most densely cratered object known. Ganymede was a geomorphic breccia of craters, grooves, ejecta blankets, and fissures. Europa—"a sort of transition body" between the solid-silica Io and the icy Ganymede and Callisto—was perhaps the most bizarre, because for all its surface crinkling, it was as smooth as a billiard ball. Larry Soderblom noted how unlike these bodies were from the realm of the inner planets, and then listed the observed superlatives: "Included in the Jovian collection of satellites are the oldest (Callisto), the youngest (Io), the darkest (Amalthea), the whitest (Europa), the most active (Io), and the least active (Callisto). Today we found the flattest (Europa)." Voyager's sequel had matched its original. Even now adulation mingled combustibly with anticipation about what Voyager might achieve. NASA associate administrator for space science, Tim Mutch, championed Voyager as "a truly revolutionary journey of exploration" and exulted that, when "the history books are written," these times would be recognized as a "turning point."[83]

Yet much remained to do even at Jupiter, and Voyager 2 was at a critical turning point of its own, a long, complex trajectory correction, timed with periapsis, that would hurl the spacecraft to Saturn, always an anxious maneuver. The episode became yet more awkward after Jupiter's radiation, far more intense than anticipated, damaged the already troubled radio receiver. On the evening of July 9, radio contact ceased, and engineers scrambled to reconnect before the scheduled course correction. Happily, the intricate commands had been previously uploaded on July 6, when contact was clear and before bursts of radiation had upset the temperamental receiver. At closest encounter, Voyager's thruster rocket began a seventy-six-minute burn calculated to maximize the gravity-assist acceleration. The combined momentum would save ten kilograms of hydrazine, and with it, the capacity to tack and steer around Uranus. Meanwhile, the instrumental recording and imaging raged on. The dark-side observations chalked up marvel upon marvel, sighting the ring, auroras, lightning, and a revealing volcano watch on Io. The forward scatter that made the Jovian ring luminous, however, was nothing compared to the foreshadowing that made nearly every mind feverish with anticipation.

Both Voyagers were sprinting to Saturn. But even as it departed, Voyager 2 left a legacy akin to Voyager 1's iconic image of Io in eruption. Careful analysis revealed a new satellite, a body closer than Amalthea and apparently associated with the ring. It was named Adrastea. Jupiter acquired its fourteenth known moon, and Voyager had discovered another world.

Others would follow.[84]

## ISLANDS

The idea of a New World was a late concept, first broached around 1511 by an Italian humanist in the service of Spain, Peter Martyr, and subsequently consolidated in his posthumous *De Orbe Novo* (1530). What Europe's *marinheiros* discovered in practice were islands. The equivalents in "this new ocean" of space were moons.

The Great Voyages between Old and New worlds was most typically a voyage between islands, or from archipelago to archipelago. Those isles were portal, way station, sanctuary, and fortress. Points of departure—Madeira, the Canaries—had their counterpart in offshore islands from São Tome to Hispaniola or in seaports that had the properties of near-islands, such as Al Mina, Goa, Zanzibar, and Mombasa. The Indies, East and West, are a concourse of islands. Even in the seventeenth century, the Spanish monarchy proclaimed the

Canaries, located at a triple junction between Europe, Africa, and the Americas, as "the most important of my possessions, for they are the straight way and approach to the Indies."[85]

So, also, islands loomed large in the imagination of explorers, schemers, and cartographers. Fantasy islands, from Atlantis to the Island of the Seven Cities, dotted the Atlantic. Real islands swelled on maps to many times their actual size. Newly discovered isles became the incentive and model for colonization. If uninhabited, like Madeira, they pitted capitalist and colonist directly against the land. If inhabited, like the Canaries, they first required conquest to tame or remove the indigenes, or to smother them by immigration. Island-hopping was progressive as new islands often drafted populations from old; perhaps 80 percent of the Canaries' population was Portuguese, many following the sugar traffic from Madeira. Islands were potentially profitable estates, less onerous than seizing holdings from the Moors, and they offered ready-made fiefdoms by which to reward enterprising knights. (Sancho Panza mockingly begs Don Quixote to reward him with the governorship of an island.) Continents such as Africa and New Worlds such as the Americas were for the early explorers barriers, not beacons.[86]

In the Second Age, as sword and cross gave way to rifle and sextant, islands testified to experiments in natural history. As had Humboldt, exploring naturalists honed their skills on Atlantic isles before venturing farther abroad, and like their voyaging progenitors, they did their business on isles as destinations. It was from their journeys among volcanic islands—the Galapagos, for one; the Malay archipelago, for the other—that Charles Darwin and Alfred Wallace independently conceived of evolution by natural selection. Interestingly, Wallace developed his idea while fever-ridden on the Spice Islands themselves, a splendid symbol of how interest in islands persisted while their cultural configuration changed. Even the exploration of Antarctica proceeded mostly from islands (and it was from Ross Island that both Ernest Shackleton and Robert Scott made their separate attempts on the South Pole).

Not least, islands offered competing visions of environmental change. As naturalists spirited around the globe, they tallied the chaotic chronicle of species gained and lost, of lands gardened and gutted, of visions still bright or hopelessly blackened. Islands returned as environmental indices: they represented semi-controlled experiments in the interplay between humanity and nature, with the circulation of species as the measure of nature's economy. For every relatively untouched isle such as Mauritius, exploring naturalists identified a St. Helena degraded into biotic dust; for every successfully colonized isle, there was one plunged into ecological chaos; for every Madeira there was a Hispaniola. A summary of an era's exploring naturalists, Alfred Wallace's *Island Life,* was a virtual compendium of colonizing ecology, documenting the often lethal competition between endemic and exotic species.

The Third Age found its isles on the continents in the guise of nature preserves (Robert MacArthur and E. O. Wilson's *The Theory of Island Biogeography* replacing Wallace), in the deep oceans in the form of seamounts and hot vents, and in space through the solar system's dizzyingly diverse satellites. Even as they were disclosed, the seamounts—lush with deep coral, hosting fisheries like diminutive Grand Banks—were being denuded by trawlers. So in the Third Age, as before, there seems to be a precarious balance between discovery and destruction, as stripped seamounts compete with momentarily spared black smokers.

And space? The giant planets resemble the barrier continents of the Great Voyages. They might be coasted or their gaseous bulks occasionally visited by a probe or two, but the real work of mapping hard surfaces and sending robotic explorers (or perhaps humans) would happen on their moons. As with islands, there were a lot of them, and they were different, enough to warrant comparative study. They formed parts of what could readily seem miniatures, talismans, and relics of the solar system. The exploration of the solar system would be primarily an exploration of moons.

For the Ages of Discovery, islands have been symbols as well as sites. They were places for imagination as much as for away teams and probes.

It was onto islands that Europe's intellectuals projected their social fantasies. From the monastic-styled Utopia of Thomas More to the bustling laboratories of Francis Bacon's New Atlantis, from the Brave New World of Shakespeare's *Tempest* to the tropical Edens portrayed by Pierre Poivre, the yet-undiscovered island was a cameo of cultural ideals. Echoing Homer, who had Odysseus narrate his tale from the isle of Scherie, Camões opens *The Lusíads* at Mozambique; but narration in medias res commanded less power than the prospect of new lands—new societies, new hopes—that could be lodged only on recently discovered or yet-unvisited islands. Columbus spoke for all such visionaries when he rhapsodized over Hispaniola—and for all scorners, when he failed to implant a society equal to that dream.[87]

By the early eighteenth century the sheen had tarnished. Daniel Defoe maroons the vagabond Robinson Crusoe on a wretched desert isle in the Caribbean; Jonathan Swift casts his voyaging cipher, Lemuel Gulliver, onto one isle after another, ending with the fantastic vision of an island metropole, Laputa, suspended in the clouds, the creation of visionary "projectors." The bright prospect of new Madeiras and Tenerifes ended with the horrors of Hispaniola and the wreckage of the Antilles; the Arcadian gardens imagined beyond the horizon collapsed into the scalped forests, goat-plagued hills, and gullied slopes that washed over the boles of felled dragon trees and the bones of extinct dodos. Instead of wistful new Gran Canarias, there were grim St. Helenas, and in place of another hypothetical Ile de France, there was Red Madagascar, destined to become a symbol of habitat degradation, ever burning and bleeding its soils into the sea. In the New World the Black Legend acquired an ecological edge. The shimmering mirage of Columbus met the bleak reality of Las Casas. Exploring utopias ended as environmental dystopias.[88]

Yet still they beckoned. And since better worlds had to be somewhere else, exploration repeatedly opened up prospects for uninhabited lands—new frontiers that might avoid the mistakes of the past. The modern founder of the genre, Thomas More, published *Utopia* in 1516. As a Platonic ideal, an updated *Republic*, Utopia did not have to be any place; but the rapid discovery of unknown lands made its literary location as a New World isle plausible, along with the use of a weather-beaten Portuguese, Raphael Nonsenso, as a protagonist and prototype for the Ancient Mariners and Old Men who would hound and hector the enterprise.[89]

The Second Age found it harder to place its utopias: there were few terrestrial locales yet unvisited. For a while Tahiti served as a tropical Teide, the first and most enduring of the Enlightenment's new Edens. Jules Verne stationed Captain Nemo on Deception Island near Antarctica, and James Hilton sited Shangri-La in the remote mountains of Tibet, but most visionaries had to turn to the past or the future. They invented modern versions of lost golden ages and former Edens, or they projected, along evolutionary trajectories, perhaps outfitted with steam power, a more perfect future. Instead of places, they turned to peoples, remote tribes as relics of a primitive virtue or exotic folk such as the Polynesians of apparent plenty and ease. Then the Third Age in space suddenly made possible a revival of place-sited utopias.

For space seers there were three possibilities. One of course was Mars, seeming to stand to Earth as the New World did to the Old, awaiting only a more benevolent terraforming. How Mars might avoid the irony of past colonization and not sink into a planetary Haiti was unclear: it would simply happen. Proponents proclaimed, in a more secular declamation to those which Franciscan missionaries had made five hundred years earlier, that scientific knowledge was great enough to guide the technology to righteous ends.

A second possibility was more Platonic, to establish a space station among the Lagrange points, specifically at L-5, at which the gravitational attraction of Earth and Moon are equal, and which would thus suspend the colony in perfect weightlessness. Here was an invented, high-tech island free from the weight of the past—a solipsistic society, though one founded on ideals and first principles. If More's ideal was a kind of monastery, Gerard O'Neill's L-5 colony Dyson sphere is a kind of space suburb absolved of the past injustices and unmarred by the pull of Earthly vices. To advocates it occupies itself a kind of historical Lagrange point: it can neither evolve nor decay. To critics it might seem like a burnished version of Laputa. Neither utopia is likely to happen.[90]

Instead, exploration will probe the third alternative, the new isles of the solar system, the moons, now revealed as diverse, fascinating, and abundant. For visionaries these places come, as they had for the Great Voyages, with a window of opportunity between discovery and full-bodied exploration. The first ignites imagination, while the second tends to extinguish those utopian flames. The places' real value lies with a more worldly if still idealistic pursuit. Here are new worlds that, if closed to prospective colonists, still hold interest for science. Here are the experiments in natural history, equivalent to those scrutinized by Wallace, Darwin, and Joseph Hooker, that might reveal fundamental truths about the origin of Earth, life, and perhaps human purpose. Here are the promises less of a New Jerusalem than of a New Galapagos. Here are the Spice Isles of an exploring science: a new Madeira among the moons of Jupiter, a new Antilles in orbit about Saturn, a new Ternate at Io.

In brief, while the moons of the outer planets could never evoke the sensuousness of the earthly tropics, nor spark an equivalent lushness of imagination, they were the best the solar system had, and a few possessed the organic molecules that might be the byproduct or forerunner for life. The Third Age's answer to the tropical Edens that so enchanted intellectuals in earlier times, enticing them to imagine other worlds, would be satellites that beat to geologic rhythms different from plate tectonics or were rich with organic molecules like methane. The new Tahiti would be a place like Saturn's Titan.

## TO THE SIRENS OF TITAN

That is where the Voyagers now trekked.

By now, Pioneer 11, having doubled Jupiter and sprinted across the solar system, was a scant four months away from its rendezvous with Saturn on September 1, 1979. Once there, it found a new ring and even a new moon (indirectly, through magnetic disturbances), and most of all it blazed a route outside that great reef of rings that would preserve the option for Voyager 2 to continue the Grand Tour.

Meanwhile, the colossal Jovian gravitational field that had quickened the Voyagers' velocity during their planetary fling now pulled them back; but they had still gained more than they lost. Even as they slowed, they sped along some twenty-five thousand kilometers per hour faster than before they began encounter, with Voyager 1 pulling still farther ahead. For Voyager 2 its long course correction put it at a velocity and trajectory that would save precious hydrazine for its maneuvering thrusters. The Voyagers reentered cruise phase. They would continue this way for twenty and twenty-five months, respectively, chatting with Earth, sampling the solar wind and cosmic rays, reorienting themselves by the Sun and stars.

On they sailed to Saturn, across a void over 648 million kilometers wide, like two gnats crossing the Pacific.

DAY 587–914

# 13. Cruise

The spacecraft now had the velocities they needed to complete their mission. Voyager 1 had entered its Jovian observation phase at 48,960 kph; it exited at 86,000 kph. Voyager 2 had begun encounter at 38,016 kph and ended at 75,600 kph. Liftoff had granted escape from Earth. Their swing around Jupiter granted them escape from the Sun. They had sufficient inertia to carry them from giant planet to giant planet.

The program, too, had a quickened purpose. The spacecraft had worked. They remained sentient. They had completed what the pilots of the First Age had termed a *gran volta*.

## PORTALS AND PASSAGES

Past exploration had launched with muscle, wind, current, and tide. An exploring party might begin with nothing more than a long stride, a horse's gait, a floating coracle, or an unfurled sail. Long voyages required cracking the codes of wind and current—understanding their constancies, their seasonal variations, their propensity for storms. Discovery unfolded as those motive forces permitted and as mariners learned to harness them to their own ends.

For the North Atlantic, this meant deciphering the southwesterly trade winds, the seasonal paths of storms, and the Gulf Stream. The process began with coasting down Africa and some ventures into the Atlantic that discovered Madeira and the Azores. It proved difficult (and dangerous) to sail up and down Africa, however, and as the magnitude of the continent became slowly apparent it was obvious that while long coasting could serve cautious probes, it could never support the high-volume far voyaging demanded for trade with the Indies.

The solution was what the Portuguese called a *volta do mar largo*. This required a journey westward until, after reaching the far isles, one could turn into the westerlies and then ride them back to Madeira or the Azores. The *volta* involved a great arc by which one sailed out and then back. Over time, this pattern was enlarged until it could cross the North Atlantic altogether. In this emerging geography of wind and water, the Canary and Cape Verde islands were ideally situated to ride the southwestern trade winds to the New World, and Madeira and the Azores well sited to capture them on return. The short *volta* between near-seas became a long one over the Ocean Sea.

Crossing the equator unsettled that lore. The doldrums were a nasty mix of thunderstorms and calms. The inherited learning of the Ancients had nothing to contribute, the winds below were reversed, and even the North Star vanished. One could not travel to India coastal port by coastal port, as one could sail across the Mediterranean. Bartolomeu Dias could inch down Africa before being blown by storms southwesterly and returning around the Cape; a successor would have to find another strategy. Perhaps outfitted with some intuition that the world had to be symmetrical, that what happened in the north should, in reverse, happen in the south, probably aided with a dose of accident and happenstance, Vasco da Gama, once past the Cape Verdes and Sierra Leone, rode the trades out to sea and then turned south to capture the austral westerlies. Those powerful gales hurled him back to Africa, just above the Cape of Good Hope, which took another week to round. When Pedro Cabral did the same with the next Indies fleet, he veered still farther west and ran into Brazil; but he also rode the winds farther south and rounded—or "doubled"—the Cape. The great turn, pivoting with the winds, made the passage to India possible by flinging fleets around the gravitational mass of the Cape and into the Indian Ocean, where they would ride the monsoons to India.[91]

That was half the problem: tracing the patterns that characterized each discovered sea. But to truly sail the world ocean required discovering the links by which to get from one decoded sea to another. The brachiated peninsulas and islands that constituted Europe defined interior seas between which straits allowed passage. The Mediterranean was, for mariners, a collection of smaller seas—the Tyrrhenian, the Adriatic, the Aegean, the Ionian, the Ligurian—broader seas east and west, and innumerable gulfs. The composite had one eastern strait, the Bosporus, that joined it to the Black Sea, and one to the west, Gibraltar, that connected it to the Ocean Sea. A similar logic applied to the north, with its own complex quilt of seas, gulfs, and straits. That experience, like the creation of the volta, expanded across the earth.

The quest for the Indies that underwrote the Great Voyages was a search for passages across seas and a search for those straits that would permit passage from one sea to another. The barrier that was Africa required a southern passage; the barrier that was Eurasia, a northeast passage; the barrier that was the *novo mundo*, a strait somewhere—at its narrow middle at Panama, through its icy northwest, and ultimately around its southern extremities, the Strait at Magellan and the Drake Passage. The chronicle of discovery arranges itself around those pursuits like iron filings around the poles of a magnet.

Continental exploration obeyed a different logic, though it still began with the obligatory seaport that linked to the metropole. Beyond the coast, however, exploration depended on unlocking the geography of rivers and interior lakes.

Those lands that had rivers capable of ready access were discovered quickly; those that did not, lagged. Its enormous rivers, though they drained to the frozen Arctic, allowed Russians to cross Eurasia with breathtaking speed, by hopping from one tributary to another, like a squirrel jumping between branches. Timofeyevich Yermak crossed the Urals in 1581, roughly twenty-five years before Jamestown was founded in Virginia; by 1632, Cossacks founded Yakutsk, two years after émigré Puritans founded Boston; by 1649, Yerofey Khabarov had sailed down the Amur and established Khabarovsk, some fifteen years before the Dutch established New Amsterdam. The drainage of the Mississippi River opened up the interior of North America like a split log, and despite its snags, seasonal flows, and shallows, exploring parties habitually made the Missouri River the entry to the Rockies.

Lands that lacked access, particularly by galliot (or later, steamboat), stalled or redirected exploration. The St. Lawrence River proved little more than an extension of its gulf, stymied by rapids and Niagara Falls, so that travel portaged to the Ottawa River in order to cross to the Great Lakes. So it was also with the Rio de la Plata, which proved a similar dead end. The one major river in Australia, the Murray, was unusable as a point of entry to the interior, causing expeditions to stagger across stony desert and sandy spinifex by foot or camel. The most curious case was Africa, which was both the first and last of the continents penetrated by Europeans. The reasons for its lag were several, not least its horrific diseases; but the absence of usable streams north and south, the endless cataracts on its major rivers, and the confusing contortions and delta of the Niger all prevented serious exploration. The world's largest river by volume, the Congo, and its longest, the Nile, were the last in each category to be navigated.

The hope for simple ways through those great landmasses surrendered to a search for a means across them. Commercial rivers and navigable lakes were replaced by cross sections of natural history; and the hope of finding Great Khans, Prester Johns, and the bullion of new Mexicos, by inventories of natural wealth and nature's wonders. As steam power overcame the motive geography of wind and currents, so railroads replaced rivers as means of transit. By 1869 the last unknown river and mountain range in the continental United States were discovered, though not fully explored, curiously the same year as the completion of the transcontinental railroad. Within another decade cross-continental surveys had traversed Australia and Africa.

For the solar system, a still different logic governed travel, for this was a journeying geography defined by gravity. It had its equivalent to seas and straits, an imprinted dynamic of forces that one could sail with or against. The solar system had its doldrums and trade winds and, in Lagrange points, its Sargasso Seas that suspended dust and asteroids. Exploration could advance only if it found a way to harness those forces to its own ambitions, to ride over gravity's waves. For that it required a new *gran volta.*

The exploration of space could begin only by overcoming Earth's gravitational field, and it could proceed in a timely way only by navigating the gravitational fields of the outer planets. In the 1920s Walter Hohmann worked out for each planet the lowest energy speed to achieve that goal, which is to say, the trajectory that would require the least departure velocity from Earth, what became known as Hohmann transfer ellipses. For the outer planets the time required to travel would run thirty years to Neptune and fifty to Pluto. No spacecraft could survive that long, no scientist could sustain a career across so many decades without results, and no institutional program could expect to thrive over such a duration. Planetary exploration seemed doomed either to sail among the inner planets or to devise some additional propulsion beyond what an Earth-launched rocket could provide.[92]

The initial thrust, then, was to concentrate on hardware—bigger rockets were an annual event. It might be possible to strap on a snazzy new propulsion system that, once beyond Earth's gravitational field, could accelerate the spacecraft to velocities that made interplanetary flight feasible. But it was also possible that better trajectories might simplify and shorten the requirements. In the early years, when getting a payload into Earth orbit, much less to the Moon or to Venus, stretched their liftoff capabilities, engineers were eager to try anything. The best one might hope for in sustained travel was a flyby mission to Jupiter.

## VOYAGER'S THREE-BODY PROBLEM

In June 1961 JPL hired a UCLA math and physics graduate student, Michael Minovitch, as a summer intern to work on some tricky calculations involving trajectories to and around planets. The work absorbed, then obsessed the twenty-six-year-old, who continued to labor over it for the next two years, not only during summers but while at school on borrowed big-computer time at both UCLA and JPL. What emerged from his analysis was the realization that as a spacecraft swung past a planet, its velocity could undergo a permanent change. Specifically, moving in the same direction as the planet could add the thrust that bigger rockets or exotic propulsion systems could not, and thus make planetary exploration accessible with existing technologies.[93]

Minovitch quickly recognized the issue as a variant of the classic three-body problem in that it involved two large objects and one small one in mutual gravitational attraction—in this case, the spacecraft, the target planet, and the Sun. A solution had bedeviled Newtonian physics since its origins; it was to celestial mechanics what Fermat's Last Theorem was to number theory. No exact solution was possible, but with the advent of computers it was feasible to generate hundreds of approximations, each of them a possible trajectory for interplanetary travel. Enthralled, Minovitch, who termed his insight "gravity propulsion," learned FORTRAN to better run endless calculations and tried to work out as many solutions as possible, defaulting to his training as a mathematician to fashion a universal set of curves. He soon appreciated the seminal status of Jupiter by virtue of both its immense mass and its sentinel location between the inner and outer planets. At summer's end he wrote up his results in a JPL technical report.[94]

The initial reaction to his study was mixed. The mathematics was formidable, and the physics, in some respects counterintuitive, since it required a shift in reference frames from Earth to the Sun. Viewed from Earth, what a planet gave by gravity to an incoming spacecraft it then took away as the spacecraft departed; but viewed from the Sun, there was a gain or loss depending on the directions in which the spacecraft came and went. With a monster planet such as Jupiter, the momentum added could be significant for the spacecraft, though infinitesimal relative to the mass of Jupiter. For Voyager 2 the impulse acquired was 35,700 kilometers per hour. For Jupiter the impulse shed was a foot of orbital velocity per trillion years.[95]

The other, more serious problem was personal. As fascination deepened into obsession, Minovitch became more secretive, speaking only to friends and his late-night IBM 7090 computers, avoiding the give-and-take of shared ideas at the lab, more and more resembling a hermetic inventor worried with equal passion that his idea might be ignored and that it might be stolen. JPL had hired him to do work to fit into existing projects, and his supervisor, with whom he did not get along, believed that his independent inquiry, subject only to his own curiosity, interfered with his project assignments, which Minovitch admitted "it did." Moreover, what Minovitch believed revolutionary the trajectory group considered evolutionary, one of many ideas about how to get spacecraft to farther planets. Other researchers (not all with JPL) were experimenting with trajectories along similar lines. The group did not immediately reconstruct its planning around what they came to call "gravity assist."[96]

The technical objections were quickly overcome, and the concept of "Free-Fall Trajectories" took its place amid the assorted inquiries of the group. In 1963, after two more summers of labor, Minovitch was encouraged to write up a summary technical report. A new, more supportive supervisor, Joe Cutting, urged him to publish his results in the *AIAA Journal*. "I hope you will be able to devote enough time to finish it up," Cutting wrote, "so it can be submitted soon before anybody else comes up with something similar." You can't patent or copyright an idea, only its expression. Still, Minovitch didn't publish.[97]

Meanwhile, Cutting decided to push the idea and see if it could apply to a real mission, not to computer-generated curves such as conic sections. He and Francis Sturms plotted a flight to Mercury—an entire profile, with real numbers and actual dates—that relied on a gravity assist from Venus. What it required, and what Minovitch had not reckoned with, was precision navigation and launch dates; others supplied them. Mariner 10 demonstrated the concept's practicality by whipping around Earth, Venus, and Mercury in 1973. Pioneer 11 then confirmed the concept at Jupiter. Well before then, however, gravity assist had become fundamental to the design of the Grand Tour. Gravity assist was both real and essential. Those who conceived it and coded it into a mission moved on to other projects and, in Minovitch's case, other institutions.[98]

Yet, even as the concept passed test after test, Minovitch brooded over what he regarded as a lack of suitable recognition compounded with imagined slights by JPL. He saw his research as a work of individual genius, unprecedented and unique. It was not enough that the trajectory group had included him on its roster, that his technical reports had entered bibliographies of subsequent studies, or that NASA had awarded him an Exceptional Service Award in 1972. He believed he had done something monumental, and indispensable, and in his estimation planetary exploration such as the Grand Tour could not have happened without him.

So when Victor Clarke, his first supervisor, applied for a joint monetary award for the trajectory group's success with gravity assist, Minovitch sensed conspiracy, an attempt by undeserving others to claim what he autonomously had done. And when a historian, Norriss Hetherington, began research for a scholarly paper on the origins of the concept by interviewing members of the group, Minovitch threatened JPL with lawsuits, ended Hetherington's study, and began a campaign to have the concept, and hence the entire planetary program since Mariner 10, follow from his lone and unfairly ignored "invention." He found collaborators in his quest; JPL invited him to return for the Neptune encounter; and he has written up his versions in several papers for the International Astronautical Federation and constructed an elaborate Web site to promote his case in the belief that, as he put it in the third person, "there may be a small group at JPL that may attempt to claim the credit for his invention after he dies."[99]

Who discovered gravity assist as a means of propulsion for planetary exploration? How the issue might be answered depends on how the question is asked. Minovitch defined his achievement in terms of a classic theme in celestial mechanics, the restricted three-body problem, and his achievement he regarded as most significant for its mathematical techniques, especially his vector equations, which when combined with the number-crunching power of computers yielded the "first numerical solution." But gravity assist was, for him, also something more. It was an "invention," like John Harrison's chronometer or James Watt's governor, that made possible a "new philosophy of space propulsion."[100]

Others saw it all differently. The personality that made Minovitch's solitary labors possible also isolated them. Upon investigation there were many antecedents for the idea of gravity perturbations; and it was almost inevitable that the idea would gain currency, if not by mathematical analysis, then by empirical observation as spacecraft ventured to nearby planets and navigational groups recorded unanticipated changes in velocity. Minovitch's claim to absolute priority stems from his sense that no one else, certainly no one at JPL, was prepared to tackle a general solution to the restricted three-body problem; and because he saw it as a mathematical topic for which only numerical estimates were possible, he had to amass all possible trajectories, with the curve of his calculations asymptotically approaching something like a proof.

But what Minovitch saw as an intellectual puzzle, the JPL trajectory group, all engineers, saw as meaningful only within the context of a mission, and its solution would not come from endless hypothetical trajectories but from scenarios in which a spacecraft had real dates, real weight, and real escape velocities, and mission control had the capacity actually to navigate around planets with the accuracy that could ensure success. In their estimation, instead of elevating the concept, his complex mathematics and endless iterations had tended to depress its pragmatic value, and other trajectory group members, who regarded those solutions as "not truly adequate," devised simpler methods, including old-fashioned manual techniques with "tabulations and graphs." JPL was not an academy of science but a laboratory for reconciling science, engineering, and politics. Gravity assist was for them not an invention but a fact awaiting discovery and then, more important, demanding confirmation, development, and expression in a traveling spacecraft.[101]

What should have shone with the brilliance of its participants has threatened to sink into mires of petty egotism. The episode reminds us that history, too, has its three-body problems. They arise with particular vehemence when someone claims priority and someone else disputes it and when third parties only perturb their interplay, and such dynamics work with special vehemence when one body is smaller than the others. In such contests there is no formal solution, only working approximations. The transfer of gravity assist from mathematical concept to working mission was one such approximation. But, then, so was Voyager.

## FIRST DISCOVERY

Disputes over priority—credit for first discovery—are a constant of science, no less than for exploration, and when the two combine, the issue can be both futile and fathomless. There is no scene more common in both enterprises than a quarrel over priority, and perhaps none more unseemly or dismaying.

Both science and exploration thrive on competition. A collectively identified problem approached with shared understanding and techniques—the co-discovery of places, species, ideas, and techniques is so common as to be the norm. Without competition, the drive to discover weakens; but with it comes the spectacle of rivals converging, and then arguing endlessly over who really got there first. Both Isaac Newton and Gottfried Leibniz invented calculus, independently and more or less simultaneously. Alfred Wallace wrote a dilatory Charles Darwin, then still tinkering as he had for a decade with his ideas, and presented a fully realized articulation of evolution by natural selection. There are precious few scientific discoveries that have not come amid multiple contestants or that would not have been announced, with a slightly different accent, by another.[102]

So, too, has it been with geographic exploration. The norm is for rivals to converge, and then for explorers and their partisans to argue endlessly and with scholastic tenacity over who deserves priority. Did Portuguese (or Biscayan) cod fishermen rather than Columbus first discover the New World? And even among his small flotilla, did Columbus deserve credit for the initial sighting? On the evening of October 11, 1492, shortly before moonrise, both he and a seaman on the *Santa Maria* thought they spied a distant light, "like a little wax candle rising and falling," and others thought they did, too, or might have, before it vanished. At 2:00 p.m. on October 12, Rodrigo de Triana, aloft on the Pinta, cried out, "*Tierra! Tierra*!" Captain Pinzón confirmed the sighting, and signaled to the flagship, where Columbus agreed. Who, then, discovered the Americas? Those who saw a flickering light? The sailor who happened to be manning the *Pinta*'s lookout? The captain general of the fleet? Those who sponsored the expedition? Besides, all of these quarrelsome concerns only pertain to those who came from afar, not to those for whom the New World was an ancient homeland.[103]

Who discovered the Great Salt Lake? Fur trappers were all around the region in the mid-1820s. Etienne Provost almost certainly saw a part of it from the Wasatch Mountains in 1824. Some months afterward, to settle a bet, Jim Bridger followed the Bear River to where it entered the lake, tasted the waters, and reported as his discovery that he had found an "arm of the Pacific Ocean." Or did David Jackson, who reportedly explored the western shores in 1828–29? Or a party of fur trappers who spent twenty-four days in a bullboat on the lake, never finding the mythical outlet to the sea, the river Buenaventura? Or did discovery lie with the first published map of the enclosed sea by Capt. Howard Stansbury of the Army Corps of Topographical Engineers, after traipsing around the lake in 1849–50?[104]

Who truly reached the North Pole first? Did Robert Peary really stand at the Pole, or should his ambiguous assertion at the time, "I suppose we cannot say we are not at the pole," be taken as an admittance of failure? Did Frederick Cook beat out Peary at the last secretive moment, or commit another obvious fraud as with his earlier claim to have climbed Mount McKinley? These are sparse facts, bitter resentments, and strong personalities. Certainly Robert Peary's character inspired more critics than it did supporters. Moreover, there are institutional claims and jealousies, demands for recognition among this patron or that, this nation rather than another, and questions about whether the person or the sponsor merits special honor.

If not Peary, then who? Richard Byrd? Did he really pilot a Ford Trimotor to the pole, or turn back short? One can sympathize with those who prefer to leave the honor of first discovery to the

USS *Nautilus* and *Skate*, nuclear-powered submarines that surfaced through the polar pack during IGY. If so, then who can claim priority? The captains? The ships and crews? The U.S. Navy? To those engaged, nothing seems more vital than clearly establishing the pedigree of priorities, and to those less committed, nothing seems so pointless.

Voyager's quarrels, such as they were, lay in the realm of scientific rather than geographical discovery. Voyager was a corporate enterprise, an exercise in big science. The mission had in place explicit guidelines by which to determine if a new moon or ring was found, and methods by which to corroborate claims. Who, for example, discovered the volcanoes on Io? The first specialist to realize that an anomaly existed? Or the second, who was already moving toward independent discovery? Or the person who appreciated the anomaly for what it was? That no such quarrel bubbled up in public speaks to Voyager's sense of itself as a collective project.

Where it most differs from predecessors has to do with the peculiar character of Third Age terrains. No one lived on Adrastea, Cressida, or Proteus, so there was no tradition to guide visiting explorers, nor old accounts to translate into a new vernacular. There were no competing discoverers with rival claims or standards of substantiation. The procedures were in place, and followed. Some disputes were inevitable: they are intrinsic to discovery of any kind, and especially to the practice of modern science with which exploration had bonded.

Except for gravity assistance. It was easy, particularly in retrospect, to identify intellectual predecessors for the concept; Michael Minovitch crystallized one of them in elegant form. But the concept could enter into Voyager only by being engineered. It had to be tested against potential trajectories, had to fuse with guidance and communication systems, had to reconcile spacecraft with rockets, had to merge politics with ambition and idea. The recognition of gravity assistance did not make Voyager: by itself it went nowhere. It no more created spacecraft than the enunciation of the second law of thermodynamics created steam engines. The idea had to become a working machine. It had to be embedded in a mission. To those personally removed from it, the quarrel might seem like arguing over who invented the lateen sail that made the caravel possible, or who first identified the prospects for the *gran volta.*

Gravity assist was one of dozens of ideas—inventions, if you will—that made Voyager possible. Who invented Voyager? Who invented the Grand Tour? Without the discovery of gravity-assist propulsion, it could not have happened. But neither could it have happened without the Deep Space Network, high-speed digital computers to project trajectories, onboard computers to monitor and reset programs, and elaborate instruments and imaging; or without those complex groups, so often tedious to ambitious souls, that did the collective task of adapting, fitting, and testing to ensure that spacecraft could actually perform over long missions; or without support from NASA and OMB; or without the cold war to pry open tax dollars to compete with the Soviet Union among the planets.

Among their payloads each Voyager carried six small aluminum plates on which were engraved the signatures of all the persons who could claim to have contributed to their trek. The roster ran to nearly 5,400 names. What, or who, was Voyager, and who invented it? Many people, and no one.

## ACROSS THE VOID

Having made their *gran volta*, the Voyagers commenced a long, lonely trek to Saturn. Even with a roughly 40,000-kilometer-per-hour boost in velocity, Voyager 1 needed another twenty months, and Voyager 2 another thirty, before they reached their near-encounters with what members of the space community were coming to call the Lord of the Rings. Meanwhile, the shuttle had throttled the U.S. planetary program into silence. Pioneer Venus, designed to do at Venus what Pioneers 10 and 11 had done at Jupiter and Saturn, launched in May 1978. Then nothing. The shuttle took it all. The Grand Tour was the only tour.

The Voyagers flew alone.

DAY 752–1,031

# 14. Encounter: Saturn

To the Ancients it was known as the Great Conjunction, that moment, roughly every twenty years, when the varying speeds of Jupiter and Saturn cause them to appear to unite momentarily in the night sky.

As Voyager 1 closed on Saturn, a Great Conjunction was under way, with those two giant planets joined by Venus and the star Regulus to create a brilliant predawn constellation. Such alignments had happened forever, interpreted as portents of fortune and fate. But this time the cluster included a new celestial body and it joined humanity to the planets not through astrological magic but by aeronautical engineering in that predawn of solar system exploration. A Great Conjunction now marked Voyager's encounter with Saturn.[105]

## VOYAGER 1

Well before that moment, as early as January 1980, Voyager 1 began recording radio bursts emanating from Saturn on a roughly ten-hour cycle. These coincided with the rudely known rotation period for the planet; the refined transmissions allowed for a sharper determination of the precise numbers. But the anticipated spectacle was the spin not of the gaseous planet but of its rings, which had mesmerized viewers from Galileo onward. Where Jupiter stunned and dazzled with its Technicolor weather, Saturn did so with its gorgeous rings.

The observation phase commenced in October 1980. As Voyager 1 barreled toward the planet at 1.3 million kilometers per day, something like weather appeared, with lighter and darker blotches on the generally opaque surface of Saturn, and the rings revealed definitions and structures previously unviewed, a drama heightened by time-lapse movies akin to those made for Jupiter's Red Spot. Voyager photographed two inner satellites, first identified in 1966 but with properties otherwise unknown. By October 24 the spacecraft had halved its distance to Saturn, enhanced imaging beyond its narrow-angle camera, and brought its other instruments to bear, particularly its ultraviolet and infrared spectrometers. Voyager 1 officially entered its far-encounter phase.[106]

The rings betrayed not only more and more detail—grooved like a phonograph record—but also inexplicable dark "spokes" that spun among the rings. The phenomenon quickly became an object of special imaging. Analysis on October 25 and 26 led to the discovery of two satellites along the F Ring. Day by day, the resolution sharpened. Saturn displayed yellowish bands and turbulence among its cloud cover; its rings multiplied into hundreds, then thousands; its now-fourteen satellites acquired substance, rendering Saturn, like Jupiter, into a miniature solar system. Among the rings the Cassini Division revealed internal divisions. Computer enhancement and false-color imaging broke down the blurry haze of surface cloud to display a score of weather belts in the southern hemisphere alone. Titan acquired some definition and loomed as what it was: the second largest moon in the solar system. Daily, Voyager imaged the rings in a grand mosaic; every six hours it directed its full battery of instruments toward Titan, while otherwise it searched for unknown satellites and swept the scene with its spectrometers.[107]

On the evening of November 6, Voyager 1 underwent a final course correction—its ninth. The maneuver was doubly complicated because the Canopus star tracker had earlier malfunctioned, but controllers were confident that onboard redundancies would compensate. The spacecraft tilted 90 degrees, fired its rockets for 11.75 minutes, and took almost four hours before recovering sufficiently to transfer control to its gyroscopes back to celestial tracking. With its new trajectory it could speed past Titan, dip below the southern pole of Saturn, and then rise back up through the ecliptic of not only Saturn but also the Sun, and race away to the edge of the solar system. For the most part, Voyager 1 also determined Voyager 2's passage; but for now its remote, trailing twin assisted its sibling by measuring the strength of the approaching solar winds (which were high), thus allowing for better forecasts of when Voyager 1 might encounter bow shock, the turbulent, filmy border between the solar wind and the Saturnian magnetosphere. Meanwhile, the plucky spacecraft took a snapshot of Iapetus.[108]

What mattered was less the Sun's weather than Earth's. Thunderstorms pummeled Madrid and blocked out transmissions to its DSN tracking station for some six hours. Not until the Goldstone station, in California's Mojave Desert, acquired a signal could data return. When it did, Voyager revealed another new moon—the third it had discovered at Saturn—and yielded more information about the rings, Titan's atmosphere, and Saturn's magnetosphere. On November 9, sharpened images only heightened the mysteries of the rings and their spokes and of the anomalous absence of upstream bursts of electrons such as happened at Jupiter. But photos of Rhea promised another gallery of new satellite worlds. And of course there was Titan, as large as Mercury. On the eve of near-encounter, the panel of planetary wise men reassembled once more, as they had for Mars and Jupiter, to discuss "Saturn and the Mind of Man."[109]

On November 10, the air was electric with expectation. Voyager 1 would pass through Saturn's bow shock and fly past Titan, but veteran scientists and journalists readied to undergo their own bow shock of novelties. In what had become a mantra, researchers repeated that "Everything we are seeing on Saturn is brand new." That, in truth, is exactly what the partisans of Pioneer 11 had proclaimed, and both were right. Voyager saw more and saw more clearly, and it encountered a differently tilted Saturn, so that even when it revisited themes, it found them altered or it reexperienced them as though for the first time. Moreover, what Voyager had found at Jupiter did not, uninterpreted, translate to Saturn. The two planets, or planetary mini-systems, more resembled each other than they did Venus and Mars, yet they differed no less than did the inner planets. The discoveries were as fresh as the Bahamas in October 1492, or Antarctica in January 1840. The images scrolled past: Tethys, Dione, Rhea; the befuddling and mesmerizing rings; the muted weather of Saturn. Still maneuvering, Voyager acquired a new guide star while its far-encounter rapidly closed.[110]

The measurements, the images, the outburst of data—all rushed forward on the eleventh with the quickening pace of the spacecraft. For Voyager 1 there were in reality several near-encounters. The first was with Titan, some eighteen hours prior to its near-encounter with Saturn proper. As it whooshed toward Titan, the giant, enigmatic moon overflooded Voyager's narrow-angle camera, demanded three-by-three mosaics, and absorbed almost all the spacecraft's scans. Yet Titan refused to part its lofty haze; its surface remained opaque, and Voyager was forced to rely on indirect instrumentation, especially a double occultation, once as it passed behind Titan, and again after it passed out the other side. With that telemetry came the calculation that trajectory deviated from the programmed schedule: the distance, two hundred kilometers, mattered less than the timing, some forty-three seconds. (Engineers scrambled to recalibrate the sequencing before closest encounter.) Even as it swept beyond Titan, Voyager glided through the Saturnian ring plane and approached periapsis from beneath and on the dark side of the planet.[111]

On November 12 the countdown to closest encounter began with photographs of Tethys, the dark side of the rings, then Mimas and Enceladus, and the co-orbital satellites, S-10 and S-11, that Pioneer hinted at but never saw, followed by an occultation of Saturn and passage to its dark side, an encounter with Dione (which it was hoped would sweep the region clear of ring debris), followed by an exit occultation and reemergence from Saturn. Telemetry twice zipped through the atmosphere of Saturn and, ninety minutes later, danced dangerously with thunderstorms at the Madrid DSN station; but the signal came through. Still within Dione's presumed clear zone, Voyager 1 passed upward through the ring plane. Soon afterward the spacecraft made its closest approach to Rhea.

Everyone watching was stunned, and overwhelmed. They were "euphoric," exhausted, but mostly "simply flooded with new data." There was too much, and there was too much new, and too much that didn't fit theories and preconceptions. In this "strange world," as Bradford Smith, head of the imaging team, commented, "the bizarre" had become "the commonplace." He spoke particularly of fresh images of an F Ring, eccentric and braided in defiance of gravitational mechanics. But almost everything about the Saturnian system had either blurred Voyager's instruments with a gauze of haze or, if it sharpened the image, only deepened the haze of understanding. Tiny Mimas had an impact crater almost half as large it was. Tethys had a near-encircling trench. Dione hinted at internal forces. Enceladus lacked mighty craters. Iapetus had contrarian dark and light surfaces. Rhea boasted a new saturation level for impacts, all the more astonishing for what was essentially an icy rock. Saturn's inner satellites, all five, constituted "ice planets," an "entirely new class of worlds never before seen."[112]

The data analysis continued, even as Voyager turned to photograph and measure a receding, crescent Saturn. That image–Saturn's shadow streaking across its luminous rings–would be Voyager 1's Saturnian signature, as Io's volcanoes had been its Jovian. When the encounter phase officially concluded on November 13, Voyager continued to scan the scene, looking for new moons, viewing Saturn from fresh angles. It now sailed above the ecliptic, headed away from not only Saturn but also the planets altogether, for a final encounter with that nebulous boundary where the Sun's wind collided no longer with planetary magnetospheres but with those of other stars.

## TITAN

Saturn's shadow had not been the expected dominant image of Voyager's encounter. That honor had been reserved for Titan, for it was hoped that Voyager 1 might penetrate the veil of Titan's atmosphere. It didn't.

From the earliest planning, Titan had loomed as a primary target. The desire for a close flyby determined the entire trajectory of Voyager 1, which is to say, the Voyager mission overall, because going to Titan meant departing the planetary ecliptic altogether. But Titan was a world in itself. Its size, nearly as large as Mars; its atmosphere, denser than Earth's and rich in methane; its proximity to Saturn, with the promise of magnetospheric dynamics between the two—all suggested that, in the words of Project Scientist Ed Stone, "an encounter with Titan" would be "the equivalent to a planetary encounter."[113]

It would, so enthusiasts noted, perhaps be equivalent to visiting an early Earth, and a world more likely than any other to have life, or at least its predecessor organics. Voyager carried those hopes and queries to the planetary borders of Titan. A week before periapsis with Saturn, the spacecraft began an imaging sequence that, with computer enhancement, allowed some hint of contrast and perhaps of surface structure. But even with false-color imaging and other tricks, Titan remained impenetrable and inscrutable. Instead of contrasting with the generally bland surface of Saturn, Titan recapitulated it. Its size overwhelmed Voyager's cameras, forcing imaging into two-by-two mosaics, and then three-by-three; there was no break in the clouds, only at best a computer-forced glaze of gloom. The cameras that had proved so illuminating on hard-surface satellites could here show close up only what telescopes had viewed from far off. The gauzy sheen screened off detail. The best that close scrutiny could discover was a haze layer, detached and hovering over the north pole, some one hundred kilometers above the impermeable primary atmosphere. The interrogation of Titan fell to instruments operating outside the spectrum of visible light.[114]

These did as they were programmed to do. The infrared spectrometer (IRIS) mapped heat, the ultraviolet spectrometer (UVS) scanned for higher-energy wavelengths, while radio astronomy tested occultation through the Titan atmosphere, and plasma wave detectors traced the anticipated interaction of magnetospheres between

Titan and Saturn. (Titan, it was discovered, lacked a magnetic field.) But there was no surface to view, and there would not be until a radar imager could be sent to penetrate the clouds. Titan shied away from offering a tangible new world, presenting only an enigmatic mist-shrouded isle to taunt and tempt.

Instead of revisiting this unpromising scene, Voyager 2 would tweak its trajectory away from Titan and toward the equally baffling but far more transparent rings.

Had those hunkered down in the windowless mission center at JPL been able to look outside, they might have found the contrast between Earth and Titan that eluded them on their monitors. As the Saturn encounter wound down, Santa Ana winds had picked up, threatening the DSN antennas at Goldstone as thunderstorms had those at Madrid and compelling JPL to rely on backup generators to ward against power failure. More strikingly, the San Gabriel Mountains were aflame.[115]

Titan's most passionate partisans were those obsessed with its organic chemistry and the prospects for life, or at least life's biochemical progenitors. Titan was Mars with methane. Voyager revealed that its atmosphere was 1.6 times as dense as Earth's and that nitrogen was, as with Earth, a major constituent. Titan offered a tantalizing prospect for a world in which methane took the place of water on Earth, with methane clouds, methane rain, and methane seas; but mostly it teased with the hope that a fluid soup of organics might reveal the cosmological foundations for life. It would offer that comparative alternative that would move life from a singularity to a statistic. What the Galapagos Islands did for the theory of evolution by natural selection, Titan might do for exobiology.[116]

That is what made those back-lot fires significant. At the time the prevailing portrayal of free-burning fire held that combustion was solely a chemical reaction shaped by physical circumstances, and that it mattered ecologically because it had been on Earth since the earliest terrestrial plants. It was a physical disturbance like hurricanes and ice storms, and affected the living world similarly. While everyone agreed that "life" elsewhere would not look like life on Earth, there had to be common features, much as the atmospheres and rings of Saturn and Jupiter, so seemingly alike, were different, and a general theory of planetary atmospheres had to accommodate both. So it would be with a theory of life. It would have to embrace both Earth and Titan. Earth chauvinism would falter before the challenges of the planetary Other.

Yet increasingly it is the conception of earthly life that has changed, a reformation for which fire may trace the emerging contours. The realization slowly grows that fire on Earth is biologically constructed; that life creates the oxygen, that life creates the hydrocarbon fuels, that life in the hands of humans is the primary source of ignition, that the chemistry of combustion is a *bio*chemistry, joining the mitochondrial Krebs cycle to the cycle of flame in chaparral. Unlike floods, earthquakes, volcanoes, or the Santa Ana winds—all of which can happen without a particle of life present, or even shards of preorganics—fire cannot exist apart from its living matrix. Fire literally feeds upon biomass.

The fires that ripped across the San Gabriels offered an insight into earthly life far more vivid than abstract appeals to hydrocarbon drizzles and smoggy screens in the upper Titan atmosphere, and they threatened not so much the electrical power that ran JPL monitors as the conceptual power that underwrote what those monitors were designed to search for. Life was indeed more pervasive and exotic than prevailing formulas allowed; but the challenges to those views would likely come from Earth. They would come when new eyes viewed seemingly old worlds afresh.

This is what exploration by Western civilization had always done. From time to time it revealed truly unknown places, but mostly it discovered places known to others or rediscovered once known places in novel ways. The Indies—the destination of the First Age—were hardly unknown: the whole point of the Great Voyages was to find a new way to get there. Charles-Marie de la Condamine surveyed the Amazon with a map in his hand produced by predecessor Jesuits, and Humboldt succeeded La Condamine in Ecuador; while across the globe the Institut d'Egypte, accompanying Napoleon's invasion, measured monuments known for thousands of years, yet saw them anew, and rendered them from obscurantist reliquaries to the measured and catalogued artifacts of science. The Second Age, after all, had begun in Europe as well as the Pacific. It was in Europe that the natural history excursion, the Grand Tour, and the new geology converged before Joseph Banks and Louis-Antoine de Bougainville took them global.

Sometimes discovery can come not by looking farther and deeper but simply by turning around.

## THE PLANETARY SOCIETY

On the eve of Voyager 1's closest encounter, as had happened with Mariner 4 and Viking at Mars and with Voyager at Jupiter, a panel, now well versed in the venue, convened at JPL to discuss "Saturn and the Mind of Man." Chaired by Walter Sullivan of the *New York Times* and the author of books on IGY and Antarctica, the panel included Bruce Murray, Ray Bradbury, and Carl Sagan, all veterans of previous encounters, and Philip Morrison, returning from a stint with the Viking landing panel. Framing the session were comments by Marvin Goldberger, president of Caltech, and Jerry Brown, governor of California.[117]

The themes identified the usual suspects: imagination, curiosity, wonder, the imperative to understand our place in the cosmos. All combined science with other themes. The physicist Morrison added an emphasis on the aesthetic qualities of Saturn—its enthralling rings. Geologist, and now director of JPL, Murray commented on Voyager as a "climax of a glorious decade of exploration" and as the vanguard of a necessary era of discovery by robots. There was no alternative to semiautonomous spacecraft: the distances were too great and the velocities too speedy to allow for hands-on guidance from Earth. Astronomer Sagan managed to mingle the scientific with the prophetic, a kind of modern astrology purged of its hocus-pocus.[118]

Afterward the panel, along with many others, retired to a fund-raising dinner on behalf of The Planetary Society, which Murray, Sagan, and Louis Friedman had just founded. Its originating purpose was to rally public sentiment on behalf of space exploration by providing an institutional focus for such enthusiasms, one that could speak to politics. Its stated mission: "to promote planetary exploration and the search for extraterrestrial life." By the time the organization celebrated its twenty-fifth anniversary, that charter had expanded to read: "to inspire the people of Earth—through research, education, private ventures, and public participation—to explore other worlds and seek other life."[119]

That private groups might sponsor expeditions or, more commonly, pressure governments to do so had ample precedent. In the Second Age it was possible for individuals or small groups simply to trek to the frontier and walk into the unknown. But the Great Voyages, and certainly space exploration, were vastly too expensive for private capital, or, if undertaken, came with generous concessions.

Somewhere the state would be involved as not simply a tithe collector but a player concerned with the shifting balances of power that exploration might catalyze. While Prince Henry could mingle public and private enterprise, most rulers preferred to outsource discovery to individuals or Companies of Adventurers. Cabot and Columbus, for example, could establish themselves as medieval barons on new lands, provided they ceded a fifth of the discovered wealth to the state. In this way private groups took the risk, and the state collected money and power. Later, scientific societies supplemented commercial institutions and campaigned for expeditions; think of the Paris Academy dispatching Maupertuis and La Condamine to measure an arc of the meridian, or botanical gardens sponsoring collecting naturalists, or the Royal Geographical Society and National Geographic Society sending exploring parties (usually under naval officers) to the ends of the Earth, from Lt. Verney Lovett Cameron to central Africa to Lt. Robert Peary at the North Pole. More recently philanthropy and publicity have combined to loosen the purse strings of private benefactors, leaving Antarctica, for example, endowed with the Beardmore Glacier, the Ford Mountains, and the Walgreen Coast. The idea that the state should undertake such enterprises for the sake of curiosity is a very recent innovation tied to the peculiarities of modern science and the perceived imperatives of national prestige as a subset of national security.

Perhaps the closest analogue to the Planetary Society was the African Association, or in its more fulsome title, the Association for Promoting the Discovery of the Interior Parts of Africa, organized in 1788 with Joseph Banks as chair. At the time, Banks was probably the best-known British scientist identified with exploration, was an aristocrat well placed in British society and scientific institutions (including Kew Gardens and the Royal Society), and was a keen promoter of geographic discoveries. The African Association targeted particularly a place of popular, even mythological lore, Timbuktu, and its associated river, the Niger. Under its auspices Mungo Park made two ventures to the region (the second proved fatal). More than sponsoring treks, however, the African Association kept the image of Africa and the putative value of its geographic exploration before the public eye and in the corridors of power. It lamented, as the Planetary Society did, the lapse in official enthusiasm after Park's disappearance, and it struggled institutionally after its most charismatic figure, Banks, died. It was succeeded by the Royal Geographical Society under Roderick Murchison, which bridged official and unofficial exploring as Britain lurched into imperial competition with France and later Germany during Europe's unseemly "scramble for Africa." (It was in turn succeeded in space exploration by the British Interplanetary Society.)[120]

The African Association resembled dozens of lobbying groups, all intent on directing the power of the state, with its capacity to mobilize truly impressive resources, toward their own particular exploring enthusiasms. Each would claim that, without such commitments, the nation would sink into slovenly insignificance. Each sought to keep its ambitions before the public. In Banks's day it was enough to interest the elite and select officials of the empire. But increasingly, in democracies, it was the public that either drove expeditions or allowed them to expire. That required a public medium, which in the heyday of European imperialism meant popular books, lecture tours, and especially newspapers. The press could make or break an exploring venture. It could browbeat an administration into sending an expedition, or ridicule one into remission, or even sponsor expeditions of its own, as the *New York Herald* did when it underwrote

Stanley's search for Livingstone. A race to the pole made good copy. Adventures into unknown lands sold papers.

By the time Voyager launched, however, the American public was plugged into television. The entire space program had grown up with TV: it was, in some respects, staged for TV. The press corps that came to JPL for the Voyager encounters watched the drama unfurl on TV monitors; the most gripping reports from Voyager science came from the imaging team that delivered graphic visuals suitable for rebroadcast. The "Mind of Man" that mattered was the TV-viewing public. Even as Voyager closed on Saturn, PBS was airing Carl Sagan's blockbuster series *Cosmos*, of which episode six, "Traveler's Tales," highlighted Voyager. It was masterly timing, as art and life intertwined into a Great Conjunction of exploration and culture.

## VOYAGER 2

Seven months after its twin, Voyager 2 began its approach to Saturn.

It had the same instruments and was visiting the same planetary world, so much would be similar. Yet it would see Saturn differently because it flew on a different trajectory, had a superior camera (50 percent more sensitive), saw Saturn when its rings were turned higher into sunlight and its weather was more boisterous, and could target specific features that its robotic sibling had noted only in passing. But Voyager 2 had always been the troubled twin. It entered Saturn plagued with faulty receivers, and it left with a crippled scan platform. Most of all, constraining its goals was the need for a trajectory that could propel the spacecraft on a 4.5-year cruise to Uranus.

The narrative of their Saturnian encounters thus differed in their fundamentals. Fraternal twins they might be, but the two Voyagers had different historical DNA, and they had different destinies.

Voyager 2's ten-week observation period commenced on June 5, 1981. This time scientists hoped for sharper atmospheric images from Saturn, more detailed photos of the satellites, and greater refinement of the ring structure, including shots of elusive shepherding satellites along with an explanation for the baffling dark spokes on the B Ring. The cameras rolled; the images of white-ringed Saturn loomed large on screens; still photos were strung together into animated movies.[121]

Voyager 2's mechanical ills continued, however. One of the memory chips in its flight data system computer failed, which forced mission control to devise a new encounter sequence—"better" than the original, they insisted. On August 19 the spacecraft executed a final course correction, a day before the fourth anniversary of its launch. Soft geography caused problems because the solar wind was high and gusty and no trailing spacecraft could issue forecasts as Voyager 2 had for Voyager 1; even more than usual, bow shock was a frothy frontier, complicated by the planet's interaction with Titan and the magnetospheric tail of Jupiter, now in rude alignment. Yet anticipation was high, less in the hope that Voyager 2 might discover whole new phenomena than that it might help clarify the raw mysteries unveiled by Voyager 1. Much as the imaging team resorted to more false-color and computer-enhanced photos to massage out details, so the mission design sought to refine the encounter sequence to illuminate some of Saturn's more baffling revelations.[122]

Near-encounter began with a close flyby of Iapetus, the outermost of Saturn's ice moons, on August 22. Thereafter, Voyager intensified its reconnaissance of Saturn's hard and soft geographies. Although bow shock proved evasive, the spacecraft undertook a complex sequence of rolls to plot out the contours of the magnetosphere and trained its UVS and IRIS instruments on everything in sight. But the hard geography had its fugitive phenomena as well: the satellites that theory demanded be embedded in the rings could not be found. Nor could instant analysis locate any explanation for the dark spokes. What the images did reveal was an ever-multiplying number of rings, now in the thousands. Compensating for the missing moonlets, however, were brilliant images of real moons, now including Hyperion. Slight perturbations in the spacecraft caused by their gravitational pulls allowed for estimations of mass. (Iapetus, it was calculated, was nearly pure ice.) As Voyager 2 crossed bow shock for the last time, those satellites would command much of near-encounter efforts on August 25.

Guidance was nearly perfect: the spacecraft was within fifty kilometers and three seconds of schedule. The images flashed by—the rings, the moon Tethys, newly discovered satellites with orbits shared with Tethys and Dione. But the mission was science, not merely reconnaissance, and for an astonishing 2.5 hours the photopolarimeter claimed exclusive attention while it measured the occultation of the star Delta Scorpii through the rings. When it ended, with a profile of 82,000 kilometers of rings on chart paper half a mile long, the spacecraft's cameras scanned the moon Enceladus in riveting detail. Then it passed behind Saturn and plunged through the rings, and completed periapsis, not to reemerge until midnight some ninety-five minutes later. Its record of occultation through Saturn's atmosphere, its inner passage through the ring plane, and its final scans of Enceladus would not be broadcast until it reconnected with the DSN. There was only minor anxiety about the physical hazards posed by the rings, since Voyager was following the rough trail blazed by Pioneer 11. And on schedule the DNS station in Australia reacquired the telemetry signal from the spacecraft, now boldly hurtling toward Uranus.

The signal, yes. But the Voyager 2 that emerged was not the Voyager 2 that had entered.

Something had happened. A few instruments transmitted odd data, and erratically; near the ring plane some control thrusters had made unauthorized firings; the scan platform now pointed not at a receding Saturn but into black space. While fresh news to mission control, this was old news to Voyager 2. Everything had happened ninety minutes earlier. The near-imaging of Enceladus and Tethys; the planned stereo views of the F Ring; a photopolarimeter-recorded occultation of the rings; the backside view of Saturn—all were gone, irretrievably lost in space. The change in temperature caused by passing into Saturn's immense shadow had also upset the balky receiver, and mission control struggled to identify an acceptable frequency and to reengage. Each new command required multiple transmissions, which even at the speed of light would take more than ninety minutes to reach the crippled spacecraft. Gloom replaced euphoria, until the mood at mission control resembled the abrupt borders of black and white that Voyager had imaged on Iapetus.[123]

The primary goal was to spare the spacecraft further harm. Mission control sought to shut down whatever was not essential and to preserve, if possible, the option to go to Uranus. New commands placed the instruments on standby, since they recorded nothing valuable and might complicate attempts to reclaim control and diagnose the disorder. They shut down the preprogrammed command sequence for the encounter, since that sequence was anyway a shambles. And they moved the scan platform away from where it had frozen, a position that might ruin some instruments by making them directly face the Sun. It then remained to wait for the download from the Voyager's tape recorder, which had held its transmissions while the spacecraft passed behind Saturn.

That broadcast arrived on Earth at 9:00 a.m. on August 26. Here was Voyager doing what it was supposed to—sighting a magnificent scan of the F Ring, and a crescent profile of an F-Ring shepherd satellite. Then the record entered its problem phase, as the narrow-angle camera sent, instead of crisp images of Enceladus, black screens. The wide-angle camera could still capture the edge view of the rings as the spacecraft crossed through, but anything that had demanded precision targeting was gone. There were no rings, there was no Tethys. Misaligned, the camera got only a marvelous if unanticipated image of the Keeler Gap in the rings. The wide-angle lens got a snippet of a moon. Worse, the scan platform's ability to move along an azimuth was clearly deteriorating. Finally, the platform froze altogether. The cause of its malfunction remained unknown, and frenetic inquiries were stymied by the faulty receiver. The startling clarity and vigor of the early images made the subsequent losses all the more poignant.[124]

The flow of science based on the information successfully sent continued, as it would for months during which analysis would replace brute reconnaissance. There was more data from plasma waves, particularly a burst of energy a millionfold higher than normal as Voyager passed through the ring plane. And there were those brave, cold new worlds, as continued study scrutinized the major satellites of Saturn and continued to find new minor ones. But what had always characterized near-encounters—the brute flood of new data, the sheer awe of first contact and shared discovery—was gone. Voyager continued to transmit empty scenes, each frame "displayed on the monitors at JPL, complete with the commanded exposure time, filter, and so forth," and they "kept coming, one after another, all day." Unless Voyager righted itself, it might go to Uranus as a blind chunk of metal rather than a robotic explorer.[125]

So engineers continued their inquiry, at once feverish and measured, to diagnose the problem and prescribe remedies. They isolated the trouble to a stuck azimuth platform, the geared mechanism that allowed the scanning instruments to rotate in a plane and point in different directions. One by one, engineers brought back online the other instruments, successfully. The soft geography Voyager could still track; it was the hard geography of visible worlds that was threatened. Then engineers began to massage the azimuth mechanism in ten-degree increments, and they discovered that they could still control the elevation mechanism. With that, and by using the spacecraft as a platform overall, it would be possible to do some basic reconnaissance at Uranus. Step by step, mission control inched Voyager 2 back to where it might at least focus on Saturn rather than continue taking its disheartening images of empty space.

As the engineering investigation continued into August 27, E. K. Davis, project manager, and Ed Stone, chief project scientist, jointly declared that the principal objective was "to recover the scan platform capability for Uranus." The mission had nearly five years to tinker and perhaps solve the problem before a Uranus encounter, and they were not willing to do anything "to increase risk by premature activity now." When asked how much had been lost, Stone preferred to emphasize how much had been gained. The mission, he announced, had been "200 percent" successful—by so much had Voyager 2 exceeded its mandated charge. The full raft of data would take years to assimilate; and even as the spacecraft struggled to return from its passage through the dark side, project science leaders trooped forward to place the latest discoveries before the world. The odd surface of Enceladus. Titan's hydrogen torus, and apparently missing ionosphere. Plasma discontinuities created by Tethys and Dione. The ever-enigmatic rings.[126]

By August 28, diagnostics had progressed sufficiently to identify the problem and suggest, as Dick Laeser, deputy project manager, put it, that "the platform improves with use." The glitch was in part an outcome of what had earlier doomed the polarimeters: overuse. The science teams had earlier demanded more of Voyager than its engineering could guarantee. Its scan platform had tried to zip too quickly among too many targets, slewing faster than its lubricant could respond, and the apparatus had gagged. The spacecraft needed a mandatory rest, followed by a cautious program of rehabilitation. The worrisome days after the crash met that need. Within hours after identifying the source of the breakdown, on August 28, mission control hoped to get Saturn back into Voyager's field of vision. They did.

Encounter officially ended on August 30 with a bittersweet celebration. Once again Voyager 2 had come back from a near-death experience. If much had been lost, far more had been gained. But the real losses were in the future: it was as though Voyager were passing through a gap in the ring plane of American planetary exploration. It was a gap in history, and this one had no shepherding satellites. Voyager *was* the American planetary program, and would be the only mission to other worlds until the Galileo spacecraft went to Jupiter in 1989. Bradford Smith anticipated a "data gap," Ed Stone an institutional gap, and others, career gaps. However hobbled, Voyager 2 was the future for planetary exploration.[127]

By now the spacecraft had recovered enough mobility that it could exercise its right to the traditional parting shot. On September 4, Voyager 2 photographed Saturn's most distant and darkest moon, Phoebe. Probably a captured asteroid now in retrograde orbit, the moon seemed to emphasize the retrograde character of the American planetary program as it looked back. Yet there was hope, too, in that new world, and there was faith, as Voyager pointed to the dark promise of Uranus.

## NAMES OF DISCOVERY

Even wonders require names, and marvels, labels. The cascade of data and images revealed worlds swarming with physiographic novelty ready for both.

Between them the Voyager twins discovered thirty-five new moons and first mapped twenty known ones; they traced the gaseous contours of Titan and the four outer planets; and they recorded hard geography surface features by the hundreds. Suddenly there were craters to name; valleys and rifts to label; volcanoes, dark splotches, and half rings to categorize; to say nothing of features for which no landscape terminology existed. They were not hollows, washes, mesas, shorelines, hills, or dales, since they expressed a different tectonism and a land sculpture purged of flowing water. They were, rather, ancient Earths, before seas and life; alternative Earths, obedient to stresses and abrasions unknown terrestrially; and simply alien worlds. Spontaneous, slangy expressions sputtered out in order for scientists to talk about just-revealed terrains, as they struggled to place odd features into familiar categories. But these would not be the names recorded on the maps and the gazetteers that would codify them formally.

Fortunately two decades of planetary exploration could build on a nomenclature and a mechanism developed for traditional planetary astronomy, which in turn reflected both centuries of geographic discovery and the rationalizing instincts of two centuries of Enlightenment exploration. As ever, there were similarities as well as differences.

All this had happened from the beginning. New places needed names, or else one could not speak of them, but also because naming was a means of possessing. The commissions dispatched under the auspices of the Spanish monarchy were thus enjoined: having arrived "by good providence, first of all you must give a name to the country as a whole, and to the cities, towns, and places." The chronicler of Juan Ponce de León's 1513 expedition recorded that "it was the custom of those who discovered new lands to give their own names to the rivers, capes, and other places; or else the name of the saint on whose day they made the discovery; or else, other names, as they wished." When Ponce de León arrived off the coast of a new land, it was Easter—"Easter of Flowers," as it was known in Spanish—and the land seemed well blossomed, so for both reasons he named it "Florida."[128]

Common sources for inspiration were legends—hence, Brazil and the Antilles, named for mythical islands in the Atlantic; California, for an island in a romance, ruled by Queen Calafía; or the Amazon River, since women there were rumored to fight like those of ancient legend. Other sources were analogues to places back home—hence, a Nuevo León, a New York, a Neuw Amsterdam—or to rivers and hills reminiscent of those from where they had come. Often some feature of the site sparked a name, as some characteristic or deed inspired a nickname. Turtles led to the Tortugas; famine, to the Starving River; snow, to the White Mountains or Sierra Nevada. Classical allusions led to recommissioned Troys, Syracuses, and Romes. Biblical terms often subjected local terms to a full-immersion baptism into Jordan Rivers and Mounts Pisgah. Mormon pioneers drafted liberally from the Book of Mormon, bestowing Deseret, Bountiful, and Kolob. Some founders, wishing to emphasize the novelty of their ambition (if not settlement), invented neologisms, perhaps from Latin or Greek, as William Penn did with Philadelphia. And not least there was the practical flattery of naming features after patrons: Virginia, for the virgin queen Elizabeth I, and Jamestown for her successor; Mount Hood, for the Lord of the Admiralty; the Jefferson River, for the sponsoring president; and Washington nearly everywhere. The explorers themselves showed varying shades of immodesty. (Cook, for example, discreetly left the inlet beside Anchorage unnamed so a junior officer might insert Cook's name.)

But few places at the time of discovery were truly unnamed because they were rarely uninhabited. Every settled place already came as thickly covered with names as with woods or flying insects. Since most explorers had native guides and interpreters, they first learned of those places through their indigenous nomenclature. Some survived, more or less intact, leaving Massachusetts, Tidbinbilla, and Rotorua. More got bent through linguistic prisms as adults struggled to mouth exotic sounds or convert them into something that sounded familiar. Nahuatl Gualé twisted into "Gualape." Algonquian *che*, meaning "big," and *sepi*, meaning "river," combined with a generic ending *-at* to become "Chesapeake," which sounded vaguely familiar, like something that might apply to a mountain. French Canadians exploring in the Plains named a river the Purgatoire, which Anglophone Americans reengineered into "Picketwire."[129]

But there were some places that seemed strikingly new, required many names for unusual features, lacked indigenous labels, and came late enough in the game that explorers invoked some system. The prime example may be the Grand Canyon. The first explorer through the gorge, John Wesley Powell, elevated the canyon from Big to Grand, and named some of the salient features of the river, particularly its rapids and their associated gorges, bestowing such memorable labels as Sockdolager and Bright Angel. But the perspective from the Canyon rim was more daunting: there were so many features, all unnamed. Clarence Dutton decided that the mesas resembled pagodas and that the Canyon deserved a spiritual aura, and so bestowed the names of Asian gods and sages. Immediately there were temples for Confucius, Shiva, Vishnu, and Zoroaster, while the rim proper got Cape Royal, Cape Final, and Point Sublime. The next major cartographer, François Matthes, favored Nordic gods, bequeathing Wotan's Throne, Freya's Castle, and Thor's Hammer. When Arthur Evans succeeded him, the Welshman insisted that figures from the Arthurian legend be immortalized, which led to Lancelot Point, Guinevere Castle, and a spire called simply Excalibur. Meanwhile the Birdseye Expedition through the gorge, an exercise in cartographic engineering, merely labeled the river's features by their mileage from Lee's Ferry (eg., Mile 128 rapid). A few vernacular terms slid into the gazetteer, such as Horseshoe Mesa and the Battleship; miners and promoters put their names on trails and select features; and the National Park Service arranged to name the main tourist overlook after its first director, Stephen Mather. Explorers ensured they got onto the map, although they found themselves clustered on Powell's Plateau, with Dutton claiming the prominent point to the southeast and army rivals such as Lieutenants Wheeler and Ives banished to less interesting western peninsulas.

---

To a remarkable extent this same pattern would govern the naming of planetary bodies and their landscapes.

One reason is that by the time the Canyon was fully mapped, the U.S. Board on Geographic Names had been established to oversee uniform place-names by federal agencies, particularly the U.S. Geological Survey, which was mapping the country. The board functioned to regulate usage much as a dictionary does ordinary language. The board was rechartered in 1947, and continues today to promulgate "official geographic feature names with locative attributes as well as principles, policies, and procedures governing the use of domestic names, foreign names, Antarctic names, and undersea feature names." It thus oversaw two of the three regions of the Third Age. It spread into space because the U.S. Geological Survey's Astrogeology Research Program, established in 1963, has responsibility for mapping and naming the features that planetary exploration discovers.[130]

Two traits especially distinguish space from the naming practices typical of the First and Second Ages. There are no indigenous peoples, and hence no existing names, and similarly there are no territorial claims associated with the naming process. The solution to the first issue is to create a past by appealing to ancient or traditional lore from around the Earth, much as Dutton and his successors did at Grand Canyon. The solution to the second is to rely on a neutral institution to govern naming, a task that falls to the International Astronomical Union, first established in 1919, which arbitrates and regularizes nomenclature for planetary bodies. At first this meant the Moon and Mars, as viewed by Earth-based telescopes. But as robotic spacecraft began their reconnaissance, the task expanded. In 1970 a Mars nomenclature working group was charged with designating names for the features revealed by Mariner 4 and its successors, particularly those anticipated from Mariner 9. In 1973 the IAU established a Working Group for Planetary System Nomenclature, which in turn created task subgroups for the Moon, Mercury, Venus, Mars, and the outer solar system. In 1982 Harold Masursky of the USGS Astrogeology Program and a member of the Voyager Imaging Team became president of the Working Group.

By then the Voyagers had recorded two flybys through the Jovian and Saturnian systems. In their wake were tens of thousands of photographs of new worlds overrunning with features. They needed names, or at least some did. In the end more than five hundred landmarks got them.

The naming process begins with a theme, approved by the proper IAU task group, followed by some names to prominent features. Improved images lead inevitably to more names. From the task group the names go to the Working Group for Planetary System Nomenclature, where they must again find approval. The protocol allows for a three-month period for objections before the names are recorded into gazetteers, maps, and the transactions of the IAU.[131]

In their search for themes, the working groups assembled a library of more than two hundred published or submitted sources, from the *Larousse Encyclopedia of Mythology*, *Kiowa Tales*, *Gilgamesh*, *The Indian Background of Colonial Yucatan*, *Giants*, *Tales of Yoruba: Gods and Heroes*, *Soviet Encyclopedia*, *Webster's Biographical Dictionary*, a list of radar scientists (provided by G. H. Pettengill), *Fairies*, *Zhenshchina v mifakh i legendakh* (*Women in Myths and Legends*), *Dictionary of Slavic Mythology*, and *Myths of North American Indians*, to say nothing of classic literature from Homer and Virgil to Dante and Milton.[132]

On November 16, 1980, some members of the working group for naming the outer planets met at JPL to identify the themes for naming features on Saturn's satellites. Hal Masursky thought that since they all had themed lists of potential names, they might match many-featured moons such as Rhea with the longest lists; but Tobias Owen, the chair, thought they might begin with the moons with the fewest features. That pointed to Enceladus and Hyperion. But a theme? Since Saturn was a Titan, most of the satellites were named for sibling Titans. Enceladus, however, was a giant that Athena buried under Sicily, so Masursky suggested that the craters on the moons be named for giants. Owen then noted that the IAU had criticized the nomenclature committees for taking too many names from Western civilization. The group considered using names of giants from cultures everywhere, which segued into a query about giants in the

*Aeneid*, and then the *Odyssey*, which free-associated into *The Tale of Genji* and then Arthurian legend. (With its sharply etched black-and-white terrains, Iapetus seemed suitable for a story of good and evil—a white crater for Lancelot, a black one for Mordred.) Then they considered naming after discoverers, but since living persons were disallowed, they could not, for example, use Rich Terrile. Instead, they decided to let him pick from a list. Returning to Enceladus, they fussed over the idea of naming features after Sicilian towns, which led back to the *Aeneid*, which then led to adjournment. In the end the full working group proposed that Tethys get its names from the *Odyssey*, Dione from the *Aeneid*, Mimas from Arthurian legend (based on Baines's translation of *Le Morte d'Arthur*), and Rhea from creation myths from around the world. Enceladus got names from people and places in Burton's *Arabian Nights*. In the end the giants lost out altogether because they tended to cluster too heavily in Nordic legends and thus found themselves buried beneath the Sicily of cultural diversity.[133]

By the time Voyager completed its flybys, the *Gazetteer of Planetary Nomenclature* recognized some fifty-one "descriptor terms" for features, almost all of which had Latin roots. These ranged from "arcus" (arc-shaped feature) to "cavus" (hollows) to "dorsum" (ridge) to "labes" (landslide). "Palus" described a swamp or small plain. "Planitia," a low plain. "Planum," a plateau. "Macula," a dark spot. "Rima," a fissure. "Rupes," a scarp. "Tessera," tile-like polygonal terrain. Matched with themes, Io's volcanoes got names from fire, sun, thunder, and volcano gods and heroes. Its catenae (chains of craters) came from sun gods; its "mensae," "montes," "paterae," and "tholi" from the Io myth or Dante's *Inferno*. Europa received names from the Europa myth or Celtic mythology. Ganymede boasted gods and heroes from the Fertile Crescent. Some small satellites, like Colorado River features, got numbers. After the space shuttle Challenger blew up, the committee resisted the impulse to name moons after the deceased astronauts.

The names rolled on, a fugue of words counterpointing the rosters of raw data. While neither names nor numbers could compete with the graphic shock of images, both were necessary means of assimilating what would never be colonized. They populated the new worlds with gods, heroes, giants, and so connected the mission with a cultural heritage that a machine could not otherwise have. They made Voyager a saga, not simply an instrument.

## EYES OF DISCOVERY

To planetary scientists, Voyager was a flying lab, outfitted with instruments for recording phenomena largely invisible to the naked eye that it then transmitted back in streams of data, which researchers would pore over for years. To the public, however, Voyager was a camera, and its startling revelations were its photos. No one could see infrared or ultraviolet radiation, much less something as recondite as a bow shock. But everyone could stand amazed at crisp images of exotic worlds viewed instantly as they were displayed on monitors. For humanity at large, and for the media, those images largely defined and sustained Voyager.

The split in perception was real, however, and it dated from the very origins of the space program. Many space partisans saw scientific research in only a supporting role; and space scientists themselves disagreed about what research warranted a place on the instrument boom. Certainly in its origins planetary science had meant geophysics; of the fifteen scientists allied with Voyager, only one, Larry Soderblom, was a geologist because, as Bruce Murray put it, "there was no geology of the outer planets." IGY had been, after all, a *geophysical* year, with special focus on the upper atmosphere; and the new astronomy was shifting away from imagery based on hard bodies and visual light and toward radio, X-ray, or infrared spectra and the objects that emitted them. The first triumph of the American space program was Explorer 1's discovery of Earth's invisible radiation belts. Simple pictures of new landscapes, moons, and galaxies held little research promise; they seemed scarcely better than postcards, or tourist souvenirs, with trivial value for scientific inquiry. As early as 1962 an august group of space scientists (including Nobel laureates) had argued strenuously against putting a TV camera on satellites as a waste of precious payload. Even as Voyager 1 hurtled toward its rendezvous with Jupiter, "nowhere," observed two chroniclers of the mission, was "there any indication of the dominant role that images would play in the exploration of these new worlds."[134]

That was the vision of professional science speaking to its own. Against its bias was the steady penetration of TV into American culture. This was an age in which the word increasingly counted less than the image, and the still image less than the moving one. From its origins, the space program had been televised; Vanguard 2 had attempted to televise Earth; and beginning with Tiros 1, weather satellites beamed back photos, both startling and self-evident, to an eager public. The eyes of discovery disseminated to anyone with a TV set: nothing created so great a bond with the citizenry whose consent made the space program possible. Still photos were effectively published on TV.

Gradually, however, geophysicists came to appreciate the value of images for interpreting the hard geography of the new worlds being discovered. Even seemingly backwater sciences such as geomorphology enjoyed a revival by being able to decode from surface images past and current planetary dynamics, from once-flowing waters on Mars to subterranean renewals on Europa. The lasting image of the Apollo program was not a posturing astronaut on the Moon but Earth rising above the Moon's horizon. The camera surrogate used by Pioneers 10 and 11—painstakingly reconstructed from scans by a photopolarimeter—had enthralled viewers; but those spare, grainy images paled before the glossy, multihued thousands that Voyager broadcast back.

Their wide- and narrow-angle cameras, not their recording of whistlers on Jupiter or the unstable magnetospheric fringes of Titan, were what most bonded the Voyagers to the wider public.

That had long been true. Before printing became cheap, explorers had indulged in word paintings, and they still did, even as the Second Age made it standard practice to carry artists with expeditions. Those paired images were the ones that electrified the public.

It was one of the great convergences of the Second Age that nature became a reputable subject for art. Vast natural history paintings replaced operatic human history canvases as a source of wonder and moral instruction; portraits of parrots, beavers, turkeys, and platypuses replaced, at least in popular lore, portraiture of aristocrats, plutocrats, and ecclesiastics; mountains and blasted trees sublimated religious allegory, and then became iconic objects in their own right. Artists such as William Hodges established indelible images of Tahiti, Easter Island, and floating Antarctic ice mountains. Karl Bodmer bequeathed a brilliant, irreplaceable record of upper Missouri tribes, as aesthetics merged with ethnography. Frederic Church reproduced Humboldt's South American adventures with tropical mists, towering peaks, and erupting volcanoes in immense chromatic canvases. Artists recorded and introduced to Europe the kangaroo, the Aborigine, the baobab tree, the Ituri Forest, the Great Plains. In America, where a sublime Nature and its monuments would substitute for the absence of Antiquity and its relics, discovery and artistic representation knotted tightly together. Thomas Moran's splendiferous canvas of Yellowstone Falls and his mellifluous watercolors of Mammoth Hot Springs, painted for the Hayden Expedition of 1871, were galvanic in pushing Congress to declare Yellowstone a national park. Immediately, John Wesley Powell persuaded Moran to visit the newly unveiled Grand Canyon, and repeat the trick–which he did.

Eventually technology changed the terms of public engagement, as the camera overtoppled the brush as the most popular medium. Photographs were what primarily engaged the public; even artists used photos to help re-create paintings later in their studios. Where expeditions had to pay for themselves by lectures and publications, as Shackleton's did, photos were critical. That makes all the more gripping the moment when, on the *Endurance* expedition, knowing they had to leave the ice floes for boats, Shackleton and Frank Hurley sat down and determined which glass-plate negatives they could afford to take with them and which they could not. Shackleton smashed the rejects to forestall any attempt by expedition members to secrete the remainder among the stores.

During the interregnum between the Second and Third ages that followed, discovery became increasingly banal, or even staged by magazines such as *National Geographic*. Everyone used cameras. Expeditions sought out niche landscapes, or simply places that lent themselves to being photographed. The photogenic mattered more than the unknown. There were a few exceptions, most spectacularly the Leahy brothers, who carried a motion-picture camera on their 1930 foray into the highlands of New Guinea, filming a first-contact record never to be repeated.

By the Third Age the camera had continued as an instrument of record, but art had abandoned nature for art, and artists had plunged into a greater unknowable, themselves. No artists accompanied the major expeditions of the era, although an Earth-based art of space did evolve, and some cosmonauts dabbled in paint and pencil while enduring the ennui of Earth orbit, like sailors on long voyages carving scrimshaw. The medium of the age, however, was the moving picture. The classic episodes of exploration were recorded on TV or recapitulated in documentaries. The dominant technological romance emerged from Hollywood.

For Voyager, in particular, its long trek paralleled a revival of space movies. *Star Wars* was released in May 1977, three months before Voyager 2 launched, and *Star Trek: The Movie*, in late 1979, when the Voyagers were halfway between their encounters with Jupiter and Saturn. The real Voyagers did not have to wait for Hollywood. Among their most memorable images were time-lapse movies made of Jovian storms, Saturn ring spokes, and fast-approaching moons. Technicians completed the trend by creating animated, computer-generated movies that tracked Voyager as it flew past new worlds.

Such productions required more than normal manipulation. "Enhancement" was already a mandatory practice, however, since Voyager did not transmit canvas oils and watercolors but only electronic daubs and brushstrokes for its pixels. The published image was assembled at JPL labs.

This, too, had a long tradition. Exploring artists rarely painted their canvases on the scene and then hauled them for months on mule and keelboat. They carried sketchbooks and journals in which they did studies and executed tricky details and indicated in words what color should go where, and then collected artifacts that they could subsequently include in a reconstructed scene. Later, they relied on photographs as well. They completed their final compositions in a studio. And not surprisingly, they indulged in a degree of enhancement.

The nature of digital images made manipulation both necessary and tempting, as what began as a means to sharpen data could segue seamlessly into artistic license. It was possible to compensate for weak and distorted signals; to lighten dark blotches; to parse bland haze into false color tints based on differing electromagnetic spectra; to highlight murky rings with sharp coloration, as though the image had been washed with psychedelic drugs. Computers became palettes, and imaging labs, artists' studios. The first survey from Mars by Mariner 4 consisted of two hundred scan lines of two hundred picture elements each, a giant printout of forty thousand numbers, which were then hand-colored with crayons to produce an image the public could understand. The volcanoes on Io were effectively discovered by manipulating data into a form that highlighted the crescent domes of sulfuric gas. As with all data, the numbers had to be processed to make meaningful patterns. The imaging teams used photo software instead of statistics.[135]

Yet the scientists who initially distrusted the cameras, particularly TV, were right in their suspicions. The romantic horizon that captivated so many nineteenth-century painters had acquired a high-tech reincarnation. Perhaps the most famous distortion was the oft-reproduced image from the Magellan mission to Venus, in which the radar-measured surface of Venus was exaggerated twentyfold to turn lumpy lava into mountains and basins. But Voyager had its enhancements, too. Some of the mission's most celebrated images were, as literary journalists might say, composites. The Voyager 1 shot of Earth and Moon began as three images with different color filters, which were then synthesized into a single composition at JPL's imaging processing lab, where the lunar scene was brightened threefold in order to yield a more pleasing and informative contrast. The images that brought spontaneous gasps from the press were almost all manipulated in some way to enhance one or another feature; and some of those images simply charmed because they were gorgeous. The outer planets were worlds of beauty as well as data.

Once again, the sublime competed with the scientific. The software engineers at the Image Processing Laboratory were the artists of the Grand Tour.[136]

Yet the Third Age, once again, broke simple continuities. It reversed the thematic arc of the Second by which art had fastened itself to exploration. The classic grand tour of the eighteenth century had been a journey to centers of art and antiquity, but by trekking across the Alps and visiting Etnas in eruption, visitors had to confront nature as well as Old Masters, and travelers appealed to the new sciences to decipher nature's hieroglyphs and illuminate its meaning beyond what classical texts could do. Art and science together pulled the cart of travel.

Yet just as space science had changed, so had art, and so had the intrinsic character of those encountered landscapes that both science and art sought to wrestle into meaningful form, which meant the Voyagers saw differently from exploring naturalist-artists of the Second Age. This was a modernist nature: abstract, conceptual, minimal, alien to the human presence. It was a world only a robot could visit and only a robot's instruments could record. The human observer was, even at the speed of light, an hour and a half distant, with the studio banks of electronics further distancing him between the real and the reconstructed. Much as those on the grand tour sought to recreate the new into the venue of the ancient, so the imagers labored to place the Voyagers into the inherited traditions of discovery in which the wonder of first contact was an expected outcome.

But the shock of the new also spoke to a different realm of art, a comparable shift for the arts to that which the sciences were making as they refocused from hard to soft geographies. Willingly or not, the Voyagers' was a modernist art. The issue went beyond obvious visual parallels that made false-colored rings into something like a Barnett Newman painting, or the blotchy surfaces of Titan and Saturn into the abstract expressionism of Mark Rothko. It had to do with the position of the artist-explorer in the scene, and it went well beyond what computer enhancement could do.

The classic views of exploring art included observers in their foreground. The painting looked over the shoulder of the explorer. It saw not only the view of the explorer but also the explorer doing the viewing. Often, too, it attempted to imagine the explorer as seen by the explored, looking over the shoulder of Fijians or Fuegians or the putatively awed gaze of Aborigines, or it found in the wondering faces of encountered indigenes a reflection of the encountered explorer. The artist could view the expedition from the outside, or at times even place himself within the scene as viewed by an outside observer.

But no one could view Voyager. The twins were too far apart for one to image the other. There was no other platform by which to see Voyager sweep past Jupiter, Saturn, Iapetus, Titan, the rings, and the rest. This was a fundamental condition of the Third Age: there would be no Other to receive, bestow awe, or even witness. There was no one on Ganymede or Rhea, no village on the F Ring, or fishing vessel on the cloud seas of Jupiter, from whose perspective it might be possible to imagine Voyager's arrival. Its only audiences were the imaging team, the press corps, and the TV-viewing public. Yet continuity with the understood heritage of exploration seemingly required a perspective that the Third Age could not offer. Voyager might hurtle blithely through the void of interplanetary space; it could not so easily leap that looming void of solipsism, the threat posed by the missing perspective outside the spacecraft.

So when the time came to create computer movies of the flybys, the producers inserted Voyager into the frame. The animation follows the plucky spacecraft as it pitches and yaws and rotates its scan platform past exotic new worlds, just as we knew it would. Much as the camera, though initially vilified by space scientists, came to define discovery for the public, so old perspectives intruded into what might have been a purely modernist moment. That the Voyagers made the compromise allowed them to bond better with a public that regarded modernist art as something best kept in a museum. An image of the spacecraft went into the scene, rather as its curiously awkward trek went into the grand narrative of Western exploration.

There Voyager sailed, while we looked over its shoulder, and while five hundred years of exploration history looked over ours.

## PARTING

The Voyagers bid farewell not only to Saturn but to each other.

So far they had been linked, with Voyager 1 breaking trail for Voyager 2, and Voyager 2 sending data on solar winds to Voyager 1, and the twins programmed to complement each other as they soared past the satellite islands and planetary continents of new worlds. Now they split. Voyager 1 left the plane of the ecliptic entirely in a rising ascent to the borders of the heliosphere. Voyager 2 careened around Saturn for a rendezvous with Uranus. They would never again share routes or pass in tandem. Almost a quarter century later they would meet termination shock separately.

They were both scarred and limping. Shortly after encounter, Voyager 1's plasma science instrument had ceased transmitting. This had happened also after Jupiter, and engineers had successfully rekindled the system by temperature cycling. This time it failed to spark. Its photopolarimeter remained broken. The scan platform stubbornly resisted slewing, although this mattered little, since its cameras were turned off on December 19. For the next decade it would transmit only data about the soft geography of the solar system.[137]

Voyager 2 also had a defective scan platform. After restoring control, engineers eventually determined that the frenzied pace of the near-encounter–the rapid rotations of the platform, almost minute by minute–had overworked the system, driven out its lubricant, and caused the mechanism to seize up. Through temperature cycling and a pause sufficient to allow some lubrication to seep back, engineers regained enough control to permit the platform to image Phoebe. Then, save for experiments to diagnose the problem, further maneuvering ceased. With enforced rest and fast-slewing prohibited, the system performed as desired. This, too, mattered little for the present, since Voyager 2's cameras were also shut down. Until it reached Uranus, it would emulate its twin and measure and map only the soft geography of interplanetary space.[138]

On September 30, 1981, the primary Voyager mission officially ended, succeeded on October 1, 1981, by the Voyager Uranus/Interstellar

Mission (VUIM). The spacecraft were well beyond their warranty, though not beyond their encoded ambitions. The goal for Voyager 1 was now to monitor the interplanetary medium preparatory to leaving the solar system and entering the interstellar realm. The goal for Voyager 2 was a flyby of Uranus, another miniature planetary system, while preserving the option to continue to Neptune.[139]

The projected mission was long and lonely, almost 4.5 years and 724 million kilometers away. Uranus was farther from Saturn than Saturn was from Earth. The gravity assists Voyager 2 had received from Jupiter and Saturn helped it pare that immense distance to a perhaps manageable stretch. But there was no way to disguise the fact that the journey was far and the risks high. The cruise phase was longer than the cycle for presidential elections. Five course corrections would be needed to steer the spacecraft through the Uranus system, a place about which only the crudest facts were known. With the euphoria of encounter over and the space shuttle hemorrhaging money, NASA demanded major budget cuts, including 60 percent of staff, which left the Voyagers to sail through solar winds and gravitational tides on something like autopilot. If a crisis developed, JPL would activate a Spacecraft Anomaly Team (SCAT), packed with Voyager veterans, to grapple with the issue. Without money, staff, and knowledge, it was daunting to plan for an encounter at a place so remote that even the simplest exchange of messages would require 328 minutes, or 5.5 hours.[140]

DAY 1,031–2,492

# 15. Cruise

Voyager 2's cruise to Uranus was longer than its trek from Earth to Saturn. That prolonged stint offered a chance for the spacecraft to rehabilitate, for programmers to study a poorly known planet in preparation for encounter, and for both Voyager and JPL to improve their ability to communicate. In the end, Voyager was only as good as the commands it could be given and the means available to give them. As time and distance increased, both messages and means became more arduous, which left cleverness to compensate for aging, and upgraded communication for remoteness.[141]

When they launched, the Voyagers' capacity for reprogramming and their power packs were revolutionary. Today, when computers sit on every desk and microchips make cell phones into miniature PCs, those software uploads seem as quaint as their transmitting power, about a billionth the wattage of a digital watch. When Voyager left Saturn, the first personal computers were just entering the market; for an IBM PC, the choice for RAM was either 64k or 128k, and program software ran on the same diskette as its output files. Voyager 2's computer command subsystem held a scant 2,500 words of memory for sequencing. For each instruction, one word specified the event, and the other, the time. Yet Voyagers' constraints were even more formidable: it would not be possible to swap out a new hard drive or install a new modem. But while Voyager's hardware could not be updated, the earthbound antennas for sending and receiving could.

The success of Voyager 2 at Uranus depended on the capabilities of the Deep Space Network. But, then, the DSN became what it was largely because it had to do what Voyager required.

## DEEP SPACE NETWORK

The Deep Space Network performed three tasks. It assisted navigation and guidance by tracking spacecraft. It sent and received messages. And its signals could serve as an instrument of radio astronomy, particularly during occultations. But the network had to do all this on a twenty-four-hour basis; it had to receive from every spacecraft in orbit, a number that swelled yearly; and it had to connect even as spacecraft, notably Voyager, ranged farther and farther afield. In this, as in other matters, Pioneers 10 and 11 had pushed technology and procedures, forcing the DSN, further hobbled by Pioneer's antiquarian hardware and its algorithms, to its limits.[142]

These were only the known challenges. There were unknowns, too, such as the intensity of planetary radiation, which at Jupiter had caused wobbles in Voyager 2's receiver. And there were known but uncontrollable variables, related to weather both on the Sun and Earth. The solar wind caused turbulence, and passing behind the Sun overwhelmed receivers with noise. Rain in Spain caused a blackout of Jupiter data. As it had for explorers trekking for a year to distant lands to measure the transit of Venus, a wisp of cloud could wipe out the observation period.

Uranus strained both Voyager and the DSN. Even as it downloaded data from Voyager 2, the DSN had to track Voyager 1, along with Pioneer Venus and the ever-receding Pioneers 10 and 11. It had to maintain contact with the Russian Vega and European Space Agency's Giotto Halley probes. It communicated with the Japanese spacecraft MST-5 and Planet A. It transmitted and received with the

Giacobini-Zinner comet probe ICE. It still followed Helios and the earlier Pioneers (6, 7, and 9). The complexity of multisatellite tracking was daunting. But none of these issues equaled the challenge posed by Voyager's Uranus encounter.[143]

The spacecraft could not perform as it had at Saturn. It was losing power at a rate of seven watts annually. It was using fuel in the RPG, and as it aged it became less efficient with what it had. It could no longer power the IRIS flash-off heater while also transmitting on both the X band and the S band; yet all those operations were required for particular experiments. The solution was to shut down those components not absolutely needed, to turn off the IRIS heater when both transmitters were working, and to prohibit the simultaneous use of X-band and S-band transmitters in their high-power mode. By meticulous sequencing, IRIS could take its critical readings, the Uranus ring occultation measurements could proceed under high-power X band and low S band, and the planetary occultation at high S band and low X band. Another fix was to upload new software that could compress the data stream, allowing more kilobits to be sent for the same expenditure of power.[144]

Still, the signal was weakening at an exponential rate. For every doubling of distance, the strength of the signal fell by a fourth. Saturn was twice as far from Earth as Jupiter, and Uranus twice again as far as Saturn. The maximum rate of data transmission at Saturn had been 44.8 kilobits per second. At Uranus, it would be 21.6 kbps and 14.4 kbps. At Uranus, Voyager's radio signal would be "several billion times weaker than the power of a watch battery." The only way to compensate was to upgrade the DSN facilities. By a quirk of celestial geometry, the Canberra Deep Space Communication Complex in Australia could track the near-encounter for twelve hours. That was where the DSN and JPL Telecommunication Division would concentrate their efforts. The long cruise phase allowed time to consider options.[145]

A straightforward solution was to upgrade the receiving dishes; but this was unlikely since it was very expensive. NASA was investing everything in the space shuttle, and there were no other planetary missions on the books until Galileo in 1986, for which the existing arrangement sufficed. DSN would have to amplify what it had.

In 1982 it commissioned a global survey of large antenna facilities, what became known as the Interagency Array Study, which issued a report in April 1983. The study recommended as the best option to expand a technique developed for Saturn by which the several receivers at a station could be electronically linked and, in effect, establish a collective range far greater than the instruments could manage individually. Uranus, however, exceeded even these capabilities; the Canberra complex would have to link with another large dish, and the obvious candidate was the Parkes Radio Telescope, which also had a 64-meter antenna. By effectively connecting all the receivers together, the DSN had the equivalent of a 100-meter antenna, an escalation in capacity of 20–25 percent. The Voyager imaging team had wanted 330 images every 24 hours. The DSN could now guarantee 320 under ideal conditions.[146]

The other way to sharpen reception was to target the DSN array more accurately, which would reduce background noise (not a trivial concern given the lower wattage and the vast distances involved). To prevent degradation, DSN had to point its sixty-four-meter antennas to less than 6/1000th of a degree. One approach, called "blind pointing," was to estimate the location of the spacecraft and direct the antenna to the forecast spot, a technique applied during occultation and signal acquisitions. Instead of visible stars such as Canopus to assist navigation, the radio telescopes relied on radio stars for fixed points. The other strategy was to scan the estimated region in a conical sweep and acquire a signal from the spacecraft, which could then be used to track its location precisely.[147]

For Voyager 2 the two techniques were complementary, and both were necessary. To complicate the procedure, however, suitable radio stars in the region were unavailable, and the spacecraft's balky receiver demanded extra tending to locate its shape-shifting downlink signal. New software, practice drills at blind pointing, the installation of a Mark IVA system, and weekly meetings between DSN and

JPL brought the apparatus to working requirements. At Uranus the DSN could support up to 29.9 kilobits per second, improve navigation, and bolster the radio science experiments.[148]

Or it could in principle. In practice, upgrading facilities, crafting interagency agreements to bind individual antennas into arrays, and writing code had consumed time and money that would otherwise have gone into training. This was the mission cost of those sharp staff cuts after Saturn; continuities, once broken, were not easily regained. When, two weeks prior to the Uranus observation phase, the flight team conducted a dress rehearsal of near-encounter, it became clear that DSN was not sufficiently fluent; it needed more training and more people. But unearthing such outcomes was the reason behind the operational tests. DSN responded by adding technical staff and using the observation phase to conduct drills, as JPL used the information received from that period to refine targeting and the final sequencing for near-encounter, and of course to correct last-minute glitches.[149]

One such glitch emerged only shortly before closest approach, as images became marred by light and dark streaks. The first thought was that the error came from computer processing on Earth. But a scurried survey isolated the difficulty instead in a faulty memory spot in the onboard image processor. Within three days new software had been written to use an alternate location, and only six days before closest approach, the package was uploaded. The streaks vanished.[150]

There remained a final course correction, the last tweaking of a trajectory that had spanned nearly five billion kilometers. But so accurate was the spacecraft's path that no further adjustments were needed. Voyager 2 hurtled within twenty kilometers of its ideal aim point, testimony to a masterful triangulation among it, JPL, and the Deep Space Network.[151]

Compared to past exploration, Uranus was unfathomably remote in space but close in time. It could substitute speed for distance.

Lewis and Clark traveled 13,700 kilometers in 28 months; Voyager

Liftoff for America's space program. After the launch of Explorer 1 in 1958, a press conference led to one of the canonical images of the ensuing space age: William Pickering, James Van Allen, and Wernher von Braun hefting a mock-up of Explorer 1 over their heads. Each man stood for a competing version of what space might represent.

Source: NASA/JPL

Voyager spacecraft being assembled at the Kennedy Spaceflight Center. The foundational bus is clearly shown, along with the dominating high-gain directional antenna.

Source: NASA/JPL

Voyager spacecraft, assembled, as it would appear in the blackness of space.

Source: NASA/JPL

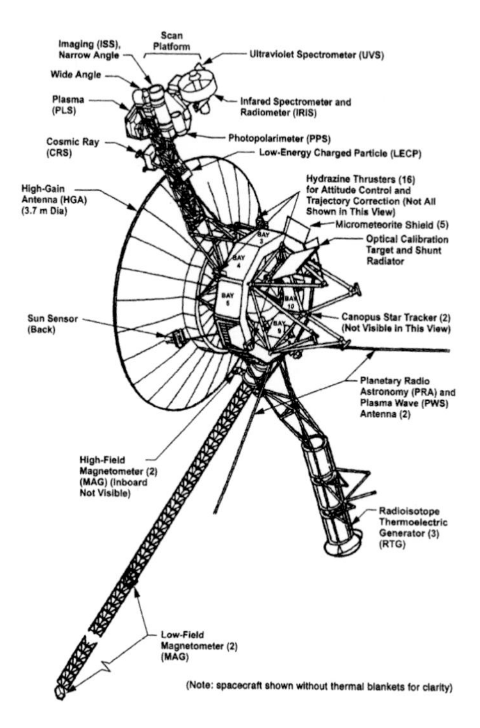


Schematic of Voyager spacecraft with instruments deployed.

Source: NASA/JPL

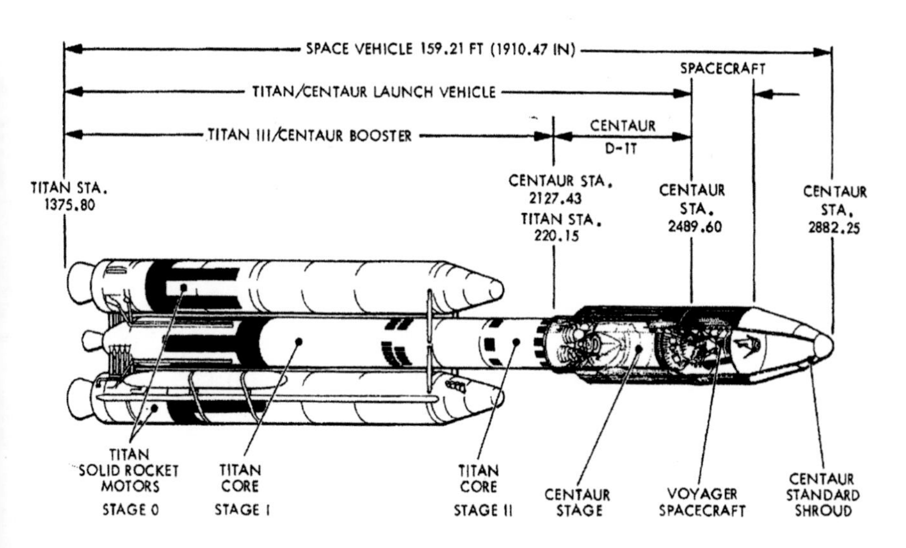


Titan IIIE-Centaur D-1T launch vehicle.

Source: R. L. Heacock, "The Voyager Spacecraft," Institution of Mechanical Engineers. Reproduced by permission.

Launch of Voyager 2, August 20, 1977.

Source: NASA/JPL

Voyager's gold-plated record.

Source: NASA/JPL

Iconic images: Voyager at Jupiter, showing technicolor clouds and orbiting moons.

Source: NASA/JPL

The Voyager family portrait: Voyager 1 looks back on the solar system.
The panels show the panorama of imaging, and the inserts, the separate images of the planets.

Source: NASA/JPL

Voyager 1 looking back on Earth and Moon. Note: the brightness of the Moon has been enhanced to sharpen the contrast.

Source: NASA/JPL

Voyager 1 looking back on Io. This is the original black-and-white image, intended to assist navigation, but which first showed anomalous features (center right) eventually identified as a volcanic eruption.

Source: NASA/JPL

Voyager 1 looking back on Saturn.

Source: NASA/JPL

Voyager 2 looking back on Uranus.

Source: NASA/JPL

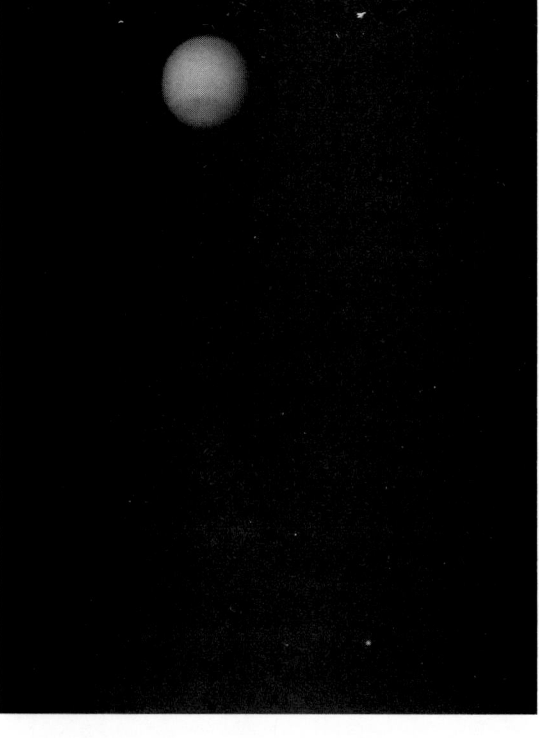

Voyager 2 looking back on Neptune and Triton, in a replay of Voyager 1's parting shot of Earth and its Moon.

Source: NASA/JPL

2 would take a few months longer than that to travel almost eight billion kilometers. The Corps of Discovery was basically out of contact for two years, and the collective journals were not published (and then in abridged form) for another century. At Saturn a message from Voyager took 86 minutes to reach the DSN; at Uranus, it would take 164 minutes, and as long to return. The first fruits of discovery were offered at daily press conferences during encounters, and published in scientific journals within months. While the ever-lengthening lag in electronic communications required Voyager to be semiautonomous, compared with historic precedents their correspondence and publication were practically simultaneous.

Such communications had always served to tell just where an explorer was and where he was headed. With space travel, however, the exchanges were the irreplaceable means needed to guide the spacecraft to its destination. This was a novelty. During the Great Voyages armadas had vanished over the horizon with hardly the prospect of a reply until they, or some fraction of them, returned. The last port of call was the final opportunity to exchange commands, information, personal mail, or warnings. While in the Canaries, Magellan received an alert that a Portuguese squadron was waiting to intercept him on his way to the New World, and so rerouted his voyage southward. One of his ships, the *San Antonio,* broke off and deserted near the Strait and returned with news (much false) to Spain. Otherwise, no one heard from the expedition again until, two years later, the *Victoria* limped into Lisbon. No camera watched as Columbus set foot on San Salvador, no radio message signaled da Gama's arrival in Calicut, no transmission measured Dias's occultation around the Cape of Good Hope. No one heard until someone returned, and if no one returned, no one knew what had happened or where. Lewis and Clark simply disappeared up the Missouri River, with no word about their whereabouts until they reappeared at St. Louis. When John Cabot disappeared in his search for a Northwest Passage, no one knew when or where.

Increased commercial traffic improved communication during the eighteenth century, and the telegraph boosted opportunities during the nineteenth. Humboldt sent letters to his sister Caroline in Prussia, who published them in newspapers. Dispatches from Russian American Company outposts in Alaska could take two hard years to reach Moscow and return. From Greenland, the 1845 Franklin Expedition sent letters home with the returning supply ship and military escort, though they might have used the whaler that spotted them afterward in Melville Bay; but they could not have expected any replies (and they themselves were never heard from again). Thomas Huxley on the HMS *Rattlesnake* could anticipate sending and receiving letters at Australian ports via the Royal Navy or commercial shipping. Henry Stanley had to wait until he returned to Zanzibar before he could telegraph his "discovery" of Livingstone to the *New York Herald*; but his more breathtaking traverse across central Africa down the Congo had him out of contact for three years. Newspaper rumors about his death preceded John Wesley Powell's successful run of the Colorado River through Grand Canyon. When Robert Scott's party died on their trek to the South Pole, it took two weeks for a search party to find their remains, another two months to inform the crew of the returning *Terra Nova,* and still another two weeks by ship before they could tell the world at Lyttleton, New Zealand. Seventeen years later Robert Byrd flew over the pole in a Ford Trimotor and then reported the feat by wireless radio upon his return to base at Little America.

Voyager had its missed, and mixed, messages, beginning years before, with JPL's error to send it an expected signal, which caused it to default to backup procedures, and continuing to moments when its cantankerous receiver would lose contact and when the spacecraft disappeared for painful minutes behind a planet. But it had what few prior expeditions did: a system of redundancy. As it pushed across the Outback of the solar system, it also had encoded in its protocols the wrenching history of past exploration. It could accept new orders. It could adapt new procedures. It could learn.

The immense distances meant that Voyager could not wait for commands from Earth. That it could communicate routinely with JPL was an astounding accomplishment; but the lengthening lines of communication meant that its managers had to allow Voyager more and more leeway. The velocity of encounters left no other option, for although communication—if compared to that available to Louis-Antoine de Bougainville in the Pacific or Nikolai Przhevalsky in Central Asia—was nearly instantaneous, "nearly" wasn't good enough. Voyager had to communicate with itself.

## THE DARKEST WORLD

Still, Voyager 2 could only report on what it was told to examine, and Uranus was a dark world, only dimly fathomed. With Jupiter and Saturn enough had been known to predict what the scientific priorities might be. But Voyager 2 now required detailed scenarios for an encounter with a planet only first discovered in 1781, two centuries before the spacecraft had left Saturn, and a body about which almost nothing was known. Its planetary plane was tilted: that anomaly was only the most apparent of Uranus's obliquities.

During its long, silent cruise phase, Voyager's staff had sought to sharpen that blurry vision and to heighten the return of Voyager information as the spacecraft blasted through the planetary plane of an eccentrically oblique Uranus. The task was formidable. There would be no Pioneers 10 and 11 to forge trails, and no Voyager 1 to do a sweeping reconnaissance. The skewed rotational plane made trajectories stranger still, and the path to Neptune trickier. Near-encounter would have to compress almost everything into a scant six hours, roughly the time it took to exchange messages between Voyager and Earth. Long distances, limited computing power, a bottomless thirst for data, and a scarcely known target—encounter demanded an unusually intricate scenario, which is to say, advance scientific scouting and planning.[152]

That began with a conference on February 4–6, 1984, which assembled more than one hundred scientists to assess the state of knowledge regarding Uranus and Neptune. The resulting proceedings revealed ignorance about even such basic parameters as rotation periods, magnetic fields, radiation, rings, and major satellites. Remoteness and a thick methane atmosphere made Uranus's surface opaque. Nothing was known of its weather, and heating at the poles rather than the equator introduced unprecedented variables in comparison with the other gaseous planets. Five satellites were known; almost certainly there were others. Nine narrow rings had been identified; again, there were surely more. (The rings had been discovered by stellar occultation only in 1977, the year of Voyager's launch.)[153]

Necessarily, much of the expected data and anticipated measurements would draw on analogies to the two gaseous giants Voyager had already visited. But some features did not translate—that oblique rotational plane, for one, and worse, Uranus's exceptional dimness, which rendered its rings and moons among the darkest objects observed in the solar system. Since sunlight decreased with the square of the distance, Uranus received 1 unit of solar radiation for every 360 at Earth. That left Uranus plenty dark, its moons nearly invisible, and its rings, reflecting some 2 percent of even incident sunlight, with half the brightness of "ground-up charcoal."[154]

Researchers were organized into three Uranus Science Working Groups, one each for atmospheres, rings, and satellites and magnetospheres. (To assist deliberations, Bradford Smith and Richard Terrile of the Voyager imaging team conducted special observations at the Las Campanas Observatory, in Chile, to gather extra data on the rings.) Between April and May the three working groups met repeatedly and independently to identify the major themes of inquiry and then to work with the flight science office to reconcile what they wanted with what the instruments could do and how to sequence the desired observations, each of which was designated as a "link," and which collectively made a chain of encounter events that could be graphed into a timeline. Each group also prioritized its scientific goals. In July 1984 they issued their final report to the Voyager Science Steering Group. Then the three groups had to reconcile their ambitions and produce a master (if still incomplete) timeline. Iteration followed iteration, with the work lasting from February 11, 1985, to July 19, 1985.[155]

One intractable issue concerned Miranda, the principal moon of

Uranus. Planetary astronomers desperately wanted to know its mass, which would affect their calculations of the planetary system's orbits as well as the path of Voyager 2, and this required real-time tracking to detect the gravitational pull of the moon on the spacecraft. But planetary geologists wanted equally to image the surface, which required image motion compensation; that is, the spacecraft had to turn along with the camera to keep the narrow-angle lens pointed steadily on target. The first needed constant contact with Earth, while the second forced the antenna to turn away, breaking that connection; there was no way for the hardware to be in two places at the same time. But it was possible for the software to pretend otherwise. During the iteration exercise, with its successive refinements of timing, planners realized that the image requirements could be satisfied best just prior to closest approach, and the mass determination during and just after closest approach, which left a difference between the two tasks of roughly five minutes during which the antenna had to reposition itself. The hardware could run only one way, but the software could be run backward, as it were; the effect on image motion compensation was the same. And so it proved: both experiments yielded excellent returns.[156]

The experience was no less a symbolic reversal of public expectations of how Voyager actually worked. The public face was rockets and robots. The reality was software and communications.

Over and again, hardware failures—faulty platform gears, broken receivers, the need for extended imaging—were compensated for by software fixes and the ability to send them across the immensity of interplanetary space. Programs for image motion compensation tweaked spacecraft gyroscopes to allow long exposures even as the spacecraft hurtled onward. Special software compressed data, and even allowed for onboard processing, which reduced the time and burden of transmission; an onboard Reed-Solomon data encoder bolstered the accuracy of data transmission while shrinking its digital "overhead." Other programs adjusted actuators for slewing, torque, and rolls.

But the grandest software, constantly rewritten and uploaded, remained in the human brain. It was the vision of the Grand Tour, and in late 1985, made real in Voyager, it sailed boldly beyond any previous encounter and toward a world darker than any other in the solar system.

DAY 2,493–2,635

# 16. Encounter: Uranus

From March 26 through November 3, 1985, preparations for encounter commenced with a routine that included regular imaging of Uranus, a recalibration of instruments, and a trajectory course correction. For almost three and a half years, Voyager had run in maintenance mode; now it had to be resuscitated, and its crew retrained for the intricate tacking and close hauling executed at velocities ten times that of a bullet. Staffing gradually scaled up. There were training sessions and trial runs, a shakedown rehearsal. As it felt the gravitational tug of Uranus, the mass of the planet altered Voyager's velocity, and the quickening events bulked up time until its history, too, seemed to acquire momentum.[157]

From October 7 to November 4, over four years after leaving Saturn and eight since it broke free from Earth, Voyager 2 put its instruments, mechanisms, and procedures through a series of checks. The power system, scan slewing, maneuvers for image motion compensation and radio occultation, likely frequency drift in the spacecraft radio reception, wide-angle photography of background stars—everything that the spacecraft would have to do in less than three months when any prospect for corrections would be lost in the blistering rush past the planet was tried now. The process culminated in a full-dress rehearsal of near-encounter. The exercise revealed assorted problems, primarily in new and inadequate staff, which JPL and DSN addressed by calling up reserves and scheduling more training. Meanwhile, engineers worked feverishly to ready final adjustments, engage the so-called target maneuver that would calibrate the image science subsystem and IRIS, and upload software patches and the all-important sequence commands.

Near-encounter would last a mere 6 hours; an exchange of radio messages between Earth and Uranus would take 5.5 hours; as Voyager hurtled around Uranus in another volta, there would be no opportunity to do anything but let the program play out. Expectation and anxiety tumbled along like wood chips through a cataract.

## CLOSE ENCOUNTER

On November 4, Voyager 2 entered its observatory phase.

Since July the extended scrutiny of the planet offered by Voyager's instruments had exceeded the best Earth-based sources. One after another the remote inventory of what JPL's *Voyager Uranus Travel Guide* called "our human quest to understand the world beyond Earth" began its third grand iteration. The ultraviolet spectrometer sought out gaseous emissions and the Uranian aurora. The planetary radio astronomy instruments searched for evidence of a Uranian magnetosphere. Of particular interest was timed imaging of the atmosphere; these photos were assembled into thirty-eight-hour movies (hence, two complete rotations of the planet) although the oddities of tilted Uranus meant that the imaging recorded the south pole, which allowed far less variety than comparable movies of the Jovian and Saturnian equatorial belts. Steadily, strikingly, the blue murk that was Uranus began to dissolve into features and movements, and the hard matrix that was the Uranian system resolved itself out of a starry background. The planet became real.[158]

Voyager's arcing trajectory around Uranus offered the usual occasion for occultation, with measurements of rings and planetary atmosphere, but halfway through its observatory phase, the spacecraft, as viewed from Earth, would pass behind the Sun. This solar conjunction allowed for occultation of the solar atmosphere as well, and because of the Sun's immense mass, for a test of the general theory of relativity, which predicted that the rays would bend (and therefore slow) under the influence of gravity, or more precisely, from the curved geometry of space around the Sun. This backside transit of Voyager yielded results consistent with prediction.[159]

Mostly, observation meant preparations: more measurements of the interplanetary medium, continuing instrumental calibrations, further testing of scan platforms, torque margins, and gyroscopic drifting. The various clocks had to be synchronized. Increasingly large images had to be edited into movies. Eventually an acrobatic Voyager underwent four yaws and four rolls that helped reset its magnetometer and furnished detailed star maps for precision guidance. The observatory phase ended with a final (of five) trajectory correction maneuver.[160]

After ninety-five days, observation segued into far-encounter on January 10, 1986. For the next dozen days the pace of monitoring matched the acceleration of the spacecraft. The pull of Uranus speeded up everything.

By now the bulk of Uranus could no longer fit into a narrow-angle lens; it took four such images to make a planetary mosaic; and soon the scope of the scene would overwhelm capabilities for whole-planet movies. Measuring the gravitational tug allowed for more precise calculations of planetary mass, which refined further trajectory corrections. Dormant instruments came back to life. The temperature of IRIS stabilized, readying it to transmit infrared data. The photopolarimeter revived and swept over the planet, satellites, and rings. The radio astronomy and plasma wave instruments added to particle and field measurements. New command packages, B721 and B723, were uploaded. A final "operational readiness test" put both the Voyager JPL team and the DSN through their paces. After years of mechanical hibernation, Voyager 2 was fully roused and ravenous, and on January 22 it began its near-encounter.[161]

In something like six hours Voyager threaded its way through the Uranian system, conducted more than ninety priority science experiments, and sped at still greater velocity toward Neptune. It was not simply the number of maneuvers but their condensation that astonishes. At Jupiter and Saturn, the Voyagers had sailed along the plane of planetary rotation, joining the flow of the rings and moons, faster than those other satellites but moving with and through a shared orbital plane over several days. Now Voyager 2 passed across them, almost at right angles. Everything had to happen in sharp, crisp succession. The spacecraft had to roll, which meant it had to shift its reference stars from Alkaid (in Ursa Major) to Canopus and then Fomalhaut in Piscis Austrinus and Achernar in Eridanus. Voyager soaked up so much data that it could not unload it, even with compression algorithms, except by replaying tapes over several days.

As it passed across the backside of Uranus, Voyager conducted almost continual occultation experiments with rings and atmosphere. The unblinking eyes of Voyager's instruments, especially IRIS and the radio science sensors, recorded atmospheric temperature, pressure, and chemical composition. Full-ring mosaics were composed both coming and going. Passage through the rings recorded particle impacts. Other instruments sampled the north polar region, invisible as Voyager had approached but now exposed as it swung behind the planet. The spacecraft interrogated the mysteries of the Uranian magnetosphere—whether it existed at all, and if so, with what dynamic geometry, since the south pole pointed into the solar wind. After passing repeatedly into and out of bow shock, Voyager could sketch a map of that magnetic shorescape. It imaged and studied the brightness of the five major moons—Ariel, Umbriel, Titania, Oberon, and Miranda—and used their perturbing gravitational effects on the spacecraft to measure more precisely their separate masses; of Miranda, the innermost moon, it made particularly close observations, passing within a mere 29,000 kilometers. But most complicated were the full-color and monochromatic photos of the Uranian satellites, all of which demanded image motion compensation, which required the spacecraft to pitch, yaw, and roll with intricate timing to hold the bleakly dark scenes steady while the spacecraft sped past. All this occurred amid constant moderate-rate slewing to capture as many objects as possible. And so it did: Voyager unveiled new rings and discovered ten new moons. For the largest, Puck, it captured a hurried but still usable image.[162]

There were a few snafus. A glitch at the Canberra DSN station lost some data, but engineers were able to have Voyager replay the stored information later. There were other minor "deviations" in expected performance. But for what Ellis Miner called JPL's "aging brainchild," they were "few in number and minor in consequences." Voyager had far exceeded its basic design parameters; it had twisted, measured, imaged, rolled, and reoriented its way around probably the most anomalous planet in the solar system; it had gone where no spacecraft had gone before; and when it hurtled past the far side of the Uranian system, it was on a dead course for Neptune, some thirty times farther from the Sun than Earth is, a trek that even with its once-more gravity-assisted acceleration would require 3.5 years.[163]

Uranus was a new world. It lay almost beyond the reach of Earth-based observation. No prior space probe had visited it, and only its most coarse contours had been traced. After Voyager, Uranus became, if not familiar, at least recognizable.

Every instrument of Voyager, and even the gravitational behavior of the spacecraft itself, added real knowledge: The mass of Uranus, and its aspherical distribution; the period of planetary rotation; the structure and chemistry of the atmosphere, including an abundance of helium, water, and methane, and their stratification; the eccentric energy budget, its reflectivity and reradiation, a pattern that leaves the planet much warmer than solar heating alone would permit; a unique meteorology, driven more by planetary rotation than solar radiation, with subtle differences in color and haze; an accurate geometry of the Uranian magnetosphere, including the wide divergence between magnetic and rotational poles; the confirmation of polar aurorae and lightning; radio emissions only weakly controlled by solar winds; low-density plasma filling out the magnetosphere; new rings and ring arcs—their composition, obscure reflectivity, particle size, dynamics. If images were "moderately disappointing," as one observer noted, that had to do with Uranus's hazy outer atmosphere, which blurred the kind of structure that riveted attention at

Jupiter and even Saturn, and with the baffling darkness of its hard geography of orbiting particles.[164]

Perhaps most spectacularly for both the public and even jaded scientists, Voyager 2 found miniature new worlds: it tripled the number of known Uranian moons. Because its passage through the system was so swift, its imaging hardware could not be adequately reprogrammed to photograph them all, nor because of its oblique rush through the orbital plane could its cameras view the outermost moons with the kind of resolution it had brought to the Jovian and Saturnian satellites; and to be authenticated, each revelation required at least two separate sightings, with enough lag to calculate a probable orbit.

But they came in a rush of near-epiphany. The first moon was found on December 30, 1985, sufficiently early that it could be targeted in the final flurry, sandwiched amid images of Miranda. Two more were identified on January 3; one on January 9 and another on January 13; another three on January 18; two shepherd satellites, flanking the epsilon ring, on January 20; and the last on January 21. Since its known moons had names from Shakespeare's *A Midsummer Night's Dream*, the first new discovery became Puck, and the others derived from various Shakespeare plays or from Alexander Pope's *The Rape of the Lock*. All this new discovery, moreover, happened while the five known moons were also imaged, with varying degrees of resolution, and their properties compared. The five outer moons were brighter, showed evidence of former melting, and were colder. The two most intriguing were also those with the highest resolutions. Ariel was lightly cratered, which revealed a period of post-bombardment surface flooding, or "cryovolcanic flows." Miranda was, simply, a jumbled mess, or more formally, a "geologic enigma." It seemed as though it had smashed apart, then cold-fused with whatever had splintered it, and there stayed, like a geologic Frankenstein's monster, full of odd lithic parts and thick stitches. Three days after its last discovered satellite, Belinda, Voyager slingshotted through near-encounter.[165]

Even to those experienced in Voyager's serial surprises, the capacity of the valiant spacecraft to amaze remained undiminished. However dark Uranus, Voyager 2 had shone.

---

On January 26, near-encounter segued into post-encounter. The Voyager team scheduled a final press conference for 8:00 a.m. on January 28, 1986, what everyone anticipated would be the celebration of a weary, wondrous quest. Instead, thirty-five minutes later, participants watched in horror as the space shuttle Challenger exploded on launch. The Uranus encounter could survive radiation, celestial mechanics, sagging power, balky actuators, particle attacks, impossibly tenuous lines of communication. It could not compete with immolated astronauts.

## BREAKUP

For almost thirty years the competing factions within the American space program had united in common cause under a fulsome NASA budget and against a cold war rival. At the onset of the space race their primary contestants, symbolized by JPL's William Pickering, the University of Iowa's James Van Allen, and the U.S. Army's rocketeer Wernher von Braun, had hoisted Explorer 1 over their collective heads, and they had continued that display of public unity, if not amity, as a downsizing Apollo program and its shuttle successor cleaved the funds that kept them all in loose alliance and forced the contestants to fight for the scraps that fell from the shuttle's table.

Both NASA and space apologists wanted to isolate criticisms within the community lest any harping harm the space enterprise overall. The official line was that both programs were necessary, and that what was good for one was good for the other. NASA tried to quarantine the issue by segregating the manned and unmanned programs except at the highest administrative level. But as the competition sharpened over the years, von Braun's vision of space colonization came to dominate; Van Allen's perspective of space as a scene for science suffered, and over the years Van Allen became more vocal, expounding his criticisms to fellow space scientists, to the public and press, and to Congress. The competition was both undeniable and subtle: the sheer weight of the manned space program could shove rivals aside. During Pioneer 10's encounter with Jupiter, von

Braun had shown up at mission control and sat on a desk stuffed with terminals to watch the show. His rump hit a switch on a video terminal and turned it off, and since all the terminals were linked, the entire system crashed. The mission staff had to remove him and reboot. That episode might well stand (or sit) as symbolic of the larger rivalry.[166]

By the time Voyager 2 reached Uranus, it was obvious that NASA's three-ply composite was delaminating. The agency had consistently favored the manned over the unmanned, the aeronautical industry over the universities and research centers, and defense over science. The manned program, with its ultimate vision of extraterrestrial colonization, might fume over the slow pace of the shuttle's development and express frustration with the lack of progress toward Mars, but the shuttle, Skylab, and the space station were at least within its informing narrative. That program, and its increasing domination by the Pentagon, cut the others off in midsentence. But while the immediate aftermath of the tragedy hushed any overt criticism—one could only voice admiration for the dead astronauts and mouth determination to honor them by reflying the shuttle–the options were out in the open.

In the January issue of *Scientific American*, which ran a few weeks before the disaster, Van Allen had publicly broken ranks and questioned the value of the manned program altogether. In May he wrote an essay for *Science* in which he systematically compared the manned and unmanned programs, noting that "many space enthusiasts blithely ignore the fact that almost all the truly important utilitarian and scientific achievements of our space program have been made by instrumented, unmanned spacecraft." He mocked the "misty-eyed concept that the manifest destiny of man is to live and work in space"; scorned the "Columbus analogy" to support a manned mission to Mars as "massively deceitful"; and concluded with a "poignant juxtaposition" of the Challenger disaster and Voyager 2's flyby of Uranus. (Some years later he suggested, with tongue only partly in cheek, that NASA sell the manned space program to China.) Instead of pursuing mirages, NASA ought to develop "space applications of widespread human importance" and sponsor "major advances in human understanding" of the universe. It ought, in brief, to be a scientific research institution.[167]

It was obvious that robots were not only cheaper and safer than people, but also capable of increasingly sophisticated semiautonomous behavior, and were improving at far faster rates than *Homo*. If NASA had committed itself to robotic spacecraft, he implied, there would have been no space shuttle to drain away funding from the truly breakthrough programs and no Challenger to explode. The science fiction that seemed to animate his rivals pointed not to a usable future but to a visionary past.

Neither Van Allen nor von Braun affected Voyager directly. Van Allen, to his and others' surprise, failed to make the scientific cut and had no instrument to place on either Voyager's boom; and von Braun died two months before Voyager 2 launched. But his ghost continued to haunt NASA and hound Voyager.

As the costs of Apollo ballooned, NASA had dumped the original Voyager mission to Mars. The reincarnation of Apollo as the space shuttle in 1972 repeated that experience and soon whittled the original Grand Tour down to MJS 77. In July 1979, as Voyager 2 commenced its encounter with Jupiter, it had to compete for media attention with the death spiral of Skylab and speculation about where it might land (and on whom) and with Tom Wolfe's paean to the Mercury astronauts, *The Right Stuff*.

Then, as Voyager 2 approached Uranus, JPL realized that the proposed shuttle manifest included a launch four days before near-encounter. Such a schedule conflict could affect support and tracking, to say nothing of public attention, and JPL asked to have the shuttle delayed. Voyager 2 had spent 8.5 years getting to Uranus; the shuttle could wait 8.5 days. The request went to James Beggs, then NASA director, who denied the request. When asked why, he reportedly replied, "The White House doesn't want the launch slipped." The Reagan administration had long sought to hobble the planetary program; this final snub, however petty and unreasonable, was in keeping. As it happened, the launch schedule for the shuttle

Challenger slipped anyway, until after Voyager 2's near-encounter. It then launched under questionable circumstances and promptly blew up.[168]

The immediate reaction at JPL and among Voyager enthusiasts was shock, and grief for their mutual enterprise of space travel. There was, as Ellis Miner put it, "a spontaneous day of silence for fallen friends and blasted hopes." But the realization quickly grew that rebuilding the shuttle program would eviscerate the planetary program, not only from financial starvation but from lack of a launch vehicle. The shuttle had taken everything. NASA had put all its eggs into one basket and watched it explode. The Challenger disaster was a "temporary death knell," Miner continued, for the "continued unmanned exploration of the planets."[169]

Others were not so forgiving. "Even then," as Bruce Murray expressed it, "Fletcher [NASA director] would not acknowledge what everyone else knew—that planetary missions were intrinsically incompatible with the Shuttle." America's planetary exploration remained "hostage to the dying embers of NASA's Shuttle fantasy." Already, to pay for the shuttle and its endless delays and overruns, the United States had declined to join the international mission to Comet Halley. Now the Galileo mission to Jupiter was postponed indefinitely. Ulysses to Jupiter and the Sun, Magellan to Venus, Cassini to Saturn, the Mars Observer, the Comet Rendezvous/Asteroid Flyby—all went into suspension as NASA struggled to make its white elephant fly again. All that remained was Voyager.[170]

What is perhaps less obvious is the challenge space science, as pure science, the search for data, posed to Voyager-like missions. This was in some respects more subtle, as Van Allen's own career demonstrated.

After all, James Van Allen had been a founder of the field. He had blazed a postwar trail to the outer atmosphere with balloons, rocketoons, and rockets; had been part of the catalytic discussions that led to IGY, and had helped ensure its commitment to the geophysics of soft geography; had sent the first instrument aloft on Explorer 1, which had led to the first great discovery of the era's space science, the eponymous Van Allen radiation belts; and had outfitted the Pioneer spacecraft, including Pioneers 10 and 11, with critical instrumentation. He had grown up with the space program, and grown famous because of it. Yet his emphasis could easily divert attention away from spacecraft and onto other platforms. In particular, astronomers might well—and did—argue for Earth-orbiting telescopes, what became NASA's Great Observatories.

If science is what you want, they delivered. Within twenty years after the Grand Tour, and after its own blinkered start, the Hubble Space Telescope (HST) and other major observatories could achieve "nearly Voyager-class imaging" of the outer planets. They could crudely map Pluto and its moon Charon; they could offer evidence that the Kuiper Belt, with some two hundred million inert comets, existed; they could improve on the Voyagers' astonishing tally of discovered satellites. Altogether the Voyagers discovered twenty-six new moons among the outer planets; the HST has since found forty-eight for Jupiter alone, most the size of asteroids. The HST can peer at a planet for long periods, return to detect seasonal and orbital changes, image the same scene over and over, and combine it with sophisticated software to detect satellites and rings. It can be patient. The Voyagers flew past once, faster than the human eye could follow.[171]

If partisans of manned programs had clawed away at planetary spacecraft, so had space scientists who wanted money and institutional attention lavished on topics of interest to them, and who might well regard robotic spacecraft as a proxy for the pointless adventuring that appalled Van Allen. The most powerful institutional voices on the NAS and NASA space science boards were astronomers, who tended to view planetary science, with its discovery of icy geysers, erupting volcanoes, and whacked-out moons, as a latter-day form of field naturalist gathering the contemporary equivalent of conch shells and orchids, and they consistently undermined the case for spacecraft in favor of telescopes and their equivalents.

Of course some measurements can't be done from observatories, great or otherwise, just as some activities robots can't do as well as people. But the pace of technological innovation continues to quicken. Much as robotic exploration was much cheaper than manned, so, potentially, was near-Earth science cheaper than far-trekking spacecraft. The arguments made in favor of robots as scientists can be made equally against robots as instruments.

The Voyagers thus had to fight on two flanks. Was Voyager, as both banks of critics implied, only an interim measure, with the robots destined to succumb either to humans or to instruments? Were such spacecraft only devices to do the work of other programs, or might they have an identity of their own? Did they do what the others could not? How might Voyager answer the charge that the Grand Tour was a Great Detour?

Like all defining expeditions, the Voyager mission was a synthesis—that's what gave it stamina and cultural power. It differed from orbiting observatories in that it traveled. That's what made it exploration, not merely extreme science. It was not the gathered data, artifacts, souvenirs, loot, logs, and journals that transformed curiosity into discovery, and discovery into exploration; it was the trek. The medium—the journey—was part of the message. It mattered not solely as a means of positioning instruments but as an event in itself, one that segued into quest. An entrée of geophysics and a dash of astrobiology might be enough for academicians scornful of spacecraft cameras and other seeming popularizations. It was not enough for the culture, which wanted to recreate that alchemical alloy of discovery and adventure, fused into a journey and culminating in an encounter. It wanted the personification of cultural identity, whether it be placed on a Nansen sledge or an adapted Mariner hex, whether by a character-testing Edwardian or a quasi-autonomous robot.

And it is the issue of character and encounter that most separated Voyager from traditional exploration. What it encountered did not require a human presence. At one level this is simply the argument for better instrumentation. Amid the geographic realms of the Third Age, there is no Other; not as a person, an intelligence, or a biota. People are simply a clever means to record data, which machines can do. The Mercury astronauts worried that they might be nothing more than "Spam in a can," that monkeys could do (and did do) what was required of a passenger. What, then, exactly, do people bring to the event? They know those hostile environs only through heavily intermediated suits and mechanical cocoons. They cannot smell or taste or touch or hear those scenes. They can only see them. Yet the "eyes of discovery" are available to everyone with access to a TV set or an LCD monitor connected to the Internet. The human "presence" is already mostly a machine presence. Exploration is becoming more and more virtual. As Robert Ballard said of manned submersibles, "ultimately" their "limits will become intolerable."[172]

What is lost is the inherited sense that exploration must be done by human explorers on the scene and the human drama that goes with flawed human actors. The argument, that is, is not that people are better at science, but that they are better at exaltation and tragedy, and that without the prospect for loss and strife, public interest in Third Age exploration will wane. Exploration will morph into something like normal science, which has its own competitions and motivating curiosity but is not organized as geographic travel. Here is one of the paradoxes of the Third Age. Because it does not involve contacting Others, it has purged past exploration of the moral dry rot of imperialism, but it has done so at the cost of its moral drama—the angst, the horror, the triumphalism, the fusion in a single personality of ambition and vision quest, the equally embedded internal struggle, what William Faulkner called "the human heart in conflict with itself"—that plagued past discovery and fascinated the public.

Apologists might argue with both reason and conviction that the reality of modern exploration was that it is an alloy of machines and men. "The truth," as Oran Nicks noted, "is that there were no such things as unmanned missions; it was merely a question of where man stood to conduct them." But equally, "crewed" missions were utterly dependent on semiautonomous machines and "flew" spacecraft that could be as easily flown on their own, without the genetically and morally flawed carbon bipeds. The proportions and perceptions might vary, but no human could venture into the realms of the Third Age except through an artificial habitat, and no robot could build, launch, and steer itself. We are of course drawn to the human, or to those creatures and machines that can be humanized. Even professional explorers such as Roald Amundsen and Henry Stanley have a complexity, a plexus of motives, surprises, and contradictions that robots cannot achieve. They are, after all, human, and it is the human story, the fusion of plot and character, powered by conflict, that drives the traditional narrative of exploration.[173]

But as the relative composition of human and machine has changed, so have the stakes. No one has calculated the attrition rate for exploration, or what trend it has shown, other than that exploring has come at a declining cost of lives over the centuries, largely as a result of better shipborne travel, medicine, and technology. On his epochal voyage, da Gama lost 120 of 180 men and two of three ships. Magellan's Armada de Molucca returned with 18 of 237 men, and one of five ships (although one vessel had deserted for home earlier), and Magellan himself died. The demands for long-voyage care and feeding were enormous. In the nineteenth century, Britain's Royal Navy recorded the percentages of shipboard fatalities as disease (60 percent), accident (32 percent), ship losses from fire or sinking (10 percent), and enemy action (8 percent). Substitute "exploring hazards" for "enemy action" and the statistics would likely compare.[174]

No such dangers attended Voyager. JPL engineers might labor feverishly to correct a flaw that threatened a mission, and advocates might lobby Congress and NASA with tireless resolve, but such efforts are still a far cry from the young Henry Stanley, himself temporarily roused from a deep fever, with the threats of Arab slavers ringing in his ears, resolving by candlelight to succeed in his quixotic search for the "Apostle of Africa," David Livingstone. "I have taken a solemn, enduring oath, an oath to be kept while the least hope of life remains in me, not to be tempted to break the resolution I have formed, never to give up the search, until I find Livingstone alive, or find his dead body." No living man, "or living men," he continued, "shall stop me, only death can prevent me. But death–not even this; I shall not die, I will not die, I cannot die!" No robot thought it might die. No spacecraft defied death in its quest. Machine failure in a robotic mission might dash dreams, but it would not claim lives.[175]

Antarctica again offers a point of inflection. Contemplating the tragedies and extraordinary exertions of the Terra Nova expedition, Apsley Cherry-Garrard characterized their endeavor as "running appalling risks, performing prodigies of superhuman endurance, achieving immortal renown, commemorated in august cathedral sermons and by public statues, yet reaching the Pole only to find our terrible journey superfluous, and leaving our best men dead on the ice." He proposed instead that modern technology—proper ships, aircraft, specialists—"will all be needed if the work is to be done in any sort of humane and civilized fashion." Then he came to the crux: that politicians and the public must "learn to value knowledge that is not baited by suffering and death," be it the death of the discoverer or of those discovered. Simple adventuring was insufficient; true exploration, he thought, was the "physical expression of the Intellectual Passion." There had to be some cultural purpose beyond simple travel and personal tests of character. He thought science might do. But his own bolt, he reckoned, was "shot."[176]

In the Third Age those appropriate technologies exist. They can allow millions to share in robotic discovery, in the only kind of sensory encounter that the forbidding geographies of ice, abyss, and space allow. The transformation of moral drama is still playing out. What modern exploration wants is not just shared sense data but shared meaning: not merely the eyes of discovery but its poetry. In some way, if only by literary tropes, the robots must be anthropomorphized. They must be agents and proxies, and a presence. They would then be, as Nicks observes, "reflections of their masters."[177]

Encounter, in brief, is not simply an event, the meeting of two entities, like the bow shock of ionic winds, but an exchange of minds, and ultimately a confrontation between moralities. How to find such moments, or their simulacra, amid ice, abyss, and space is the issue that confronts the sponsors who dispatch machine explorers. And it is what Voyager 2 enigmatically, brilliantly, poignantly displayed at Uranus.

## FIRST CONTACT

For Voyager, planetary encounter was a programmed event. It was a prescribed ritual, written into the software commands that told the spacecraft, minute by minute, often second by second, what to do. But that did not make the event idiosyncratic to the mission, or to space exploration. What Voyager codified was five hundred years of experienced encounters.

When Columbus made landfall, the practice already had its prescriptions. The first task was to look to the safety of the ship—wait for the dawn's light, watch for shoals and rocks, find a protected harbor or lee anchorage. Then the admiral and captains would take the longboat ashore, kneel, and give thanks. As soon as possible they would try to locate natives (who typically were already watching the newcomers, and often came to the ship in boats), display emblems of authority, communicate by signs and perhaps share gifts, begin to train interpreters, and collect directions to the headmen, and, always, identify the way to wealth. Amid so many isles, the crews had ample time to drill and adapt, but behind them stood more than seventy years of Portuguese probing among Atlantic isles and African coastlines.

With further centuries of exploring, the protocol evolved, reaching something of a pinnacle in Capt. James Cook, as he made contact with island upon island. His landing party consisted of him, his chief scientist, a physician, an interpreter, a security guard of marines, and sailors to row to shore. (Eerily, this is exactly the composition of the "away teams" so beloved by *Star Trek*, save that Hollywood scriptwriters could dispense with interpreters, having disseminated universal translators, not to mention other plot apparatus deemed necessary to go where, outside Hollywood, no one was.) As Cook understood, encounters typically went bad not because of natural circumstances such as high seas but because of social ones. A native stole, a sailor was killed, a chief was seized or hostages taken, the exploring expedition retaliated; or a captain failed, a crew mutinied, an exploring party fell apart. Conflict, character, choice—these are what metamorphose adventure into drama.

What explorers most valued was the founding encounter, what popular culture has come to call first contact. Here, potentially, was exploration stripped of the qualms, morasses, and banalities into which, historically, discovery had morphed into imperialism and normal scholarship. Here, physical adventure underwent a transubstantiation into moral drama.

—

Few first-contact narratives speak as today's partisans might wish. They were not recorded as part of a genre they knew the future desired. But there is one extraordinary exception, because it happened in recent times, and that is the saga of the gold-seeking Leahy brothers when they trekked into the unknown interior of New Guinea in 1930. Not only were they seeking new lands, but they also were doing so as explorers, and they even carried a motion-picture camera to document what they saw and experienced.

When they passed over the summit of the Bismarck Mountains, they looked down on immense highland valleys, densely cultivated, flush with fires, awash with people unknown to them and who themselves knew nothing of the world beyond their tribal borders. The prospecting expeditions, under Mick Leahy, continued into the mid-1930s, extending discovery into valley upon valley, beyond the vales of Goroka and Asaro, and meeting tribe after tribe, pushing on to the putative source of the gold, much as Hernando Cortés kept moving inland until he found the great depository of Aztec wealth. That first expedition ended up crossing the island, a veritable microcontinent. Here was raw first contact of a sort not seen since the Great Voyages and the entradas of the New World conquistadors.[178]

In truth, it is unsettling how fully the Leahys' encounters echoed those of previous centuries. Without a common language, exchanges were limited to signs, pantomimes, and demonstrations. The Australians wanted food, information, safe passage, and, later, women and workers; the highlanders wanted shells (their equivalent to bullion), steel axes, and, later, weapons. The explorers demonstrated their firepower and their superior technologies, and in later years flew select indigenes by airplane to see coastal cities; they took youngsters who could learn their language and serve as interpreters. The highlanders sought to fit the strangers into their existing economic and political dynamics as well as their prevailing cosmology; they wanted the unbelievable wealth of shells the strangers could distribute, and they sought to exploit the newcomers to advantage in the complex balance of endless wars among neighbors. Quickly, the Leahy party evolved protocols for contact.

Yet the encounter was profoundly asymmetrical. Like Pizarro or Stanley, the explorers had come without permission or prior notice. They had simply appeared and, by their sheer presence, broke the old order. They stood outside the existing etiquette of exchange, were not subject to taboos, did not warrant traditional courtesies to travelers or pilgrims, were neither friend nor foe, just a pale Other. Their very identity was upsetting. Given the options, the highlanders labeled the newcomers as the spirits of former tribal members now returned. So, too, they sought to incorporate the cornucopia of shell wealth and the firepower of the interlopers within their existing economic and political contexts. Unsurprisingly, where communication was limited to crude signs and barter, violence was almost inevitable. The explorers were determined—believed it essential to their survival—that they demonstrate their lethal weapons. The indigenes wanted, first, to seize the wealth of the intruders and then, once they understood the folly of direct force, to steer that violence against their hereditary enemies.

In the single-mindedness of their gold lust; in claiming special spiritual powers by predicting events from eclipses to the arrival of a Junkers transport plane; in reliance on indigenous labor and local lore; in their sexual relations with native women; in their violent retaliations to theft; in the cultivation of interpreters and a common language of trade and travel (pidgin); in the white-hot rivalries occasioned by competing prospectors and the bitter quarrels over priority that followed; in their awkward relations with missionaries and colonial authorities; in their destabilizing presence; in their growing weariness over the endless violence around them and the sheer strangeness of an Otherly morality; in an exhaustion that could lead to either submersion into that order or a desire to exterminate it—the Leahys were a throwback to First-Age exploration. They recapitulated a historical scenario of the civilization they represented.

Each group was a novelty to the other. But the Leahys coped more easily. Why? They knew what they wanted, where they wished to go, and how they proposed to get what they sought. The New Guineans did not. The Leahys were surprised, but not stunned. They carried in their baggage train half a millennium of cultural tradition based on encountering new peoples in new lands. The New Guineans knew no one outside their hostile neighbors, all of whom worked to keep one another strictly in their place. They learned quickly enough; but the momentum lay with the explorers, as it had so often in the past. It was when the explorer returned that troubles so often boiled over. That had happened with Columbus, Stanley, Cook, and it happened with the Leahys. Familiarity bred not only contempt but covetousness, as each party sought to turn the other's strengths to its own advantage.

In this, however, the exploration of New Guinea marked perhaps the last hurrah of discovery before the Third Age. While the march of exploring prospectors across the highlands had its oddities, it shared with those other ages some great constants: they all had to deal, instantly and unavoidably, with the indigenes. Survival skills had less to do with hunting, tracking, making lean-tos, wrestling with grizzlies and shooting rhinos, or with more rugged nineteenth-century versions of camping out and backpacking, than with cross-cultural politics and negotiations. Explorers would succeed or fail according to their ability to deal with local peoples. They needed them as guides, interpreters, collectors, laborers, porters, assistants, and soldiers. With few exceptions—those stray desert islands such as Diego Garcia and Midway, those patches of continental terrain too empty for people and nearly for life, such as the inner Gobi, the Barren Grounds, and the interior ice sheet of Greenland—exploration was about encounters with people and, through people, places.

That was the great divide among discovered places—and it is the major divide between the Second and Third Great Ages of Discovery. When Verrazano cruised off the coast of North America, and Cook off New Zealand, they saw smoke that they interpreted as evidence of people; but the Voyagers saw nothing that spoke to life, much less to exotic hominids. When Voyager 2 zipped past Uranus, there was no one on the planet or its moons to marvel at the strange spacecraft or wonder if it signified a returned ancestor, a forecast god, or a marooned machine ready for sacrifice. There was no one to threaten harm or to offer friendship. There was no one to enlist as a guide or impress as an interpreter. There was no one to reflect back a self-image of the explorer or serve as a prism to refract visions of the future. There was no one to stand, simply as himself, for an alternative moral universe. There was no one at all.

Encounter was the great set piece of exploration, in which something happened that placed the journey beyond the routine of travel and through which narrative moved beyond tedium and banality into the essential drama of discovery. Yet it was less a simple process of reaching a goal—the source of the Nile, the North Pole—than it was an exchange between peoples whose aftershocks could affect the discoverer as much as the discovered.

While Columbus might exalt, "How easy it would be to convert these people and to make them work for us," the reality was that conversion might go either way. Europeans might themselves be forcibly converted, or simply choose to go native. The threat was present from the beginning, as Cortés discovered at Cozumel. Through an Indian interpreter named Melchior (who "understood a little Spanish and knew the language of Cozumel very well"), Cortés learned of two Spaniards who were held as "slaves" by Indians farther inland. One, Gerónimo de Aguilar, was freed, while the other, Gonzalo Guerrero, chose to remain. Aguilar told their story. They were the refugees of a voyage from Darien to Santo Domingo, brought on the wrecked ship's boat by currents to Campeche. Some fifteen men and two women had escaped the downed ship. Upon the refugees' landing, the indigenes they encountered sacrificed some "to their idols"; some died of disease; the two women had perished "of overwork"; but Aguilar and Guerrero had escaped, and now they alone survived. So far this was almost a parody of those narratives in which the indigenes found themselves under Spanish control. The climax came when the captives had to choose which society to follow.[179]

When Aguilar appeared, he squatted "in Indian fashion" and wore no more clothing than the natives. He had taken holy orders, was grateful to be rescued and reclaimed by Spain, and willingly accepted clothes and relearned Spanish. But Gonzalo Guerrero, a sailor from Palos (where Columbus had first embarked), did not wish to return to Spain. Tattooed and pierced, he was married now, a father; a Cacique and a "captain in time of war," he had been absorbed into indigenous society. Aguilar reminded him that he was a Christian and "should not destroy his soul for the sake of an Indian woman," and if necessary could take his family with him to Spanish settlements. Gonzalo refused; and "neither words nor warnings" could persuade him otherwise. He had converted.[180]

And that was the threat, that exposure to other cultures might subvert loyalty to the true faith, be it of religion, ethnicity, or enlightened science. The discoverers could not simply slide newly discovered peoples into thematic pigeonholes. Their presence challenged those categories, and more powerfully, they did so by experience and felt belief, not simply by myth and codes of conduct. Over the coming centuries the proliferation of discovered peoples and the elaboration of their cosmologies disturbed Western civilization's assumed values as fully as the bones of *Megatherium* and *Archaeopteryx* did the great chain of being.

Because so many explorers thrived to the extent that they adapted native technologies, emulated native practices, or even assumed native identities, their experiences gnawed at the ideological and ethical roots of their self-identities and those of their sponsoring culture. The experience could be corrosive. Isolated, unfettered from their own social norms, they might lose the restraints of either society, as the officers of his rear column did on Stanley's last expedition across Africa in 1887. They kidnapped or bought Manyema women for sex, became recorders (and perhaps enablers) of cannibalism, and sank into the kind of debauchery and brutality that Joseph Conrad would later characterize in *Heart of Darkness*.

More experienced in the temptations, and so tempered against their ugly seductions, Stanley offered an explanation without apology. The men had, he believed, been changed by their "circumstances." "At home these men had no cause to show their natural savagery," but suddenly transplanted "to Africa & its miseries," they became "deprived of butcher's meat & bread & wine, books, newspapers, the society & the influence of their friends." Fever "seized them, wrecked minds and bodies," while "anxiety" banished their "good nature."

The breakdown was relentless. "Pleasantness was eliminated by toil. Cheerfulness yielded to internal anguish . . . until they became but shadows, morally & physically of what they had been in English society." As exploration segued into empire, the same dark unravelings threatened their sustaining society as well. [181]

Yet there was no alternative. If exploration was to happen, someone had to go, and that someone would be as affected as those he met, and would become a carrier of a cultural virus, whether malign or benevolent, back to the society to which he returned. The sustaining culture dispatched explorers to discover what it wanted—wealth, souls, allies, trade goods, beetles and butterflies, knowledge that would expand the existing cosmology of religion and science, but all ultimately an ordering of the world that was at heart moral.

Yet what the explorer often acquired—what the sensitive observer could not help but get, whether pinning it like a beetle to a collection box or getting it into his blood like malaria—was an empathy for alternative cosmologies, for understanding and often admiring the perspectives and codes of his native companions even though that appreciation might be profoundly unsettling of his own culture's existing order. The very nature of contact made such an exchange inescapable. If the discovered peoples had not wanted to be discovered, neither did the discovering peoples wish to see their own beliefs frayed and unwound. Yet that is what, inevitably, exploration did.

## POST-ENCOUNTER: MODERNISM AS EXPLORER

The nature of the encountered geography and the character of its purposes as science and national prestige pushed the Third Age more and more toward machines. Remote sensing acquired its own motive power. Antarctica had robots descending into the caldera of Mount Erebus, and robotic snowmobiles traversing crevasse fields and empty ice sheets. The deep ocean had a flotilla of unmanned craft, some capable of autonomous voyaging for weeks or months. And interplanetary space had its spacecraft. The first contact with the outer planets, the first traverse through the solar system, came with robots.[182]

This went beyond naming expeditions by their transport vessels, which already had a long pedigree. The voyage of the *Beagle;* the *Challenger* expedition; the *Discovery, Nimrod,* and *Terra Nova* expeditions to Antarctica—these were signature labels that summarized the complex societies and stories they carried. So it was with spacecraft. Mariner 4 or Viking distilled an institution, a population of engineers, technicians, scientists, and bureaucrats, and a culture. A trek as long as Voyager's spanned generations. It was an easy step to go from encapsulating to anthropomorphizing, just as mariners had long invested their ships with personalities.

The culture at JPL openly disdained anthropomorphism. Engineers treated the spacecraft as they would a glitchy computer or a cantankerous vacuum cleaner. Edward Stone characterized Voyager as only "a tool," like "using a telescope; it's the same," and refused "to personalize" the spacecraft. Preparing for the first of the planetary encounters, Mariner 2 to Venus, Oran Nicks dismissed the spacecraft as "a machine that had no real consciousness." It didn't "'know'" what it was doing or why. "At the time," Nicks observed, "few of us thought about the similarities of the spacecraft to ourselves or to other living creatures."[183]

But of course anthropomorphism crept in. Voyager was not simply a prosthesis, but a projection of the personalities of its creators. They saw themselves in its history: its conduct was theirs, its triumphs and flaws their own. The Voyager mission took on a life of its own, as a work of art might: a *Moby Dick*, a *Mona Lisa*, a Ninth Symphony, distinctive of its author (in this case its complex society of authors), but somehow possessing an existence of its own. Inevitably, "Voyager" as a label went beyond the traditional shorthand of characterizing crews and support institutions by the name of the vessel.

Particularly as commentators sought to explain its trek and purpose, they did so in traditional tropes, genres, and narrative styles that had evolved out of centuries of human explorers. Whether or not Voyager was a person, it behaved like one. Voyager became a character.

Carl Sagan explained its encounter at Jupiter as another in the long legacy of "travelers' tales." Oran Nicks, speaking for those who worked on Mariner 2, "conscious of the precariousness of the enterprise and the unpredictable behavior of that historic spacecraft," concluded that it became "not so much a rudimentary automaton" as "a beloved partner, feverish and slightly confused at times, not entirely obedient, but always endearing." So, similarly, did Voyager appear to Bruce Murray: "I certainly think it has a personality." That's "perceptual," he confessed, and "in the mind of the beholder," and something that scientists, ever reaching for analogies to help explain, were more disposed to adopt as figures of speech than were engineers, whose numbers had to measure against empirical designs. It was also a tendency of those who had to interpret the mission to the public. Still, he believed that Voyager did have a "personality," a "very nasty" one for a while, "then lost it," and behaved itself.[184]

As the saga grew, so did the tendency to attribute to Voyager the character traits of its designers, operators, and interpreters. As it pushed on to newer, outer worlds, it acquired such epithets as "plucky," "valiant," "indomitable," and "enduring." It became the Little Spacecraft That Could.

The grand gestures of the three ages had each embodied what Cherry-Garrard called the intellectual passion, each attuned to its times. The voyage of the *Victoria* was to discovery as the Renaissance was to learning and the Reformation to religion. The travels of Humboldt to the New World were of a piece with European Enlightenment and imperialism. So the Voyagers were, in peculiar ways, emissaries of a Greater Modernism and its postmodern progeny. That was a second reason why people didn't need to go.

Without an Other, exploration became about the self. Encounter turned inward; dialogues simplified into soliloquies; discovery slid into self-disclosure. All this was not peculiar to exploration. It was an essential trait of modernism, which thrived on paradoxes of self-reference and self-scrutiny. Russell's paradox, Heisenberg's principle, Gödel's proof, Bohr's concept of complementarity, all built on the apparent contradictions and topological riddles that made the self into a Möbius strip. Does the set of all things include itself? Can a set be both consistent and complete? Can you know both the location and velocity of a particle with equal precision, or does the act of measuring, the presence of an observer, introduce irreducible uncertainties? Efforts to study the self yielded odd distortions of logic and meaning. And when the self had no Other, the boundary conditions dissolved into the circularity of a Klein bottle, whose inside and outside surface are the same.

Yet this is precisely what the geographies of ice, abyss, and space created. In previous ages, an explorer might stare thunderstruck, like Defoe's Robinson Crusoe, at the footprint of another man in the sand. In the Third Age, the explorer was more likely to photograph, as Apollo astronauts did, their own footprints on the lunar dust. The protocol went a step further when the Viking lander sent, as its initial image of Mars to Earth, a photo of its mechanical foot on the red sands. The constant discovery of new peoples had forced scholarship beyond its inherited taxonomies. New tribes and customs overwhelmed ethnography as new mountains did geology and new flora did biology. It demanded a distinctive scholarship, and got it with anthropology; and when the flow of exotic cultures ceased, that scholarship turned inward and shriveled. The anthropology of the Third Age is the psychology of the explorers themselves, or their virtual facsimile.

Perhaps nothing so conveys the paradox, however, as that canonical photo from Apollo 11 taken as men first walked on the Moon. The image is a portrait of astronaut Buzz Aldrin staring at a camera held by Neil Armstrong. Here, it would seem, was at least a vestige of a classic encounter, if only of one member of the party with another. Yet with the reflective visor of Aldrin's helmet lowered, the image conveyed of his face is of the photographer taking the photo. The Möbius loop is complete.

No Voyager took a photo of its twin, and no one photographed Voyager on its flybys. But Voyager took images for Earth and of Earth, and its decades-long communication might well stand as a conversation with ourselves. The Third Age was different not only because of its weird terrains, but also because of the cultural syndrome that determined what it saw and how it spoke. Without an Other, there was no reason to send a human self. That paradox any good modernist would have understood instinctively.

Even as euphoria alloyed with exhaustion and as the Voyagers' triumphs sharpened the contrast with Challenger's tragedy, they indefatigably continued their journey. Voyager 1 rose unveeringly upward toward the remote heliopause, routinely sampling the interplanetary medium. Voyager 2 still had plenty of observations to complete, and they remained squarely within the realm of geophysics, not metaphysics, even as another jolt of gravity-assisted acceleration flung it away from Uranus.

Its post-encounter phase lasted from January 26 to February 25. There was new data to gather and course corrections to make. Passage outward, after all, complemented passage inward. For the first three days Voyager 2 crossed bow shock seven times. It tracked the rotation of Uranus's magnetosphere. Its instruments sampled the dark side of Uranus. It reported more occultations. It attempted a ring movie, though the images proved too dark to discern. Its interminable record of fields and particles continued. Yaws and rolls and other maneuvers recalibrated instruments. Most critically, Voyager executed two playbacks of its near-encounter data—twice, to ensure a successful transmission; briskly, before the tapes were overwritten by further activity and because the European Space Agency's Giotto spacecraft was scheduled to rendezvous with Comet Halley within a week after Voyager's near-encounter and the DSN had to redirect its antennas. For four months Voyager 2 had enjoyed priority: Uranus was unique, and the prospects for another visit, remote. But now the other fledglings in the space science nest squawked for attention.[185]

On February 16 Voyager 2 executed a trajectory correction that propelled it to Neptune. It was a slow burn, the longest in the lengthening history of its interplanetary traverse—some three hours. When it ended, the spacecraft was sprinting toward an encounter with the outermost planet, scheduled for August 1989. The Voyager twins returned to a state of quasi-hibernation.[186]

DAY 2,636–3,829

# 17. Cruise

For the next forty-two months Voyager 2 cruised at 70,000 kilometers per hour toward a cold, dark planet far beyond the realm of what the naked eye could see from Earth, a world first intuited only mathematically by the quirky perturbations its gravitational presence caused in the motions of Uranus and whose first sighting was confused with a comet. It was an abstract world, posited by astronomers, outside the realm of ancient mythology, beyond Newton's classic model of the solar system. Of Neptune's particulars very little was known. It had a moon, Triton; it had some kind of ring; it was gaseous and experienced a variety of weather conditions. Much of this information came from intensive observations amassed after Voyager had launched (and conducted to better inform Voyager 2's encounter). Neptune was as little known as the abyss that its namesake ruled. At the time of encounter, Pluto's skewed path would bring it closer to the Sun than Neptune, which left the murky blue giant as the outermost planet of the solar system.[187]

Both spacecraft were well beyond their specifications and warranties. Ellis Miner likened them to a "vintage automobile" whose accumulating eccentricities its owner understood and could tweak to advantage. Yet while its mechanical aches never lessened, while its power supply relentlessly faded, and while distance magnified communication needs geometrically, the capacity of its computers for reprogramming meant Voyager 2 was in some respects a better spacecraft at Uranus than at Saturn, and would be better still at Neptune—and so were its human handlers. Without competing projects—the Challenger disaster had shut down all the others—the Voyager mission kept more of its experienced staff, and JPL understood better the demands for preparing, the procedures for crafting a master sequence for encounter, and the protocol for rehearsing. Between them the Voyager team had experienced five planetary encounters.[188]

Now they prepared for Neptune. It would be, its handlers liked to say, Voyager 2's "last picture show." It would, in effect, perform an occultation on the entire mission.[189]

## OCCULTATION OF THE GRAND TOUR

Once again, readying Voyager meant managing a complex staging by which science objectives evolved, trajectories were tweaked, communications upgraded, and new software uploaded. A change in one meant a change in the others. The process of establishing guidelines for the final uplink of commands consumed over three years. But doing it previously was not the same as doing it at Neptune.

Too little was known, the planet was impossibly far, and while Voyager 2 could receive new software, it could not rebuild its aging hardware. Its power output was particularly troubling. It would broadcast at "a billionth of a billionth of a watt." At Jupiter, Voyager could transmit a maximum of 115,000 kilobits per second; at Uranus, even with clever compression coding, 21,600 kps; and at Neptune, perhaps 14,400 kps. Some new algorithms helped, but the only option was to improve Earth's capacity to hear Voyager's whisper against the background static.[190]

Fortunately, the Canberra DSN station would once more serve as primary antenna. So, again, the DSN arranged to contract for the Parkes radio telescope, and NASA agreed to allow the Voyager team to delay the spacecraft's anticipated arrival by five hours in order to maximize the time of Canberra reception. So vast were the distances, and so attenuated the signal, however, that effective communication would need much more, a virtual array the size of the Pacific Basin. The DSN thus reached northward to Japan's Usuda 64-meter tracking antenna on the island of Honshu, and to western North America as well, first to the Goldstone complex, which now included one 74-meter and two 24-meter antennas, and then to the very large array of radio telescopes at Socorro, New Mexico, operated by the National Radio Astronomy Observatory, which added the equivalent of two 70-meter antennas. The actual synthesis of data would have to occur after the flyby rather than in real time, but Earth could now hear what Voyager had to say.[191]

The other worry was identifying the roster of targets and the means to navigate to them, particularly since so little was known. Planners made their best guess, and then hedged by actually building into the scenario places for the acquisition of new data and recoding on the fly. The scientific working groups had already begun identifying targets even as they planned for Uranus, since the route past Uranus had to include a path to Neptune. And in one sense, the spectrum of interests was already hardwired into Voyager 2's instrument package, a medley of hard- and soft-geography scans. Specifically, Voyager 2 would interrogate the Neptunian atmosphere, notably its composition, energy budget, structure, and aurorae; it would map, delineate, and identify the constituent parts of the planetary rings and ring arcs; it would search out and characterize satellites, especially the planet's largest moon, Triton; it would survey the magnetosphere and its dynamics; and it would conduct an experiment close to the science working group's heart, four occultations and a close passage over the north pole that would deflect Voyager into a course as near as possible to Triton. Along the way it would record such fundamentals as the planet's rotation period, the orientation of its axis, and its precise mass along with those of its moons. And the spacecraft's master sequence would not only allow for fresh input but even anticipate it.[192]

The Neptune groups held a series of meetings between August 1986 and July 1987 to agree on a roster of wishes and to establish priorities. It submitted its final recommendations, pending further discoveries, on July 15, 1987. The process would continue up to encounter, a total of three years.[193]

---

Of all the Voyager encounters, Charles Kohlhase concluded, the planning for Neptune's was the "most challenging," and he likened the task to guiding the spacecraft "through an imaginary needle's eye about 100 km wide, while the spacecraft is going a blistering 27 km/sec—and they expect to predict when this will happen to within *one* second!" The final trajectory was the outcome of nearly eight years of planning, editing, refining, tinkering, a ceaseless juggling between what scientists wanted, what engineers could provide, and what the spacecraft could bear.[194]

The one ameliorating factor was that Neptune ended the Grand Tour. Voyager 2 would not have to tweak its trajectory in order to continue to another planet, so its passage around Neptune could be maximized for the values that planet offered. But the only hope for anything like the precision demanded was a good sequence of course corrections, constant refinements of trajectory based on improving data about Neptune's hard geography and its gravitational effects, and a stream of successfully uploaded commands to tell Voyager 2 how to perform its acrobatics.

Nothing in the Grand Tour, perhaps, was more daring than the hope—the planned expectation, really—that the spacecraft could in fact acquire the additional data it needed for precision guidance. The engineering term "critical late activities" disguised the audacity behind the ambition. The particular issue was that the spacecraft's navigational system could not work reliably from Canopus alone; it needed other bodies, preferably Neptunian moons, objects that were not, at the time, known to exist but whose identities were programmed blankly into Voyager's computers. If—when—such satellites were discovered and their orbital parameters measured, the additional information would be uploaded into the code. Voyager 2 needed at least one such moon; it found six. It then needed to tweak its trajectory to thread the Neptunian needle. It planned for six corrections, the first coming on February 14, 1986, a scant three weeks after the Uranus encounter. It needed only four.[195]

All this of course required smart sequences uploaded to an onboard computer command subsystem that dated to the Paleolithic age of computing by a cobbled-together array of DSN dishes that spanned from Japan to Australia to California and New Mexico. By 1989 Voyager 2's computers were nearly as creaky as its scan platform rotors, and its onboard memory was an order of magnitude less than that of even the lower-end personal computers then available. Still, JPL planned ten uploads of command sequences, the product of "high-fidelity timelines" translated into machine language and the outcome of nearly thirty-six months of intense labor.[196]

In the end, the reality was that robot and handlers would have to rely on each other–Voyager 2, that JPL could tell it what it needed; and JPL, that Voyager 2 could operate autonomously, for once the spacecraft commenced near-encounter, it would be on its own. The distance between Neptune and Pasadena was such that even a simple transmission would take 246 minutes, or over four hours, and for a full exchange, 492 minutes, assuming that a response was instantaneous. So among its programs were "failure [or fault] protection algorithms," which would allow Voyager to detect and repair many problems, or shunt into a safe mode. There were, for example, routines to have the spacecraft begin a sequence of pitches and yaws to relocate the Sun, and then roll to refix with Canopus should something break navigational contact. The computer often tested itself. Critical instruments could be shut down and then reignited in the event of voltage surges. And a "backup mission load," frequently uploaded, programmed Voyager to execute a basic survey of Neptune should something prevent the final near-encounter transfer of commands. Only at the most frenetic moments, when every fragment of memory space was needed, was the backup mission load removed.[197]

What remained was to test the system under simulated conditions. JPL and NASA had learned a hard lesson at Uranus. They could not scatter staff across other projects, or even lay them off, and then expect an instant return to service, nor could they expect that upgrades in hardware, software, or operating procedures could synchronize without practice. Beginning in October 1987, nearly all project teams underwent "operational readiness tests" to "validate and calibrate" the DSN and prepare for occultation experiments.

The Voyagers themselves experienced special "capability demonstration tests," with Voyager 1 serving as a "test-bed." Between 1988 and early 1989 both spacecraft and those communicating with them confirmed protocols and calibrations, one involving a roll-turn course correction maneuver. In May 1989 the entire operation undertook a dress rehearsal.[198]

## THE LONG TREK

From near-encounter at Uranus to near-encounter at Neptune, some forty-two months passed. It had taken the spacecraft twenty-three months to reach Jupiter, another twenty-three months to reach Saturn, and a whopping fifty-three additional months to reach Uranus. Neptune was almost as far from Uranus as Uranus was from Saturn, but the additional velocity Voyager had acquired from its sling-fling around Uranus had hastened its travel. From Earth to Neptune, however, had taken almost exactly a hundred and forty-four months, or twelve years. The Grand Tour was a notably Long Tour.

Interplanetary space is in many respects a benign environment for a machine. Voyager did not experience that long passage as astronauts might: it did not become bored or restless or overcome with anxieties. It shut down most of its instruments, shrank power and hydrazine requirements, and cruised. It entered a mechanical hibernation. Unlike an earthly vessel, it suffered from no physical waves or shearing winds; nor did it have to navigate around unknown shoals. The usual stresses that assault exploring craft were absent. There were no acoustical blasts, no aerodynamic tugs and shears, no gravitational forces pulling it to extinction, no salt to corrode or rain to rot, and no organisms to chew on it. It sailed through no sargasso weeds, acquired no barnacles, endured no ship worms. And it had no human occupants for which it had to sustain an artificial environment. It felt fields and particles, but this was a soft geography, and apart from the planets and asteroid belt, or the rare meteorite, it could sail serenely on autopilot, unworried, and unafflicted by lethargy and ennui.[199]

But this was not true for its human servants. While Voyager had trekked to Neptune, the United States had experienced four presidential elections and was ten weeks away from a fifth. The Voyager mission had gone through six program managers and was preparing for a seventh. Twelve years was an academic lifetime—two full tenure-review cycles. Staffers had grown up with Voyager. They graduated, married, divorced, had children, transferred and traded jobs, retired, got cancer and injuries, healed, died. Candice Hansen remarked that people "actually remembered when their babies were born relative to encounters." Ellis Miner noted wryly that the group had been "considerably younger" in 1965, when the Grand Tour was conceived, than in 1989, when it reached Neptune. The long cruise phases were particularly trying for scientists who had to publish constantly to advance careers. Voyager could experience a rhythm of comatose cruising broken by frenzied encounters without a change in life or career. Its human tenders couldn't.[200]

While space science seemed a promising career during IGY, in truth it became risky several times over. The spasms of discovery were infrequent and too scattered to sustain regular positions and funding. It always proved easier to create a new project than to continue paying for an old one, more appealing to sponsor a new voyage of discovery than to subsidize the publication of results from returned ones. The missions themselves were too sparse to sustain a self-supporting community. A successful encounter could make its staff into celebrities; but such fame was fleeting, and a failed spacecraft or instrument could be a career-ending blow. Planetary science was a bet against long odds.

Yet the sense that scientific productivity had to accelerate continuously in every field is a recent aberration. In the past, exploration had often taken years, and its scholarly write-up could consume decades. Charles-Marie de la Condamine spent ten years measuring in the Andes before returning down the Amazon; the last member of his expedition did not return to France for thirty-eight years. Alexander von Humboldt devoted his young manhood to preparing for an exploring expedition, spent five years in the field, exhausted five decades and his inheritance distilling it into his fifty-four-volume

*Voyage to the Equinoctial Regions*, and died still writing his popular summary, *Cosmos*. The Institut d'Egypte, the corps of 151 savants whom Napoleon assembled to survey his conquests along the Nile and who, in the words of Gaspard Monge, would "carry the torch of enlightenment" in its *mission civilisatrice*, lost 31 members directly, a disaster that distorted the scientific establishment of France for decades afterward; the record of their findings, *The Description of Egypt*, took its members twenty-six years to complete. The U.S. Exploring Expedition–America's belated answer to Cook's voyages–soaked up ten years from first approval to actual departure, spent some four years at sea, and for nineteen subsequent years crawled toward the publication of nineteen volumes (out of twenty-eight originally conceived). By the time the expedition's findings got into print, the public had more or less forgotten about it, and the country was headed into civil war. Lewis and Clark had taken two years to cross North America, and never did see their vaunted journals published. Beginning in 1867 the Russian general and geographer Nikolai Przhevalsky conducted five major expeditions in Asia with the ultimate ambition of reaching Lhasa. Eighteen years later, he died suddenly while in the field, Lhasa still a mirage over the mountains.[201]

A successful expedition could not only make a career, it could be a career.

For oceangoing expeditions, long voyages with short encounters were the norm. Still, few could rival the saga of Vitus Bering and the discovery of Alaska.

The motive began with Peter the Great's 1724 order to Bering, a Dane in his service, to determine whether the far eastern lands of Russia's expanded empire met the western extent of North America. It took Bering four years to travel to Kamchatka, build a vessel, and sail to the straits that now bear his name. Dense fog prevented a direct sighting, but the nature of the strait and its currents strongly suggested that the two continents were not joined, and native Chukchis confirmed that fact. When he returned, Peter had died, and his successors decided that Bering had not fully answered the question, so in 1731 they sent him back, this time with the naturalist Georg Steller. He finally sailed in the spring of 1741 with two vessels, again constructed in Kamchatka, the *St. Peter* and the *St. Paul.* In July 1741 the St. Peter sighted the southern shore of Alaska (at Mount St. Elias), made landfall at Kayak Island, and subsequently wrecked off the coast of Kamchatka, where its crew spent a wretched winter at Fox Island and where Bering died in December. In August 1742 the survivors reached Petropavlovsk and began the long trek across Asia to report their findings.[202]

The second expedition had lasted over ten years and granted Steller a scientific "encounter" with North America, at Kayak Island, for something like ten hours. The rhythm was eerily like that of planetary exploration. The difference was the immense tedium, frustrations, and suffering that Steller and crews endured to earn those frenzied few hours ashore.

Perhaps the closest historic analogue is the fabled HMS *Challenger* expedition of 1872–76, generally recognized as the origin of modern oceanography and a model that anticipates Third Age conditions as La Condamine's did the Second Age.

The *Challenger*'s complex voyage around the world lasted forty-two months, exactly the length of Voyager 2's traverse from Uranus to Neptune. The essential rhythm was one of long hiatuses broken by spasms of encounters with islands or ports. In many respects the cadence was also a foretaste of what Antarctic discovery would be like, as explorers broke through the barrier ice at the end of the austral summer, and then wintered over under confined conditions until spring allowed for sledging forays. This was of course equally the pulse of interplanetary exploration. What made it tolerable was the presence of a large cohort of like-minded comrades, and the anticipation of adventure with exotic peoples or unknown lands. Even with other cultures and strange ports of call, this is a formula for a robot. Without the prospect of those interludes, it is hard to imagine how an expedition could thrive.

Of the *Challenger*'s crew of 243, a quarter deserted out of boredom, numbed by the tedium of sounding and dredging ("drudging," as the sailors called it), and out of restlessness on the converted (and confining) corvette; 7 men died, and 26 were invalided out of service or left at hospitals. The scientific corps was marginally better. The "great adventure" across 69,000 nautical miles came at the cost of an "industrious boredom" that nearly drove even the most zealous mad. After they disembarked, the vast tedium continued as some 50 volumes, involving 100 scientists, saw print over the subsequent 20 years, concluding in 1895. Interestingly, as with so many proposals for manned space exploration, the scientific quest involved a search for life in forlorn niches; and this the *Challenger* found. But so long was the ordeal, and so complex its production, that the public and treasury lost interest, and by the time the *Challenger* expedition formally completed its task, general enthusiasms were poised to explore new lands.[203]

Still, the enterprise loomed large in its day—everyone recognized it as a vast and important undertaking, even if it could rouse only sparse popular excitement. It has remained a landmark, a historical point of reference for exploring expeditions much as Tenerife remains a geographic one. The last of the lunar modules (Apollo 17) bore its name as did the first of the space shuttle disasters, and the most successful of the deep-drilling oceanographic research vessels. Yet it has endured perhaps better as an ideal than as a practical guide. In the 1870s there was no alternative to crewed ships and their onboard "scientifics" and "philosophers." By the 1970s, when the RV *Glomar Challenger* was coring deep-ocean sediments, there was.

Among the *Challenger*'s lessons was the difficulty of sustaining research once the adventure ended. The fact is, science is long, politics short, and public interest capricious. The insistence by the "scientifics" that the expedition meant only research, and that research had to be thorough and be published according to the standards of the guild, meant that the drain on the public treasury was bound to be bottomless. Politics and public interest, however, had their own cadences.

In the past, as often as not, scientific societies or wealthy patrons had to intervene to complete the task, which could easily absorb the careers of those involved. Accounts of adventure in the guise of personal narratives were popular and could be published by subscription. The latter often helped underwrite private exploring expeditions in ways that shelves of dense reports could not. Politicians, publicists, and even sciences moved on. Where cruises lasted for years without tangible consequences, where investments were swallowed in upgrading instruments and improving staff performances for brief encounters, it was difficult to sustain much fervor.

Two years after Voyager launched, five months after encounter with Jupiter, the first *Star Trek* movie was released. It posits a later, fictional, Voyager mission that disappears into a black hole, finds a machine civilization, and then returns laden with knowledge it seeks to transmit to its "creator." Ten years later, after Voyager 2 had encountered Neptune, the fifth movie in the series, *The Final Frontier*, opened with a bored Klingon commander blasting a dead-metal NASA spacecraft, now dismissed as "space junk," for target practice. The tempo of modern society had quickened, and public expectations seemed to shorten. Even as the Galileo mission to Jupiter at last launched, after a decade during which Voyager was the only exploring vessel in action, there was only so much passion that a nominally spacefaring society could muster.[204]

Many staffers considered the Voyager mission the adventure of a lifetime. For anyone present from its conception—and they were more than a handful—the mission *was* a lifetime.

## THIS NEW OCEAN

The character of the Earth's oceans set the tempo for the HMS *Challenger*'s cruise, and the properties of what partisans eagerly called "this new ocean" of interplanetary space set those for Voyager. In the Third Age, however, the densest discoveries came not from the analogous ocean of the solar system but from the true new ocean of Earth's abysses.

The deep oceans and interplanetary space were the two commanding realms of the Third Age, and their exploration proceeded in an eerie fugue. Bathyspheres were the high-altitude balloons of the sea; the bathyscaphe *Trieste* descending to the Challenger Deep in 1960 was the wet twin to Mariner 2 flying past Venus. The knowledge of the Laurentian Abyss was no better than that of the crater Daedalus on the far side of the Moon. There was putative bullion in the abyssal plains. There were schemes for undersea tourism and colonization. There was even precedent in science fiction; after all, Jules Verne had written *20,000 Leagues Under the Sea* as well as *From the Earth to the Moon*. The cold war rivalry played out beneath the waves as fully as it did beyond the clouds.

Yet a profound paradox emerged. Space captured the public and political imagination; the deep oceans did not. Space had a mystique that the abyss lacked. Arthur C. Clarke's sea novels sank; his space fiction soared. Much of what happened in the seas occurred in bureaucratically secret "black" programs, not publicly promoted and not known until the cold war ended and declassification began revealing what had happened and the technologies of submersibles and robotic probes became generally available. The high ground for military rivalry was not thermonuclear bombs in orbit but nuclear-armed submarines. The search for life that drove the post-Apollo planetary program, particularly to Mars, found nothing, and might discover it only at immense cost and in token fossils. The exploration of the oceans discovered startling realms of previously unknown life, from the largest ecosystems on Earth to exotic niches powered by chemical rather than photosynthetic metabolism. Likewise, the long-anticipated revolution in geosciences came not from other planets but from Earth's oceans, as remote sensing unveiled world-girdling mountains, faults, and seamounts. The submersible *Alvin* visiting black smokers in the East Pacific Rise (the same year Voyager launched) and plucking rocks from the Mid-Atlantic Rift provoked more change than Moon rocks brought back by Apollo astronauts. In his critique of the Apollo program, *The Moon-doggle*, Amitai Etzioni shrewdly contrasted the panicked response to Sputnik with the overlooked prospects for the oceans. The paradox was that while attention looked up, the bigger payoff came from below. The promises of the Third Age were unfolding less in the depths of space than in the depths of the Earth's seas.

---

The reasons are many. The oceans are closer, and cheaper to explore, and however alien to humanity's quotidian existence, the seas are continuous with humanity's world both geographically and historically.

The "void" of the deep seas is in reality a tangible medium, a watery matrix and a more plausible nursery of life than microbe-carrying meteors from Mars. While light fades below 2,400 meters, while pressure can crush anything dropped from the surface, and while nutrients either concentrate in niches or diffuse through chunks of salty seawater the size of continents, this is the place where life originated on Earth, and where life has adapted in forms more fantastic than any worlds conjured up by writers of science fiction. Organisms abound at all levels. There are biota on the abyssal plains; vast throngs of micro- and macroorganisms gliding upward and downward on daily and seasonal cycles; giant squid, jellyfish-like siphonophora, sleeper sharks, viperfish, lantern fish, a menagerie of bioluminescent browsers, and predators that thrive according to alien metabolisms and obey a different ecological logic. The migrations of pelagic species are the largest on Earth, dwarfing those on African savannas. There are ecosystems clustered to hot vents organized around chemosynthetic bacteria and symbiotic tube worms. There are deep corals and fisheries attached to seamounts, submerged versions of volcanic islands and atolls. There are worms and benthic bugs that burrow into the ooze that blankets the abyss. There are opportunistic communities that feed on fallen whales and other macrofauna.[205]

Far from being abiotic vats, only capable of supporting life at the surface, the oceans are the dominant habitat on Earth. However hostile to terrestrial life, the sea is an abode for earthly life and is continuous with it. Increasingly it appears that the rifts, submarine volcanoes, hot vents, and black smokers that parse and stitch the solid floor of the sea are fundamental to a grand geochemical cycling of planetary water, and with it the nutrients and minerals essential to a living planet. Not least, it is possible to step from land to sea and back again. The shore is a familiar boundary between the two, literally ebbing and flowing with the rhythms of earthly existence.[206]

The void of deep space is a vacuum. Nothing lives within it. It may be that organic molecules travel through the void within particles blasted off planets, but there is no ecosystem possible, and no connectivity between those spores or bio-shards and a Gaian Earth. Passage between planetary bodies is a mix of tiny violence and immense void; the fiery impact of meteorites or the flame-powered launch of rockets. The solar system may loosely resemble Earth, with its interplanetary space assuming the role of oceans, the planets that of continents, and planetary moons as islands. But contact between them requires years of travel across a vacuum. It demands power; and behind power, will; and behind will, belief. One can jump into the ocean by stepping into the surf, but into space only by a leap of faith.

Another continuity is historical. The oceans have led each age of discovery. The exploration of the world ocean virtually defined the First Age. The Second began with that outrush of circumnavigators of which James Cook, Louis-Antoine de Bougainville, Alejandro Malaspina, and Thaddeus Bellingshausen are its informing captains, and which ferried the fabled naturalists of the era, from Humboldt to Joseph Hooker, to their destinations, and made islands into scholarly ports of call. The age even attempted to extend its powerful Enlightenment science to the sea, making it an object of inquiry and not simply a means of transport. Matthew Fontaine Maury's *Physical Geography of the Sea* (1855) adapted Humboldtian techniques to map the oceans. A century after Cook, as the Second Age was completing its final continental traverses, the *Challenger* expedition applied the accumulated lessons of the Second Age to a global survey of the world ocean, even as its mountainous efforts seemed to yield a molehill of fresh science. Other nations followed, sporadically, with outcomes that seem impressive mostly in retrospect.[207]

The Third Age commenced seriously in the oceans. After World War II, naval vessels were commissioned as research ships and began mapping the solid floor of the world's seas. Submarines replaced capital ships as nuclear-armed Poseidon missiles became one leg of the American strategic triad; the need grew to know better how to navigate the deep oceans and to communicate with undersea vessels.

By the time of IGY, the USS *Nautilus* and *Skate* had surfaced at the North Pole, the Trieste was bobbing in the Mediterranean, the Mid-Atlantic Rift was rudely mapped, and it was apparent that the deep oceans were not the lifeless chemical tombs and geologically inert sinks early research and prevailing theory had indicated. By 1962 Harry Hess had published his "History of Ocean Basins," which inverted the standard geophysical model of Earth, and conceptually catalyzed what became the theory of plate tectonics within a handful of years. No less than comparative planetology, oceanography became a cold war science. As the space race heated up, the U.S. Navy found a desperate need for submersibles and poured money into their development. The *Alvin* became the naval counterpart to Apollo. The deep oceans were the first of the Third Age landscapes to be mapped, and the one that catalyzed a scientific revolution.

Not least is the similarity of debates about purpose, especially the wedge issues of human versus robotic exploration, and the ultimate vision of colonization beyond Terra. On this the oceans once again appear to lead. Costs, dangers, the relative merit of human senses versus instruments, the diminishing value of a human brain and hand on board are all propelling deep-sea exploration toward remotely operated vessels or even autonomous submersibles that can wander for weeks. The only sense available to an aquanaut, as to an astronaut, is sight, which is replicated with a richer spectrum by instrumented cameras. The pilot of the *Alvin,* Robert Ballard, a pioneer of submersibles, having used both crewed vehicles and robots for decades, has concluded that "the robots are better." So, too, early prospects for a frontier of ocean settlement died in the depths. Ocean exploration can advance unencumbered by techno-utopian colonizers and aspiring homesteaders.[208]

After roughly sixty years, the evolving contours of the Third Age suggest that the solar system will not be the primary arena of discovery, that the pattern of exploration among the planets will more resemble the pattern of exploration of the First Age, when complex armadas undertook long voyages, an era of relatively few expeditions but huge returns. Rather, the deep oceans will likely claim the variety and vivacity of exploring, perhaps even the swarm effect that characterized the Second Age. But if so, why does "space" continue to claim pride of place?

Secrecy is one reason. Much of the oceanic work had military sponsorship and was not broadcast. Nor was there any overt, broadcast-over-TV, soap opera drama for an abyss race as there was for a space race. There was no literature for deep sea exploration as there was for planetary. Robert Heinlein did not write a novel titled *Submarine Troopers*, nor Ray Bradbury a *West Mariana Basin Chronicles*. Arthur C. Clarke did not imagine *Childhood's End* happening on the East Pacific Rise. Stanley Kubrick did not film *2001: A Sea Odyssey* or place alien obelisks on the Valdivia Abyssal Plain. There was no Tsiolkovsky or Goddard to imagine a Great Migration to the Laurentian Abyss, or a Percival Lowell to sketch the contours of a dying civilization on the Loihi Seamount. There was no Carl Sagan to fantasize about cosmic connections with galactic intelligences, or rhapsodize about chemo-spiritual liaisons with "salt stuff" shared between people and black smokers.

Over and again, the most publicly ardent proponents of planetary exploration said they saw their endeavor as part of a larger mission to colonize, and an astonishing number traced their enthusiasm to an adolescent literature of technological romance in which worlds were found and lost in space. While the oceans had their lore, and champions of the sea held fiercely to their distinctive sagas of exploration, that tradition stayed on the surface. The utopians wanted colonies on other worlds, by which they meant worlds that looked like planets. What the deep oceans showed, however, was that exploration was neither confined to space nor defined by it. Geographic exploration and space occupied separate cultural realms, though they could from time to time intersect.

The deep oceans will likely claim the lion's share in terms of numbers of Third Age expeditions and discoveries. The yet-unvisited oceans may reveal more animals than the terrestrial Earth, answer fundamental queries about the origin and character of life, and tell us more directly than analogies to Mars or Europa about how to care for an abused Earth. But neither numbers alone nor the robustness of oceanographic science can determine cultural clout. Space will retain its partisans, and its promise, and it will offer in the prospects of a sweeping journey, a trek beyond, something that the deep oceans cannot. And that distinction may explain why abyssal exploration has failed to imaginatively inform the coming age. It lacks a Grand Tour. It lacks a Voyager.

DAY 3,830–3,949

# 18. Encounter: Neptune

On June 5, 1989, some 42 months after leaving Uranus, and nearly 142 after leaving Earth, Voyager 2 began the end of its Grand Tour.

This was the sixth time the Voyager team had met a planet, and Voyager 2 would do what it had done at each prior encounter: it would direct its instruments toward a full-body, geophysical scan of the planet and its satellites. But if the scenario had become ritualized, it had lost none of its wild alloy of awe, anticipation, and anxiety.[209]

## LAST CONTACT

JPL uplinked the first of new commands that would prepare the spacecraft to do its gamut of tasks. To scan for ultraviolet emissions that might identify atmospheric chemistry. To search for radio signals birthed where Neptune's magnetic field met the solar wind and for long-wave radiation that could measure more accurately the opaque planet's rate of rotation. To measure the brightness of the Sun and of select stars pertinent for navigating through near-encounter events. To ready its infrared instruments by turning their flash heater on and then off. To continue, as it had throughout its long trek across interplanetary space, to sample fields and particles, from time to time recalibrating its instruments by rolls and yaws to allow researchers to account for the distorting influence of the spacecraft itself; to commence a series of four occultation experiments. To search for rings and satellites.

This was what discovery meant in the popular mind: the revelation of new worlds. The Voyager twins had found a covey of moons at Jupiter, Saturn, and Uranus, and there was every expectation Voyager 2 would discover even more exotic moons here at the outer fringes of the solar system. So its cameras imaged each side of Neptune, and in early July, Voyager found its first new satellite, unimaginatively dubbed 1989N1. This discovery gave the navigation team the needed coordinate by which to guide Voyager through near-encounter maneuvers. Other satellite discoveries soon followed. By the end of the month, Voyager had detected four new moons, subsequently named after figures from Greek mythology—Proteus, Larissa, Galatea, and Despina. On July 30 it captured all four in a single, mesmerizing image.

On August 6, Voyager 2 left its observatory phase behind and raced toward full encounter. A world that before Voyager had no sharper identity than that of a hazy blue Smurf ball, the third largest of the planets now beckoned at 67,000 kilometers per hour.

Everything began to surge. Commands—three new computer packets were uploaded. Imagery—Neptune was now too large to fit within a single aperture, so mosaics became the norm. Data—the extra antennas enlisted by the DSN came into play. Less than a day into far-encounter, the last dress rehearsal, a complex execution for an occultation experiment dubbed Radio Science ORT-4, ran for ten hours and identified soft spots in performance, which were quickly corrected.

The expected discoveries acquired their rhythm. The mosaics were assembled into movies of Neptunian weather; the search for moons intensified; rings, ring arcs, and shepherd satellites came into focus; IRIS recorded temperatures; PRA and PWS tracked radio waves, emitted where Neptune's magnetosphere met the solar wind. As Voyager 2 quickened its pace, estimates of the planet's hard parameters sharpened. "The uncertainties that everyone fussed over for so many years," Charles Kohlhase noted, were "dropping precipitously." Neptune's mass and position would be known three times more accurately, Triton three to six times better, and Voyager 2's time of arrival a third better. The Voyager staff would no longer find Triton's mass "a mystery." With the new data, engineers refined parameters for the final, complex roll-turn course correction that would climax near-encounter. Everyone waited for bow shock, the final hurdle before closest approach. It arrived on August 24, 1989.[210]

From August 24 to 29, an aging Voyager 2, now zipping at more than 71,000 kilometers per hour, hit its target trajectory, performed a riot of acrobatic maneuvers, amassed mountains of data, and executed nearly 90 high-priority optical and remote-sensing observations. At 8:56 PDT on August 24, Voyager passed within 29,240 kilometers of the planet's center.[211]

It photographed the surface and its wild weather. It sped through Neptune's ring plane and behind the gaseous giant, sending and receiving radio signals vital for occultation experiments. It used the star Nunki (δ Sagittarii) to occult the rings. It photographed the rings and arcs directly on both approach and departure, as illuminated on the foreside and as backlit, and merged those images into movies. It measured particle impacts. It flew over the north pole of rotation, recording auroral emissions and identifying the magnetic pole. It absorbed the soft geography of Neptune's ionic and electromagnetic fields, mapping the spongy borders of the magnetosphere. It measured the brightness, pressure, temperature, chemical composition, and cloud structure of the planetary atmosphere. Just before periapsis, Triton occluded the star Gomeisa, and then the Sun, and Voyager seized both opportunities to add to its inventory. Then, while in Neptune's shadow, it reversed its instruments and surveyed the planet's dark side. Some 110 minutes after closest approach, Voyager undertook a series of pitches and yaws, realigning from Canopus to Alkaid as a navigational referent in order to position its instruments to interrogate Triton. Relying on image motion compensation, it proceeded to create a high-resolution mosaic of the satellite's surface. At 2:10 PDT on August 25, Voyager 2 made its closest approach to Triton, some 39,800 kilometers from the moon's center.[212]

The Triton flyby—much sought by scientists—took Voyager 2 out of the plane of the Sun's ecliptic, and hurled it obliquely southward at forty-eight degrees. With that effort, Voyager had shot its bolt: it would encounter no further worlds. In fact, the odd trajectory actually caused a gravity-assisted slowdown. Some thirty-eight hours after entering bow shock, Voyager exited the magnetosphere, and continued to move in and out of bow shock amid the filmy fields for another two days. Finally, seventy-nine hours after closest approach, it turned its cameras back to the planet and its moon.[213]

The gesture had become a valued parting ritual begun by Voyager 1 when, hurrying to Jupiter, it had turned and captured Earth and the Moon in a single image. This time Voyager caught the double crescents of Neptune and Triton in an unforgettable scene: haunting, austere, radiantly lonely. It was the Grand Tour's last work of art.

Post-encounter had its usual anticlimactic aura—"like the cleaning crew that does its work the morning after an all-night party," Ellis Miner thought. The phase began officially on August 29 and didn't end until October 2, 1989, almost exactly thirty-two years after Sputnik 1 launched.[214]

This time post-encounter did not have the anxieties attendant on the need to target another planet. But Neptune's remoteness added an anxiety of its own. The immense distance, low wattage, and feeble computing capabilities meant that much of what Voyager 2 had sensed it still had to send to Earth. Its digital tape recorder had to play back the drama of near-encounter. It did so twice in order to reduce noise, fill gaps, and substitute redundancy for lost transmitting power. Even so, critical instruments continued to record and image. The far side of Neptune had its interests and its partisans, and the fields of soft geography busily reclaimed the place they had temporarily yielded to hard geography. Then the spacecraft undertook a final series of rolls and yaws to recalibrate instruments and navigational systems before beginning its final, uninterrupted cruise to infinity.[215]

## TRITON

Triton was the last of Voyager's surveyed worlds, and with a fitting sense of closure, perhaps the strangest.

Almost everything about Triton proved exceptional. It combines variety with size. It has "perhaps the most diverse geological land forms found anywhere in the Solar System," the result of a complex and dynamic history. It has an atmosphere flush with ices of nitrogen, methane, carbon monoxide, carbon dioxide, and water. Among icy satellites, it has the "most spectrally diverse surface." It is the seventh largest moon, with a volume more than five times that of Titania (the eighth largest), but it is the only large satellite with an inclined, retrograde orbit, an orbit almost perfectly circular. Unlike other captured moons, it shares no features with asteroids; it more closely relates to objects from the Kuiper Belt and best resembles Pluto. How it originated–whether by capture or by co-evolution with Neptune or whether it might share some eccentric history with Pluto–is at present indeterminate. What is known is that Triton is the coldest body identified in the solar system, a scant thirty-eight degrees Kelvin above absolute zero.[216]

That made the discovery of geologic activity astounding. Studying images revealed what appeared to be transient streams or plumes of gas. In all, Voyager 2 photographed four eruptions–nitrogen geysers–that lofted emissions some eight kilometers high before shearing winds carried them one hundred kilometers beyond. The proposed explanation was that insolation, however feeble (900 times lower than on Earth), could still penetrate Triton's clear icy cap, create a greenhouse effect underneath, and build up a gaseous pressure that then vented violently back to the atmosphere.[217]

The scene offered a fitting sense of closure: the cold eruptions on Triton bracketed the hot ones on Io. The unforeseen discovery of volcanoes on Jupiter's Io had announced convincingly that Voyager would find new worlds, and now at the end of its trek, one had come to expect the unexpected. At Triton the Grand Tour's alpha had found its omega.

## INFORMATION AND IDEA

That was what the Third Great Age of Discovery had done overall. Not only had it revealed unknown lands but it had intellectually remade an Earth that many believed already discovered and old. Like the First Age, it had found new lands, and like the Second, it had seen old lands with new eyes. Knowledge, after all, was relative. It was contingent not only on fresh data but on novel means of interpreting.

The Second Age had bequeathed a legacy of geographic surfaces that left the solid geography of Earth and the planets obscured by ice, seawater, and gaseous atmospheres. To probe beneath those contours required special sensors and innovative means to get those instruments to the scene. Until then the terrains of ice, abyss, and space were unknowable. Antarctica did not have its perimeter fully mapped until Operations Highjump and Windmill accomplished the task by air in 1946 to 1948, and no one had traversed the continent until the Commonwealth Trans-Antarctic Expedition did so under loose affiliation with IGY. The deep oceans had been sampled but never seen, and only vaguely understood. They were conceived as a geochemical tomb and a geophysical void, a vast sink removed from the dynamics of the planet. As for the solar system, photos of the planets and the larger moons from Earth-based telescopes existed, but the gritty inner planets and gaseous outer ones were the dregs of astronomy, which was far more fascinated by distant pulsars and quasars, white dwarfs and red giants, spiraling nebulae, cosmic rays, dark matter, and traces of the Big Bang.

The Third Age revolutionized those perceptions. Field-testing had occurred in Antarctica, and then leaped outward, much as the proposed third polar year had bulked up into the International Geophysical Year. It was, J. Tuzo Wilson wrote, "in the Antarctic, which was least known [of ice sheets], that recent efforts have produced the most marked changes in our knowledge."[218]

Before IGY the basics of Antarctica as a continent were barely appreciated: the pastiche of ice shelves and sheets, the underlying solid-earth surfaces so different between East and West Antarctica, the interconnections of south polar weather with the rest of Earth. Modern technology such as Sno-Cats and DC-4s, seismic profiling, and remote-sensing instruments gradually unveiled the nearly extraterrestrial world that was the Ice. It became both more familiar and more alien. Unlike abyss and space, this was a place where people could walk, breathe, and, along its edges, survive by hunting. With permanent bases established under the aegis of IGY, Antarctic discovery made the transition from exploration to a normal, if extreme science. It made a suitable point of intellectual departure for the discovery of those new worlds, now being exposed, where humanity could not live and would not go.

The deep oceans were, at the dawn of the Third Age, a deep mystery. A century after Maury had synthesized the known ocean in his *Physical Geography of the Sea*, Rachel Carson summarized the state of knowledge in her prize-winning *The Sea Around Us* (1951). Maury's geography appeared two decades before the *Challenger* expedition; Carson's, two decades before the revolution in earth science was distilled into the theory of plate tectonics. If her book was prophetic, it was because she believed that Earth's seas, its collective Oceanus, were the grand synthesizer of geology, climate, and life, "the beginning and end"; for "all," she insisted, "at last return to the sea." Besides, she intoned, it is "always the unseen that most deeply stirs our imagination." Her prophecy was based on an aesthetic sense. The real revolution came with hard data.[219]

A decade later, as Carson revised and reissued the book, Wilson published his personal account of IGY as viewed from the perspective of his tenure as president of the International Union of Geodesy and Geophysics. "The history of the exploration of the sea floors is brief and simple," he observed. Not until IGY was their study "first faced in an adequate manner." Preparations for IGY had forced a consolidation of existing evidence, and an IGY-inspired Special Committee on Oceanic Research perpetuated the experience. Even as Carson wrote her paean of wonder, a flotilla of war-surplus research vessels with their sparkling instruments—their hydrophones, echo sounders, fathometers, piston corers, cameras, explosion seismometers, magnetometers, dredges, and thermal probes—was stripping the veil from the abyss. That year Lt. Don Walsh and Jacques Piccard rode the bathyscaphe *Trieste* to the bottom of the Marianas Trench. The imagination no longer needed to rely on analogies, poetic tropes, and appeals to "ultimate causes": it could feast on hard data. Ideas expanded to match that empirical bounty.[220]

The modernist revolution that had swept one field of inquiry after another like rolling thunder through the early decades of the twentieth century at last rumbled across geology. As IGY closed, Wilson could ponder that "No one knows with any certainty how the earth behaves, why mountains are uplifted, how continents were formed, or what causes earthquakes. We know some anatomy of the earth, but no real physiology." Earth sciences "await" a revolution. As discoveries poured in, reformation followed. A quiet Earth became dynamic. Its sciences were reborn—renewed like its crust. Earth became the first of the planets surveyed by the remote-sensing instruments of the Third Age, and plate tectonics appeared as the founding theory of planetary science.[221]

There was little opportunity, and less need, to indulge in rhetorical tropes or incantatory musings about the "unseen" and the unfathomable. The unseen abyss had been mapped. The unfathomable had been plumbed and measured. An elevated tone, if it matches its subject, works if its voice comes well before an episode of massive change or well after one. In either setting, a few select data points, stories, anecdotes, and personalities, be they people or marine worms or fishes, can be abstracted, granted an epic aura and a poetic cast, and hold up as a narrative. But during a revolution, with new discoveries rushing like a turbidity current, they cannot. Rachel Carson's ode to Oceanus could hardly be possible after the Third Age had commenced its serious exploration: there was too much data, too many new species, too great an abundance of geologic features, a crowded, jostling, spilling-over-the-desk thesaurus of natural-history and geophysical exotica. By the 1960s there was little need to elevate rhetorically the study of the deep oceans; discovery was pouring out of the abyss and onto the continents. Thanks to the Third Age the oceans were moving to the center of Earth history. The silent abyss joined moons and distant planets as new worlds.

Something similar happened with space. Before the advent of exploring spacecraft, the solar system seemed a tired, even clichéd subject, a relic of Newtonian physics in an age of relativity and quantum mechanics. Planetary astronomy, in particular, was a backwater, and worse, one contaminated by the fantasies of Percival Lowell and techno-romance novelists. The labors of planetary astronomers, poring over photographic plates, spoke the scholarship of classicists analyzing variants of obscure texts. Here was the laboratory turned library.

Prior to the Third Age the study of the planets—of Earth, for that matter—seemed moribund, committed to ever-greater musings over untestable theories and refinements in its numbers, with august authorities such as Henry Norris Russell and Sir Harold Jeffreys recycling old themes and issuing ponderous pronouncements. When the newly created U.S. Air Force funded a summary series of books on planetary astronomy in the early 1950s, Gerard P. Kuiper of the University of Chicago's Yerkes Observatory was the world's sole professional planetary astrophysicist. When Nobel Laureate Harold Urey delivered the Silliman Lecture in 1951 on *The Planets: Their Origin and Development*, the same year Carson published *The Sea Around Us*, he devoted fifty-five pages to the "terrestrial" (inner) planets and a scant five to the "major [outer] planets and their satellites."[222]

There was some movement as astronomers occasionally directed their new instruments, from spectrometers to radio waves, toward the planets to firm up rotational periods, atmospheric chemistry, and densities; and better photographic media allowed for a trickle of discovery in planetary satellites. But the subject seemed as opaque as a gaseous giant and as dead as its. Before Mariner 2 arrived, Venus's thick clouds had screened the planet from close scrutiny; even its rotational period was uncertain. Before Mariner 4, it was believed that Mars had no craters. The major college text of the time, Robert Baker's *Astronomy*, could assert as late as 1964 that "the times of the Martian year when the dark markings change in intensity and color are such as would be expected if the changes are caused by the growth and decline of vegetation." In a volume of 557 pages, Uranus and Neptune claimed a page each, half of that devoted to grainy black-and-white photos. Mostly the text spoke to their discovery, not their properties. Kuiper lamented the lack of "reciprocity" between geosciences and planetary astronomy; the latter could only marvel at the "incredible richness of the data" the former possessed.[223]

Then exploration blasted off and the data streamed back. IGY required three world centers to hold the rising stream. Explorer 1 found the Van Allen radiation belts. Tiros 1 began imaging the dynamics of Earth's atmosphere, and even before the Apollo program was announced, it had photographed all the continents save Antarctica. Ranger spacecraft went to the Moon. Mariner 2 flew by Venus. Pioneer 6 amassed tape recordings, "shipped daily, big 9,600 foot, 17-inch reels"—"truckloads of tapes." Then came the major missions, culminating in Voyager. Writing in 1981, when the Grand Tour had just begun, and focusing only on the inner planets, three prominent geoscientists remarked that their field was "immersed in a planetary information explosion." As Voyager 2 rushed toward Uranus, JPL estimated the volume of data the mission had so far dispatched to Earth as four trillion bits, enough to "encode over 5,000 complete sets of the *Encyclopedia Britannica*."[224]

The exploring spacecraft had sparked a revolution. They not only amassed fresh data but also prompted astronomers to redirect their instruments (if not their minds) to the solid-bodied new worlds of space; and more than raw digits, the missions established a context for their comparison. Earth science became planetary science.

All this—the fevered incantation of information, dazzling scenes, novel experiences, the rhythm of trek and encounter—was the cultural and psychological drive that made planetary exploration distinct from scientific observation or technological adventuring. Here was Cherry-Garrard's Intellectual Passion leaping from Earth's miniature ice-world of Antarctica to other planets and moons. Make it new, Ezra Pound had demanded; the robots did.

So little had been known; so much was revealed. Looking back from 1985, Oran Nicks observed that "it is not easy to recapture the extent of our ignorance a quarter-century ago; *everything* we learned was new." Everything about spacecraft, everything about interplanetary voyaging, everything about the worlds the spacecraft discovered. That was true for Mariner 2 at Venus, Mariner 4 at Mars, Pioneers 10 and 11 at Jupiter and Saturn ("everything we found out at Saturn was totally new," Van Allen declared), and it was true for every planetary visitation by the Voyagers. Writing after the Neptune encounter, Ellis Miner declared simply that "no other experience is likely to come close to matching the excitement of anticipation and discovery that accompanied the Voyager Mission." Those sentiments were the hallmark of a golden age.[225]

To that astounding era, Voyager came as a climax. The twin spacecraft went to more places, sent back more data, did more varied things, and continued for the longest time. "No other mission," explained Edward Stone, "explored so many different worlds" and revealed "such unexpected diversity." After Voyager no one could see the solar system in the same way, or see Earth, or for that matter themselves, as they had before. "You only discover the solar system for the first time once," observed Larry Soderblom. "Voyager did that."[226]

## GOLDEN AGE

A golden age.

All those who participated in the American planetary program from Mariner 2's flyby in 1962 to Voyager's embarkation in 1977 agreed that this was a privileged time. Spacecraft visited virtually every planet, and revisited the closest; almost every year saw a launch; and when, after the hiatus imposed by the Challenger debacle, planetary exploration revived, it built on the legacy and hibernated ideas of those epic years. It was all "the stuff of legend and myth," as space historian Roger Launius put it. Proponents differed only in their sense of the era's tempo and their reckoning of its capacity to persist. They fretted because the age depended less on engineering cleverness and scientific purpose than on the whims, wealth, and mores of its sustaining society. Voyager, Bruce Murray noted, was the last mission in which "the technical challenge was dominant. Since then it's been politics."[227]

A golden age is to history what a utopia is to geography. It is a time of the good, the just, the ideal. Good people do good things. The world works as it should. If the past can't supply a golden age, the future might. For space advocates, that forecast future lay always just over the horizon; and after World War II it seemed it might happen in their lifetimes. Then it did happen. For NASA, the golden age of manned flight was the sixties, the age of Apollo; and for planetary exploration, the seventies, when its spacecraft visited every planet save Pluto or were on their way to do so. The ideal became the expected.

At the time, those in the political trenches recognized what a close-run thing it was. Even Voyager had faced cancellation, or rechartering, and had downsized dramatically. The era seemed golden mostly in retrospect. The space shuttle siphoned and then hemorrhaged funding and energies away. The Reagan administration was hostile—wanted space, but in the hands of private companies or the military, and was eager to force the cold war to a conclusion, but not through proxy expeditions to other worlds. Office of Management and Budget director David Stockman sought to shut down planetary exploration altogether. Voyager—in some ways "really a product of the 1960s"—was, as Murray expressed it, "the last hurrah." As the twins sped across the solar system, the sentiment could easily unfold that they climaxed a golden age, now lost.[228]

Yet there was also a sense at the time, often a quiet euphoria, that space exploration would continue because it had to continue. The space program was more than an event: it embodied a movement of evolutionary importance on a planetary scale. It would transcend its sordid origins in the cold war. NASA could commission studies that likened the space program to the advent of the railroad, movies could be made of *Childhood's End*, and otherwise sober observers might declare that outer space would revolutionize humanity more than the industrial revolution. The golden age, once arrived, would stay. That was part of the promise, and the appeal of historical utopias of all kinds across all ages. It characterizes exploration no less than other cultural endeavors.

---

Yet it is the nature of history's golden ages to be fleeting. To those living amid them, their significance becomes apparent often only at the time they are primed to implode. Before then, too much is happening to stand aside and contemplate, but it is just as the climax comes, with a softening of urgent tasks and a fading of vision, that the recent past begins to glow.

Great outbursts of exploring enthusiasms are rare. They are possible because of distortions in the normal routine of their societies; and for that very reason they cannot be sustained. They feed off their larger nurturing culture; they can, from time to time, feed back into that culture like a self-reinforcing dynamo; but the larger dynamic comes from society, not from exploration. Geographic discovery can continue only insofar as it creates ongoing wealth. Planetary exploration expended surplus wealth; it did not create it. The great outbursts of exploring that punctuated Western history rather resemble the gravity-assisted acceleration granted by passage around a giant planet like Jupiter. Viewed from the planet, the velocity gained on approaching periapsis is also lost when the spacecraft recedes. It is only from another, more remote reference frame that the event can be seen to yield a long-term gain. So it is with the golden age of planetary exploration.[229]

Many proponents refused to accept that the golden age might end, or have urged, with a progressive urgency hedging into hysteria, that it be revived and expanded. That, too, is typical, and can segue into parody. It is present in the spectacle of Hernando Cortés's stumbling around Guatemala, Hernando de Soto's bulling through Georgia, bold knight Francisco Coronado's wandering through Kansas, and the Portuguese *degredado* (convict turned conquistador) António Fernandes's blundering around the interior of Africa—all in search of another Mexico or Peru. After the Second Age, it resurfaces with Roy Chapman Andrews's leading an expedition to Shiva Temple, an isolated mesa in Grand Canyon, to search for lost worlds and with Richard Byrd leaving Little America to live by himself at Advance Base on the Ross Ice Shelf, and nearly perishing from carbon monoxide poisoning. Today it appears in proposals to sponsor another Apollo program or a national commitment to imminently colonize Mars.

Such willful pursuits have their costs, not all dismissable as foolish vanity or misplaced idealism. Such obsessions can distort public discourse, and cabals dedicated to them can capture public policy and lead into costly misadventures. These can go beyond satire into darker parodies. If Don Quixote, a knight-errant mounted in a haze of imagined ideals, is a benign response to a lost golden age, the footloose *bandeirante*, an overland buccaneer opaque to anything other than slaving, raiding, and plundering, is his malign double. Yet it is hard to know when a golden age has crested, and when to adjust ambition to possibilities.

Most enthusiasts shun even asking such questions, because to do so suggests that the recent golden age, our age, might not revive and exceed itself. But in the late 1990s, Bruce Murray did ask.

While director at JPL, he had noted the "terrible contrast" between the technical successes of Viking and Voyager and the social failures to "reinvest" in the future. The space shuttle, in particular, he regarded as "the greatest threat to space exploration" since pre-NASA days. In April 1980 he wrote Arthur C. Clarke, several of whose books he had just finished reading, thanked him for providing "a full supply of much-needed nourishment for our imaginations and spirits in this difficult period when Man seems unable to keep up with destiny," and invited him to JPL for Voyager 1's encounter with Saturn. He also planned to invite Freeman Dyson and others for a session on interstellar travel. As that theme suggests, Murray, like many of his contemporaries, thought the trek would continue to the stars.[230]

Then he realized it wouldn't; not in his lifetime, not outside science fiction novels. The spectacular supernova of exploring that Voyager had "epitomized" he came to regard as "a very brief anomaly," one that expressed a "combination of politics and economics and technology" that had allowed American society to rush through a phase at "an extraordinary rate" before hitting a limit. He raised the possibility that there would be no further escalation to the stars, that "this was the end," that it was "the end of adolescence for humanity," that "we'll have to learn to live within the space we have, physically and intellectually." In the unmanned exploration of the solar system, he thought "we've already reached the limits of what we're going to be doing. We're not going to the nearest star." Voyager, he believed, was "a symbol" and an "extraordinary historical event." But it also raised questions about whether such rates of activity were "unsupportable." Murray concluded they were: it was not possible to "keep exploring like that." There would be a falloff, which he found both depressing and impossible to stop. He wondered what such a lapse might mean for a society that "has always been expanding."[231]

Space proponents hated the meditation. Even close friends like Philip Morrison and Carl Sagan found that such prospects "violated their intuitive beliefs." But the longer the American space program continues, the longer exists the empirical record of what the public will support, and what out of the general "space" budget it will expend on genuine exploration. In 2007 the NASA budget for planetary exploration was some $3 billion. This was more than the American college textbook and resale market ($2.3 billion), and less than the estimated sales from fast food restaurants in Los Angeles ($3.4 billion). That same year, the National Football League and Major League Baseball each accrued roughly $6 billion in revenue. Those who claimed in the early 1960s that the space program would inaugurate an economic revolution had been proved wildly wrong.[232]

Planetary exploration, in particular, was a significant but niche enterprise, larger than handmade soap and self-publishing, smaller than professional sports and movies. It was, in brief, a cost that society justified as it did much scientific research and art museums. Observatories in space were on par with opera houses, and for similar reasons: they were a chosen, discretionary cultural activity, and to a large extent, an elite activity. They satisfied social needs and yearnings. They might vanish if alternate or surrogate activities could meet those desires. Society might choose instead to colonize cyberspace. It might prefer virtual exploration and extreme sports.

How does a society respond to the passing of a golden age? Some would reverse the stance of King Knute trying to halt the rising tide and stem its ebb. The danger especially exists of the quixotic turn, the determination to pursue the old glories and chivalry into new times. The scene abounds with fancied Isaiahs and aspiring Cassandras, alternately warning and threatening, all abuzz with prophecies and jeremiads. But an exploring age integrates too many activities to respond to rhetoric and vision alone.

If the social order doesn't collapse entirely, golden ages segue into silver ones. These tend to be more broadly based, less given to heroic posturing, more prone to irony and a sense of constraint. In place of new initiatives, its participants fill out the old agenda, stabilize and firm up its institutional foundations, and replace raw if grand gestures with more polished manners and elegant phrasing. The edge is off. Steady work replaces inspiration. Rude narratives become textbooks. Epic poems become eclogues and elegies. For exploration, bold traverses into the wild give way to trading posts, land surveys, and intercultural exchanges. It is not that participants are less hardy or courageous or clever; they just perform in a different context, in new settings of time and place. Besides, massive programs create a kind of wake turbulence after their passage. The institution pauses; a sequel requires a caesura.

Something like this has happened with planetary exploration. After a hard crash, programs have recovered equilibrium. Complex single missions have taken Magellan to Venus, Galileo to Jupiter, Cassini to Saturn, and Huygens to Titan, and a small fleet of rovers and orbital observers to Mars. These are brilliant, successful missions, but they necessarily lack the zest of the vision quest and the drama of first-time encounter. Perhaps from a vantage point in the distant future, everything from Mariner 2 to the New Horizons mission to Pluto and other expeditions yet to launch will all merge into a single, heroic panorama. But from the perspective of those present when Voyager completed its Grand Tour, a golden age had declined into a silver one and struggled not to sink into bronze.

## PARTING, TWO

As the adage goes, it's better to be lucky than good. The Voyagers were both. They were lucky in the providential alignment of a Grand Tour and a golden age, lucky in their handlers and their determination to rescue them from a shaky launch, lucky in that their revelations beat those from orbiting telescopes. It was good in that they didn't just hit their designated target as Viking did at Mars, or orbit a planet endlessly like Pioneer Venus. They spanned the solar system.

Voyager was, in Bruce Murray's words, "a concatenation of really good people, good ideas, support by society, luck in the engineering sense and luck in the scientific sense." The mission was "always successful." And more than any other space enterprise, it has lasted. It has outlived all its competitors. It has renewed itself. It has retained the capacity to discover. "History is what you write when a mission is over," Murray observed in 1997, but "this mission keeps on going."[233]

It continued after Neptune. Voyager 1 still arced upward toward the heliosphere, one final task remaining for the Grand Tour. Voyager 2 spun downward, also on a trajectory to termination shock. Both continued to sample the interplanetary medium and report back to Earth, still going. They are still going today.

DAY 3,950–∞

# 19. Cruise

Once again, the Voyager staff began to turn off instruments. Those most closely linked with hard-geography science went first, along with those most prone to guzzle power and hydrazine. But this time some would never be revived. Infrared sensors, imaging systems, the photoelectric photometer–these were instruments designed to inventory worlds, and the Voyagers had no more worlds to visit.

Even the exploration of the solar system's soft geography was reaching a terminus. Time and again, planet after planet, encounter had begun and ended with bow shock, that fluid, twisting field where the planet's magnetosphere collided with the solar wind. Now there remained one last such border, the heliopause, a thick fringe of ions and rays, where the Sun's magnetosphere met the interstellar medium. "Termination shock," its inner edge was called, and that phrase might stand not only for the clashing of electromagnetic fields but for the Grand Tour itself.

This time, too, there were no further burns or gravity assists. The trajectory each spacecraft had was the trajectory it would always have. The Voyagers' pasts set their futures.

## TRIANGULATING THE THIRD AGE

Why does a society explore? For many reasons, but at base because it chooses to, and it chooses to because the enterprise appeals to felt needs. What makes exploration powerful is that it bundles those longings into compelling packages and sends them on a journey. Over time some pieces are lost and some added, and the assemblage bonds to its culture with weaker or stronger valences. When those mixings give rise to qualitative change, they can spark great ages of discovery; the Third Age is such a moment.

But how to parse the significant new from the enduring old? The space program's instinct to look back as well as ahead may help. The program had begun by looking back. It looked back to Columbus for an exploring legacy, back to America's westering frontier for a past that it could project forward, and back to the memory of a murdered president who first set the lunar lander on its way. Its most celebrated crewed and robotic missions framed their images of the future with glances back to Earth. Apollo 8 gave us earthrise, and Apollo 17 the whole Earth. Voyager framed the Grand Tour with a first look back to Earth and its Moon and a last look back at the solar system. Along the way it framed its discoveries by backsiting on the eruptions of Io and Triton, by backlighting Jupiter's storms and Saturn's rings, and by sculpted backviewing of Uranus and Neptune that transformed their cloudy murk into haunting crescents of light.

In a similar way Voyager has looked back on the long chronicle of geographic exploration by the West, and in so doing it has highlighted those features of the Third Age that are most distinctive. What they share is that the Third Age is going where no one is or ever has been. In past ages, geographic discovery *had* to be done by people. There was no other option by which to learn the languages, to record data and impressions, to gather specimens, to meet other societies and translate their accumulated wisdom. It is impossible to imagine the great expeditions of the past without considering the personality of individual explorers who inspired, collected, witnessed, fought, wrote, sketched, exulted, feared, suffered, and otherwise expressed the aspirations and alarms of their civilization. But it is entirely possible to do so now. Not only is there no encounter between people, there need not even be a human encounterer.

The geographic realms of the Third Age are not places where people learn from indigenes or live off the land. These are environs that offer no sustaining ecosystems; their geographies remain, for all practical purposes, abiotic and acultural worlds. If the Third Age has propelled exploration beyond the ethnocentric realm of Western discovery, it has also thrust it beyond the sphere of the human and, with regard to space, perhaps beyond the provenance of life. No one will live off the land on Demos, go native on Titan, absorb the art of Venus, the mythology of Uranus, the religious precepts of Mars, or the literature of Ceres. There will be no one to talk to except ourselves.

This is a cultural barrier to exploration, in comparison to which the limiting velocity of light may prove a mere technological inconvenience. The reason goes to the heart of exploration: that it is not simply an expression of curiosity but also involves the encounter with a world beyond our ken that challenges our sense of who we are. It is a moral act, one often tragic, that bonds discovery to society. It means that exploration is more than adventuring, more than entertainment, more than inquisitiveness. It means that exploration asks, if indirectly, core questions about what the exploring people are like. In the past this happened when one people met another. In the future it won't, and the explorer is most likely to be a robotic surrogate.

The good news is that the coruscating ethical dilemmas of so much earlier exploring and empire building will disappear. No group need expand at the expense of another. Ethnocentricity will vanish: there is only one culture, that of the explorer. As long as other life or cultures are not present, there is no ethical or political crisis except whatever we choose to impose on ourselves. Beyond Earth there may well be no morality as traditionally understood, that is, as a means of shaping behavior between peoples. The morality at issue is one of the self, not between the self and an Other.

The bad news is that exploration's moral power—the tension, awful and enlightening both, that is involved in a clash of cultures—also vanishes. The price of ethically sanitizing exploration is to strip it of compelling *human* drama and the kind of narrative and poetry and epic tragedy that have historically joined heart to head. Planetary probes become technical challenges, to make machines to withstand the rigors of space travel, a technological equivalent to extreme sports, like white-water kayaking in Borneo or racing in NASCAR's Daytona 500. The space enterprise may become a kind of national hobby, a jobs program, or a daytime TV soap opera.

People do not have to be physically present at the discoveries of the Third Age, and there are sound reasons for arguing that they should not be. Even if colonization is attempted, successful settlements historically followed after long gestation periods of reconnaissance with aid from indigenes. More likely is an era of space tourism or historical reenactment. A risk is that partisans will try to force the Third Age into an earlier model that no longer connects to the lived world, or that they will choose to indulge in digital or chemical surrogates, or that they might simply give up on the project altogether.

## COURSE CORRECTIONS: THE FUTURES OF EXPLORATION

As the Grand Tour faded, space exploration, too, seems to have entered a long cruise phase. Although the Voyagers have no more course corrections scheduled, some are possible—and likely—for the tradition of exploration they embody, and would point that larger enterprise to its future, which is less a spectrum of choices than a constellation of options from which various patterns might be traced. Three figurations, in particular, seem plausible. One points to cyberspace; one to a Third Age that, in closing, might also close the Great Ages; and one to an untraditional narrative in which the journey becomes its own purpose.

The first prospect assumes that exploration is a response to humans' need to know bonded to the stimulus of adventure, and asserts that it will not be possible to shut those instincts off. Technology will continue to make them possible. From its origins, in fact, space travel has exhibited just such a hybrid of people and machines. Humans must rely on an artificial habitat; and spacecraft and rovers must possess a degree of autonomy. Designers have even come to speak of an "embodied experience" with regard to Mars rovers such that robots not only behave more like people, but that people identify with, and behave more in conformity with, the robots.[234]

But this might lead to a virtualized exploration. After all, curiosity can be satisfied by many means. Literature, art, laboratory science, invention, travel, collecting–individuals and societies have filled, and overtopped, their urge for the curious (and avoided boredom) in many ways. Scholars have spent lifetimes pondering Shakespeare without exhaustion. Researches in field and lab have proved unbounded if not infinite. Geographic exploration is only one anodyne among an infinite set. Already both planetary and oceanographic discovery have apparatuses by which distant observers, far removed from the operational staff, can share in discovery by watching in real time through the same cameras. Geographic exploration is no longer something that an individual alone must do or experience; it can occur among millions simultaneously. It is only a short mutation of technology and a hop-skip of faith to have exploration venture into cyberspace altogether.

Engineers might take the process further. Science might find what physiological triggers exploration satisfies, and technology might discover ways to satisfy them without the bother and boredom of trekking, just as vitamin C can be distilled from oranges into a pill or amphetamines can bring a rush that in the past had to come from danger and adventure. The old quarrel between robots and astronauts might pale before the opportunity for virtual exploration. One could experience the ecstasy of first-contact discovery without the annoyance of first-contact mosquitoes, fevers, troublesome natives, unreliable collaborators, and a nature reluctant to yield its secrets.

Each of the motives bundled into exploration might be separately satisfied by other means. The reductionism that allows science to express itself in machines might equally be applied to disaggregate the experience of exploring and to address each urge independently of the others. It would be but a series of short steps to go from remote sensing to remote exploring to virtual exploring. Already NASA has Web sites devoted to the task. The massively multiplayer online game Second Life has an exploring option, an island named CoLab, operated by NASA Ames Research Center. Here, it is hoped, a new generation can develop a "vision for space exploration" where "participatory exploration" could emerge and where earthlings' avatars might join astronauts as they return to the Moon. The island could serve as a "portal" to new, virtual worlds. A cyber Ocean Sea has again been populated with imagined isles awaiting discoverers.[235]

Will such exercises consume the varied cultural urges that exploration has, for half a millennium, satisfied for Western civilization? Virtual reality is rolling over one frontier after another. A generation that has grown up immersed in high-tech gadgets may respond less to their ambient environment than to their digital one; their community is what they connect to electronically. Already nature enthusiasts are vocally worrying about the future of conservation as children come to experience nature not from personal contact but by digital simulacra. Will exploration be next? Could designer drugs trigger the ventral striatum to release neurostimulators such as dopamine in the absence of tangible stimuli? Could a tamed exploration-designed LSD allow "trips" through space? Could cyberspace supply the wonder without the bother? Perhaps. Almost certainly the exploring tradition will adjust. And if it does not expire, it might, like big game hunting, morph into surrogates such as nature tourism or digital voyeurism, or become ceremonial, as with reenactments of Civil War battles.

In the end exploration will remain a cultural creation, and it will assume the character of its sustaining society. How that civilization hybridizes with its technologies will shape the future of geographic discovery and decide the future cadences of cruising and encountering.

A second option points to the rhythms of great ages, and suggests not only that the Third Age, too, will wax and then wane, but that its ebb might take the ages with it. In this conception, if the Voyagers could see beyond the termination shock of the golden age, they would find a future exploration that may look a lot like the past, only different. Exploration will go back to the future, with the Third Age recapitulating the First, and then perhaps reverting to its pre-exploring ancestry.

The possibility for a rekindling of rivalries that will translate into geographic discovery exists. Political contests may boil over into space, for example, if China declares a colony on the Moon as essential to its prestige and the European Union or Japan joins the fray. There is a prospect that the search for life will take on an imaginative, even a theological cast, sufficient that a significant fraction of the culture wants to pursue it among the planets. It may happen that extreme arts, brash new sciences, an as-yet-undeveloped commerce, an astropolitics, and some critical personalities will combine to yield a Third Age echo of the Second Age. In some form or another, a virtuous cycle is possible. But it is not likely. As Damon Runyon advised, the race is not always to the swift nor the battle to the strong, but that's where you place your money.

The most plausible prognosis is that the future will resemble the past, that the Second Age's monadnock of activity will mark an axis around which the evolving contours will unfold with rough historical symmetry. The Third Age will resemble the early Second, though in reverse. Expeditions will slide toward a new steady state—for space, perhaps on the order of one or two a year. These will be complicated probes, requiring years of preparation, not unlike the expeditions launched during the Great Voyages and quite unlike the brawling swarm of state-sponsored and individually motivated treks that so inflated the Second Age. They will be targeted to some particular purpose—commercial, scientific, technological, national prowess and prestige. They are unlikely to spill out from colonization: they will rather resemble those expeditions that established trading factories on islands or episodically visited coastlines for barter or sought out new routes. If the process thrives, there will be several competitors, not some collective United Earth Space Agency; and that institutional unrest is what will keep the pot simmering. Steadily, more and more of the solar system will be visited, catalogued, mapped, assessed. Perhaps, here and there, an outpost will appear, staffed for a few years.

The Third Age will then burn itself out, as other ages have before it. Its narrative will end as exploring expeditions traditionally end: it will return to its origins. It will have a beginning, a middle, and a conclusion. It will advance from launch to journey to splashdown. Reversing this trend would require an immense, global commitment that could come only from some dark necessity or irresistible rivalry, say, the discovery amid the asteroids of some mineral absolutely vital to national existence—the equivalent of the Potosí mines of Mexico, perhaps—or from Venutians announcing that they intend to colonize Mars and the moons of Saturn, and daring earthlings to stop them.

As the Third Age winds down, it may perhaps carry the great ages of discovery with it. They were created; they may expire. The conditions that sustained them may cease altogether; they may no longer inspire interest as a tradition worthy of institutional support. One can even imagine a robotic Columbus ceremoniously announcing an end to the enterprise. If the late nineteenth century marks a bilateral middle in this saga, that passing may happen some four hundred years later, the early twenty-third century, where *Star Trek* now resides in the popular imagination. Exploration, even of space, may then exist only in literature, history, film, and popular imagination, and in a past where no one, boldly or otherwise, wishes any longer to go.

Or there is a third figuration of the future. We might create an alternative to the narrative that has characterized previous eras, as different from its cadence as Third Age terrains are from the isles and lands of previous ages.

A counter version might hold that the traditional narrative structure might not apply. Like the Voyagers, the process will simply go on and on, constantly redefined and redirected. What makes the Third Age distinctive will allow it not so much to endure as to ultimately morph into a successor. It will retain some features—enough to have it be recognizable as exploring—and discard others. That sorting process will permit it to persist. It will survive by becoming different. It will simply continue.

In such speculations the Voyagers still have something to say. Their journey may yet write a different kind of narrative, pushing through the heliosphere of a greater modernism. The Grand Tour did not end with Neptune. The Voyager mission will not end with the solar system. The Voyagers continue to surprise.

DAY 4,083–4,084

# 20. Last Light

Looking back.

Before the Grand Tour ended, Voyager had a chance to peer back for one last time and reflect in its lenses something of what it had accomplished. Since Saturn, Voyager 1 had cruised with its cameras shuttered, while Voyager 2 continued to assemble its brilliant photo portfolio of new worlds. Now Voyager 1 was roused from hibernation to perform a task that would, for many observers, distill the entire enterprise into an enduring, defining image.

The post-encounter phase of every planet had included backlit shots of the planet and its satellites, and often of many such scenes into a single image. The practice had begun when Voyager 1 turned around and photographed Earth and the Moon together on September 18, 1977, two weeks after launch. Now Voyager 1 was asked to turn back on behalf of the entire Grand Tour and photograph the planets as part of a grand ensemble. On February 13 to 14, 1990, it did just that.[236]

The idea emanated from Carl Sagan, who conceived it after the Saturn encounter. He wanted a space-based version of the celebrated earthrise images from Apollo. These, he thought, had revolutionized humanity's sense of itself. We could see ourselves as others might see us. Now it was time to expand that horizon, to see how we might look "to an alien spacecraft approaching the Solar System after a long interstellar voyage."[237]

But mission officials balked. Some dismissed the gesture as a stunt, far removed from engineering necessity or scientific duty. Most believed it was better to wait until the programmed tasks were done; better not to point those lenses toward a still-vibrant Sun and risk scorching sensitive instruments; better to see the whole once than to juggle commands and slew scan platforms. The hard-science crowd scoffed: this was public entertainment, more postcards, a stunt that might embarrass a real astrophysicist before his peers. But these were the same arguments by the same people who had scorned having cameras at all. They never realized that the power of Voyager lay not in its scientific role but in its cultural context. The American public did not spend $600 million to record the magnetosphere of Neptune and the rotational period of Titania. They sought the wonder of new worlds and a reflection of their own.

After Saturn, Voyager 1 had no further duties for its cameras, and after Neptune, neither did Voyager 2. JPL's Voyager staff faced a dramatic downsizing. The DSN had more urgent tasks. If the scheme was to be done, it had to be done soon. Like the alignment of planets that made the Grand Tour plausible, an alignment of planets, particularly Earth, made a parting gesture possible. NASA administrator Rear Admiral Richard Truly ordered the photos taken.[238]

On February 14, Valentine's Day, 32 degrees above the ecliptic, traveling at over 64,300 kilometers per hour, almost 6 billion kilometers from the Sun, Voyager 1 shot 39 wide-angle views and 21 narrow-angle ones that encompassed 6 of the 9 planets. Mercury and Mars were too close to the Sun to show, and Pluto too minute amid the stellar glare. Neptune and Uranus were so dark that the long exposures caused smearing. Earth, freakishly, caught a bounce of sunlight off the spacecraft that allowed it to appear "in a beam of light." Over the next three months, as the DSN schedule allowed, Voyager 1 transmitted the 640,000 pixels of each image from its magnetic tapes to Earth, then some 5.5 light-hours away. Each photo required 30 minutes. The Voyager Family Portrait, as it came to be called, thus complemented the record of Earth that the spacecraft carried.[239]

NASA spokesperson Jurrie van der Woude and Sagan quickly dismissed any special status for Earth, which only occupied 0.12 of a pixel and had to be magnified sixfold to show at all. The point of the exercise, Woude insisted, was to show "how insignificant we are." Sagan elaborated: the Earth was a mere "pale blue dot." That phrase became the title for his final book on space exploration, published in 1994. The story of Voyager's farewell photos provided the opening chapter in what Sagan proclaimed as "a vision of the human future in space." Such pronouncements, however, were the flip side of those that dismissed the exercise as merely a publicity stunt. Most participants probably echoed Candice Hansen's judgment that this was "the picture of the century." It fell to Ed Stone to craft the right alloy of poetry and science. He turned to the practice of astronomers, who call the initial imaging by a telescope "first light." This, the final use of the spacecraft's lenses, Stone labeled Voyager's "last light."[240]

It was a scene, the self-portrait, often found in exploration history.

In the past, when the portrait couldn't be photographed or painted on site, it might be done later in a studio with the explorer outfitted in full expedition regalia. It was not the place that mattered so much as the explorer positioned within it. It was a ritual act: the quest achieved, the hero returned. By the end of the Second Age, such staged events were expected. Think of the triumphant photo of Robert Peary at the North Pole, or the haggard image of exhaustion evident in the self-portrait taken by Robert Scott's sledging party at the South Pole, each in its way a concluding record of the Second Age.

In the Third Age it was not possible, either geographically or philosophically, to position a camera or an artist to record the explorer in his milieu. There was no way to include Voyager within the frame of its accomplishment. Often Voyager would appear to the side in composites, but this was a studio confection. Voyager 1's "family portrait" was in truth a self-portrait, a record of the places it had visited and of the people who had sent it. Much as its golden record was not a message to extraterrestrials but to Earth, so its gallery mosaic was less a vision of how the solar system might appear to visiting aliens than an invitation to earthlings to see themselves as others might, or perhaps more accurately as they would like themselves to appear to others. The self-reflexive event testifies that in Voyager the sojourner and the journey had become one. The encountered Other has become the exploring self. In that gesture the Grand Tour found a way to bracket its journey.

The Grand Tour, yes. But the Voyagers' journey was not yet over.

# part 3

## BEYOND THE UTMOST BOND: JOURNEY TO THE STARS

And this grey spirit yearning in desire
To follow knowledge like a sinking star,
Beyond the utmost bond of human thought.

—Alfred Tennyson, "Ulysses"

# Beyond Bow Shock

# 21. Voyager Interstellar Mission

On January 1, 1990, the Voyagers commenced the Voyager Interstellar Mission (VIM), a cruise without a hard encounter at its end, only the soft frontier where the solar wind buffets against and finally yields to the interstellar media.

The instruments so vital to interrogate planetary hard geography were turned off. The infrared interferometer and spectrometer and radiometer. The imaging cameras. The photoelectric photometer. While the rest could still perform usefully, they did so at greater costs, which Voyager could less and less afford. The amount of hydrazine, needed by thrusters to perform rolls for magnetometer recalibrations and to keep antennas pointed to Earth, was finite, and consumed at a rate of six to eight grams per week. The ultraviolet spectrometer remained on in order to sample hydrogen in the outer regions of the heliosphere, but then it and the scan platform were powered down to save energy. Supplemental heaters were turned off. Throughout, the power yield from the radioisotope thermoelectric generators steadily decayed; by 2001 overall production had dropped from 470 to 315 watts; by 2008, to 285. And if they were to reach bow shock, the Voyagers had a long way to go.[1]

The mission's budget was draining even more dramatically. From

May 1972, when planning began for MJS 77, up to the encounter with Uranus in 1986, with an inflationary storm in between, the Voyager mission had cost $600 million. In the favored analogy, this amounted to the price of a candy bar each year for every American citizen. Hard as it was to scrounge money for a new mission, so much harder was it to find funds to maintain an old one. The political regime that authorized a program would be long gone when the program succeeded, though it might be remembered at encounter. No one would honor those who sustained a cruise. It became harder and harder to argue for VIM when termination shock would occur years away and no new worlds might flash the event across TV screens.

As the second Bush administration projected a crewed mission to Mars (or sought to pay off cronies in aerospace, as cynics argued), money became even tighter, and the persistence of old missions such as Voyager seemed like leaches that sucked the lifeblood out of new ones. In 2005 the Voyagers still had a staff of ten (down from three hundred in 1989) and an annual budget of $2 million when NASA sought to shut down the operation along with five others. But once again Voyager survived a political threat to kill it. It would end only when its internal power finally drained away. For now, it continued to send back reports from new settings. It was doing what no other spacecraft could. Its narrative simply defied closure from Earth.[2]

The border between the solar system and the stars is broad, sloppy, and untraced. The bubble of gases, or heliosphere, that contains the solar winds has a complex structure. Its outer perimeter, the heliosheath, has two edges, one facing the solar wind and the other, the interstellar winds. At its inner border, or termination shock, the supersonic solar wind slows to subsonic speeds. This was believed to occur around eighty to ninety astronomical units (AUs) from the Sun. To reach it from Neptune would take Voyager 2 as long as it had taken the spacecraft to reach Neptune from Earth, or roughly twelve years. Within the heliosheath the pressures of solar and interstellar winds roughly balance. Its thickness varies because the entire solar system moves, which compresses the heliosheath in the direction of that movement and extends it on the downwind side, much as comets have compact heads and long tails. The outer border, or bow shock, thus ranges in thickness from 10 AUs on the windward side to 100 AUs on the lee. That geography defines the three phases of the Voyagers' Interstellar Mission. When they finally pass through the outer bow shock, they will be sailing amid the winds of stars.[3]

Voyager 1 was fastest and farthest, and at 85 AU, in August 2002, it recorded fluctuations that were reported as possible evidence of termination shock. This could not be confirmed, however; the billowing border is unstable, full of pulses and pauses from gusty solar breezes and their turmoil with interstellar winds, and the magnetic disturbances that shape them; and the plasma science instrument on Voyager 1 no longer functioned, which forced researchers to infer shock from indirect sources. But on December 16, 2004, at some 95 AU, it unblinkingly passed through termination shock and entered the heliosheath. Voyager 2 crossed that bar at 84 AU on August 30, 2007, and then reexperienced passage at least five times over the course of several days as the shock boundary twisted and bubbled.

Since the Voyagers entered the heliosheath at different places some sixteen billion kilometers apart, their encounters helped map the geography of that frontier. As predicted, the heliosphere was "squashed" on top and bottom. But the amount of variance was more than expected, and so were the temperature differences on both sides of the border; the heliosheath was far cooler than expected. Voyager, as its then program scientist Eric Christian noted, "has once again surprised us."[4]

But surprise had been a Voyager specialty since launch. On January 5, 2005, the mission surpassed 10,000 days of operation. In August and September 2007, the Voyagers reached thirty years. Measured in robot lifetimes, they were Methuselahs. They were leaving the solar system with computing power inadequate to run a cell phone, and electrical power insufficient to animate a clock radio. Yet they had much yet to survey: the dynamics of the solar wind, sunspot effects on the interplanetary medium, low-energy cosmic rays, reversals in the Sun's magnetic field, interstellar particles, radio emissions from various sources within and beyond the heliosphere, and of course the interstellar medium, if all went well. Of particular interest are their

UVS instruments in that they observe spectra no other spacecraft can.[5]

Their trek continues.

They have enough energy that, if conserved, they might still be able to transmit after passage through bow shock at the outer heliosheath. By 2015 Voyager 1 will have to shut down its data tape recorder, and Voyager 2 its gyros. A year later, Voyager 1 will cease its gyros, and after another year, Voyager 2, its data tape recorder. Steadily, in a creeping paralysis, the Voyagers will trade dead instruments for only the most basic functions—ultimately, the measurements of interstellar wind and the communication of that information to Earth. By 2016 for Voyager 2 and 2018 for Voyager 1, those few instruments still online will need to share power. By 2020, barring a sudden collapse, the robotic equivalent of a stroke or heart attack, their RTGs will no longer be able to support even a solitary instrument. The spacecraft will wither, like an organism wasting with scurvy. They might still be able to send signals, useful for tracking, but even that capacity will eventually fail. Sometime in the mid-2020s they will cease to function and become inert time capsules in a realm beyond time.[6]

For now they cruise, settling into trajectories that will propel them toward the fluid edge of the solar system and to the very borders of the Third Age.

# 22. Far Travelers

There are now four spacecraft headed beyond the solar system. Pioneer 10 is moving downwind toward termination shock, roughly heading to the star Aldebaran in the constellation Taurus, and at 43,000 kilometers per hour (relative to the Sun) might reach it in some two million years. Pioneer 11's long reverse trek to Saturn has pointed it toward the closer, upwind border, cruising at 41,500 kph; it will pass, within the next six light-years, by Proxima Centauri, and some 6,500 years later it will sail within four light-years of the star Ross 248. Both spacecraft have gone silent. All contact was lost with Pioneer 11 by November 1995. Shortly after its thirtieth anniversary in 2002, Pioneer 10 blinked off. The two Pioneers now sail onward, ghost ships from Earth.[7]

By comparison the Voyagers are still vigorous. They crossed termination shock with their field instruments and communications intact. And they are faster. On February 17, 1998, Voyager 1 surpassed Pioneer 10 as the most distant human-made object in space. By January 1, 2008, relative to the Sun, Voyager 1 sailed at 61,600 kph and Voyager 2 at 56,000 kph. Voyager 2 was slower because its peculiar targeting at Neptune had turned the magic of gravity assistance against it and cost it velocity—that was the price of Triton. At that date, Pioneer 10 was 94 AU distant from Earth; Pioneer 11, 75 AU;

Voyager 1, 106 AU; and Voyager 2, 85 AU. The Voyagers have become Earth's farthest explorers.[8]

In all this there is a certain symmetry. The Voyagers had begun their journeys by riding enough energy to escape Earth's gravity, and they had acquired sufficient additional energy on their travels to escape the Sun's. But there is also a profound asymmetry. They will not return. That of course has been true for many exploring expeditions, and the ships that carried them; it was a risk every explorer took; it was part of why their sustaining societies granted them special honor. But it was never an intention of design. For the Voyagers, it was.

What has been the fate of famous vessels? In the past they returned to sea and plied the waves until they sank or were scrapped. Columbus's *Santa Maria* foundered in the New World, while the *Niña* rejoined him for the second voyage, and later took him back yet again to Hispaniola. The Victoria, which first circumnavigated the Earth, was refitted and sent out for transport, until it eventually sank. That was the destiny of vessels generally: they were reconditioned, rechartered, and reworked. Very few had been originally designed for exploration; they had been refashioned, ice-strengthened, copper-plated, supplied with living quarters in place of cannons, fitted out with heaters and libraries for polar campaigns, given laboratories and collection cabinets; and when finished, they were reworked again further down the food chain until they sank or were sold for scrap. A few—the *James Caird*, lifeboat of Shackleton's *Endurance*, is an example—found themselves accidentally placed into service, yet achieved such fame that they seem to have been designed for their destiny. The *Caird* has become a museum piece, now displayed in traveling exhibitions like a painting from the Old Masters.[9]

A few might find a second wind and explore further. After serving as a flagship for the U.S. Exploring Expedition, the USS *Vincennes* was dispatched on a second tour (the North Pacific Expedition) before running out its life cycle. The HMS *Erebus* and *Terror*, having sailed with James Ross to the Antarctic, joined John Franklin's quest for the Northwest Passage, where they sank. The HMS *Beagle* was recycled into three expeditions (Darwin's was the second), before being refitted from coastal surveying to coastal patrols against smuggling, ending in the marshlands of the River Roach.

Ships were expensive: in order to endure, they had to mine a deeper vein than their legacy to exploration alone. Who frets over the fate of the *Astrolabe*? The *Mirnyy*? The *Terra Nova*? The keelboat used by Lewis and Clark? La Salle's *Griffon*? But where a richer cultural bond exists, the desire to preserve or re-create (and reenact) can thrive. Enthusiasts have rebuilt and voyaged in Norse longships; they have resurrected Columbus's *Santa Maria* and reenacted its voyage to the New World; they have reconstructed and sailed John Cabot's *Matthew* to Newfoundland; they are rebuilding the *Beagle*. Land vehicles could survive junking as well. Douglas Mawson's pared-down sledge is a museum artifact at the University of Adelaide. The *Discovery* expedition's hut at McMurdo Bay, Antarctica, is a protected historic site.

The HMS *Endeavour* began as a north-country collier, was refitted to suit Cook's first circumnavigation, and then slid back into the obscurity of the Admiralty's rosters. It was again refitted, this time as a store ship, and sold in 1775 for £615. Its subsequent history is murky, but the likely story is that, renamed the *Lord Sandwich,* it worked the Baltic trade and then was accepted into the Transport Service to carry Hessian soldiers to Rhode Island during the American Revolution. At Newport, Rhode Island, it became a prison ship and was later sunk to help blockade the harbor. In the 1960s the anchor and cannon that were lost when the *Endeavour* struck the Great Barrier Reef were recovered and put on display in Australia. In 1988, as part of Australia's bicentenary, a replica was constructed using the drawings from the 1768 refit; the project was completed in 1994, and a few years later reenactors sailed the ship around the globe before it came to rest at the Australian National Maritime Museum in Sydney.[10]

The HMS *Endeavour* was special, not only for its legacy to exploring but for its role in the national story of Australia, and it was the Australians who oversaw its partial recovery and reconstruction. The HMS *Challenger* had no such national narrative. The fifth Admiralty ship to be named *Challenger,* it was converted from a corvette in 1872, abandoned in 1880, and finally sold in 1921. It was reincarnated only in name; the seventh and eighth ships were designed for survey and research (the last sold in 1993). By then the name had been transferred to an American research vessel for deep drilling, and to a space shuttle.[11]

Only toward the end of the Second Age had exploration become sufficiently professional and self-conscious that it designed vessels for its own purpose, and considered saving them as museum pieces after their task was done. Perhaps the most famous is Frijthof Nansen's *Fram,* which survived the frozen Arctic Ocean it was designed for and then went to Otto Sverdrup for another round, and finally to the Antarctic and back with Roald Amundsen. It then dry-rotted in storage until it was reconditioned into a museum in Oslo.

The Third Age has been different—and the same. It could beat rockets designed as swords into plowshares to launch spacecraft. But the spacecraft themselves had been custom-built for their missions, and their societies have sought to memorialize them (or their facsimiles) after they completed those missions.

All this reflects the wealth of the societies that have sponsored exploration, the difficulty of recycling spacecraft, the self-consciousness with which modern exploration builds on its legacy, and the self-interest that seeks to parlay historical continuity into political conviction. No sooner do capsules return to Earth than they head to museums. Mercury Friendship 7, Gemini IV, Apollo 11—all grace the halls of the National Air and Space Museum. In place of the lost robots, curators have substituted replicas, many constructed from the spare parts left over from the originals or the reserve craft that remained on the ground, where they functioned as working models to help resolve engineering glitches, and were then available for display rather than scrapping. So the National Air and Space Museum's Milestones of Flight includes Explorer 1, Mariner 2, Viking, and Pioneer 10. Voyager resides within another exhibit, Exploring the Planets. Only the Stardust probe, returned to Earth from comet Wild 2, may be recycled into another mission.[12]

But while the Voyager simulacrum has gone into the NASM, the real Voyagers are inverting the expected relationship. Instead of fitting into an earthly museum, they are carrying a museum within them. Attached to their frame is a miniature of Earth, a distillation of the planet's sights, sounds, and stuff. It is an odd, yet oddly apropos, act of self-reflection that once again bears witness that the Voyagers are truly mechanical doubles of their creators.

# 23. New Worlds, New Laws

Once beyond Earth, the Voyagers would seem to have left behind earthly concerns with law and politics; and once beyond Mars, obsessions and qualms over colonization. After all, while the spacecraft carried plaques from Earth, they had neither the capacity nor the intention to deposit them on the new worlds they surveyed. Now, as they pushed beyond the Sun, they would seem to have shed such concerns altogether as mere metaphysical distractions. Even an abstract rule of law had to fade at termination shock.

But while the Third Age might go beyond Earth, it could not go beyond history. In the past, geographic discoveries had led not only to new ideas but also to new relations among peoples. Notions of sovereignty, of just wars, of the rights to rule over distant lands or remote others—all were upset and reformed as much as ideas in natural history, text-based scholarship, and the literature of travel. Exploration stretched, deformed, and remade inherited legal and political regimes. In particular it kept in constant turmoil two sets of governances. One sought to mediate among the explorers themselves. The other sought to mediate between the discoverers and those being discovered.

The Third Age skimmed over the latter. But it could not avoid the former. Like all ages, the Third was powered by rivalries, and its keenest political energies arose out of the cold war. It had to govern those passions so that exploration did not mutate into more lethal expressions. The revelations of the Third Age forced inherited institutions and ideas to evolve, and much as natural selection must act on what exists—turning a mammalian digit into a fin or a thumb, for example—so novel legal and political orders appeared that still bore the legacy of their hard-thought past.

From the onset of the Great Voyages, explorers had simply gone where they were sent or wished to go. They rarely sought permission from those being discovered.

When Vasco da Gama arrived at Calicut, for example, he was asked why he had come, and famously replied, "for spices and Christians." But the question had been posed by a Muslim trader who was neither astonished nor rapt with wonder, but dismayed and angry at the arrival of an unwanted commercial competitor. The Armada may have been a grand adventure for Portugal, but it was a destabilizing presence for the peoples it contacted. It meant fighting, treaties, forts and trading factories, a reconfiguration of commerce, and a realignment of political alliances. Was this havoc warranted so that, in modern terms, the Portuguese might exercise their inalienable curiosity and right to wander? They weren't wanted. The Indians didn't accept as justification that they were only responding to ethnic imperatives or the call of their DNA. Their choice to explore did not affect them alone. All too often trade, especially monopoly, required force, and force segued seamlessly into empire.

No one put it more succinctly than Henry Stanley at the close of the Second Age. "We went into the heart of Africa self-invited," he confessed, "therein lies our fault," and conceded that Africans certainly had an "undeniable right to exclude strangers from their country." Once having tramped into new lands, explorers might then have to fight their way out, or be rescued, or incite rivalry with another country that had also sent in exploring teams, which removed local matters into international politics. Similarly, the Institut d'Egypte's appeal to science hardly justified Napoleon's invasion, nor did the discovery of the Rosetta Stone warrant a forcible translation of institutions onto a society that didn't want them. A Stanley critic from the *Saturday Review* summarized the situation succinctly: "a private American citizen, traveling with negro allies, at the expense of two newspapers" had precipitated the deaths of indigenes "with no sanction, no authority, no jurisdiction—nothing but explosive bullets and a copy of the *Daily Telegraph*—into a country where he and his black allies are intruders and natural enemies."[13]

Of the justifications offered—science, proselytizing, the Enlightenment of commerce—only the cause of knowledge could persist into the Third Age. Instead of the call to missionize, advocates substituted a genetic hardwiring, and in place of empire, the need for national rejuvenation. Without indigenous peoples and without obvious sources of wealth, the classic competition sublimated into rivalries over science and status, or the paranoid desire to keep someone else from claiming something whether or not it had any value to the claimant. Instead, techno-romances brood over the possible contamination of newly discovered worlds; and the adventures of *Star Trek* pivot around a "prime directive" that specifically forbids interference in the affairs of discovered cultures and that prescribes protocols for contact. Apart from forcing sterilization procedures on spacecraft, these concerns have remained in the realm of fiction.

In the Third Age, as earlier, the first set of relationships, those among the exploring nations, still require regulation. There is clear value to all participants in having rules and in signing treaties that can prevent incidents from spiraling into ruinous competition or outright war.

This, too, can trace its genealogy to the Great Voyages. Between 1479 and 1494, Portugal and Spain negotiated three treaties that partitioned the discovered new worlds between them. The first divided Morocco and the Atlantic isles—Spain got the Canaries, and Portugal the rest. The most critical determinations involved the primary realm of the First Age, the ocean. The Iberians naturally wanted to rule the seas as they did isles and discovered lands; and as other nations entered the scramble, Spain and Portugal also sought to extend their monopoly over sea as well as land. In practice, however, they were unable to enforce such an edict, and the more nations that joined the contest, the more difficult it was for any one to assert unquestioned ownership.[14]

Instead of the right to a *mare clausum*, or closed sea, challengers proposed a *mare liberum*, or open sea. In 1609 the Dutch jurist Hugo Grotius codified the arguments for a mare liberum into a book of that title, establishing a body of customs and prescriptions that evolved into a generally recognized Law of the Sea. It was not an easy sell. That the Dutch were, at the time, keen competitors demonstrated the working alliance of politics, commerce, and jurisprudence. Self-interest figured as fully as idealism. Grotius was himself equally engaged in justifying various Dutch intrusions, such as seizing everything from vessels to ports, particularly against the Portuguese and English. Nations had a right to their coastal seas out to three miles (roughly, the range of a cannonball), but the idea that the high seas should be freely open to all became an enduring legacy.

Over the centuries a corpus of international law elaborated on what constituted discovery and on what basis a nation might claim rights to discoveries. When islands and coastlines were the prime realms, first sighting was deemed sufficient until other explorers demanded a more tangible act, a landing along with a formal declaration and a monument of some kind. Trading posts and colonies escalated the requirements into "effective settlement," and then into various concepts of "hinterlands," seeking to establish just how far a presence might justifiably extend into the backcountry, a judgment that had to factor in the character of the land and its inhabitants (if any). These issues were still under legal definition as late as 1933, when the international court at The Hague ruled that Norway could not claim eastern Greenland, since Denmark's two settlements on the south and west were sufficient to extend its rule over the uninhabitable ice sheet.

The Third Age required new protocols for its discovered realms. Unsurprisingly, the process began in Antarctica. In the aftermath of IGY, the nations active in Antarctica met in Washington, D.C., to establish a workable arrangement for governance.

At the time seven nations had territorial claims over the continent, three of which overlapped significantly. A large chunk of West Antarctica was unclaimed but widely recognized as the American sector, although officially American policy was to advance no formal claim and not to recognize the claims of others. During IGY, the Soviet Union carefully placed bases in each of the claimant territories. The prospect that the cold war might extend to the Ice was real. Yet IGY also demonstrated an alternative: in the name of science, it allowed for the free movement of all parties across the continent. In effect, for the duration of IGY, it treated Antarctica as high seas. The Antarctic Treaty that emerged in 1959, and entered into force on June 23, 1961, sought to continue that regime.

The treaty applied to all land and ice south of sixty degrees south. Among its articles were provisions to eliminate military activity, weapons trials, and nuclear tests or waste disposal; to allow the free movement of personnel and information; and to resolve disputes through the International Court of Justice. By adding further protocols and accords, these provisions have established a system of governance. Most interesting, perhaps, is the way the treaty has finessed the question of national sovereignty. It neither recognizes such claims, nor does it insist that by signing the treaty nations are renouncing claims. The treaty can thus mean one thing to domestic audiences and another to international, a kind of legal equivalent to Bohr's principle of complementarity. Sovereignty becomes symbolic, not operational. Modernism finds a geopolitical expression.

The deep oceans, likewise, increasingly found themselves outside a legal regime. The world sea was collecting problems as it did garbage; its concerns embraced the oft-oil-rich continental shelves as well as pollution, military traffic, nuclear waste disposal, even schemes for colonization. The more the Third Age mapped its terrain and identified its constituents, the greater the incentive for potential rivals to announce claims. The mare liberum threatened to become a submarine mare clausum. The case went to the UN in 1967, when the Malta ambassador, Arvid Pardo, warned about a cold war in the abyss, and called for a regime to govern these newly accessible but lawless regions. They should belong to no one, he insisted. They were rather the "common heritage of all mankind."[15]

In response the UN sponsored a series of conferences on the Law of the Sea. The last, convened in 1973, produced a convention ready for adoption in 1982. Unlike Antarctica, the oceans had a long history of human usufruction that was unlikely to be set aside. The resulting compromise extended the territorial seas from three to twelve miles, established an exclusive economic zone of two hundred miles, and otherwise granted the deep oceans the same freedoms as the surface seas. With one exception: adopting the "common heritage" principle, it established an International Seabed Authority, under the auspices of the UN, to oversee exploitation of the ocean floor. That principle the United States refused to recognize, and accordingly it has yet to sign the convention.

These same principles, and problems, have extended to that other realm of the Third Age: space. Again, the political promptings came from exploration, specifically the space race ignited during IGY; again, it looked to the Antarctic Treaty for inspiration. Unlike the ocean, there was little heritage of national economies connected to space at the onset. There was no fishing of the upper atmosphere, no military presence through the satellite equivalent of nuclear-armed submarines, no dredging of the solar wind or drilling for cosmic rays. As with other realms, the active parties sought to avoid a "new form of colonial competition."[16]

Addressing the UN in 1960, President Eisenhower urged that the principles enunciated in the Antarctic Treaty be adapted and extended over the solar system. Subsequently, the Limited Test Ban Treaty of 1963 removed the question of orbiting nuclear weapons from consideration; neither the United States nor the USSR argued to allow their presence in Earth orbit or beyond. Discussions continued, with one objection after another overcome, mostly along the analogous lines devised for Antarctica. In 1966 the General Assembly recommended approval of what became known as the Outer Space Treaty. In 1967 it was opened for signature and entered into force.

The treaty declared that the "exploration and use of outer space" should be conducted "for the benefit of all peoples," that such realms "shall be the province of all mankind." That phrasing looked generally to the useful ambiguities of the Antarctic Treaty and to the heritage of the open seas. An alternative, the Moon Treaty, restated the principles into the language and institutional framework of the Law of the Sea in that it relegated jurisdiction to the UN, particularly over the commercial exploitation of resources. With regard to the deep oceans, the institutions of exploitation were a matter solely of principle; almost nothing of real value then existed (save perhaps manganese nodules). But near-Earth space was rapidly filling up with satellites for commercial use or military reconnaissance, and other schemes were likely. In this case not only the United States but all the nations with launch capabilities have refused to sign the Moon Treaty. Completed in 1979 the treaty entered into force in 1984 for the thirteen nations that have acceded to it.

Complicating the picture are the inevitable ideologues, or the Third-Age equivalent to privateers and freebooters—the Cecil Rhodes and William Wallaces—who want untrammeled access to new lands and unfettered rights to claim them as private fiefdoms. Antarctica already suffers from unregulated tourism; the oceans are becoming gyres of garbage; near-Earth space is cluttered with debris from commerce and the clamber from well-heeled tourists, and its near-militarization threatens to unhinge existing treaties. The prospect of a scramble for Africa on the Moon or Mars should be enough to frighten even the most fanatic partisan.

At issue is the same question posed by Stanley except that it involves lands instead of peoples. Why should individuals have access outside any legal regime? The concern is that what people do has social consequences, that private profit will come at social costs. When an Air New Zealand DC-10 carrying tourists crashed into Mount Erebus in 1979, the U.S. Antarctic Research Program had to shut down in order to attend to the wreckage. An adventurer might go on his own, but some apparatus of government would have to rescue him. Stateless entrepreneurs among the Moon and planets might easily have the destabilizing presence of stateless terrorists. All this the Third Age has set into motion.

---

For Voyager, such concerns were not even hypothetical. It carried no weapons, and served no geopolitical reconnaissance. It landed on no planets. It issued no manifesto, deposited no tokens of sovereignty, proclaimed no imminent demesne, and implied no manifest destiny. It encountered no one, so was not encumbered with the moral qualms of cross-cultural contact. It made no landfalls, so was not plagued with questions of biotic contamination, state sovereignty, or corporate ownership. It was an alloy of adventure and exploration, a voyage of discovery that was in its narrative trajectory and purposes stripped of the moral and legal barnacles that encrusted and burdened the heritage that helped launch it. It did not seek new worlds to loot, colonize, or lord over.

It was an old world made new by its journey and became a new world unto itself.

# Beyond Narrative

# 24. Voyager's Voice

The Voyagers spoke to the public primarily through images, for which words served more as captions than as stand-alone texts. Unlike past explorers, the Voyagers would write no personal narrative upon their return, nor would they address thousands on extended speaking tours, or receive medals and be toasted by kings and academies. NASA and JPL could find spokespersons to address their mission, and press conferences were important ceremonies of planetary encounter, but what they said accompanied what they showed. Without those photos, the public was left with mumblings about bow shock and occultations. What Voyager needed was someone who could speak publicly to its meaning.

That task fell to Carl Sagan. His biography and that of the space program had co-evolved: he had proposed, explained, boosted, and preserved in print, on TV, and even in gold-plated records the reasons and necessity for humanity's exploration of space. For much of the American public, he was the public face of planetary exploration, and he became, with a tenacity both particular and paradoxical, the public voice of Voyager.

The golden age of planetary discovery and Sagan's career had evolved in tandem, like binary stars. He received a master's degree in astronomy and astrophysics from the University of Chicago in 1956, the year before Sputnik; a doctorate in 1960, a scant two years before Mariner 2's flyby of Venus. He spent a summer with Gerard Kuiper at the McDonald Observatory specializing in planets and their atmospheres when those topics interested very few people. Even more revelatory, he studied biology, including work with H. J. Muller; met Harold Urey and Stanley Miller; and saw planets as arenas for life beyond Earth. These interests granted him a special niche in the natural sciences. What brought him to the public was a flair for presentation and writing. In 1966 he authored a popular-audience book titled simply *Planets*, for Time-Life, and co-authored with I. S. Shklovskii a general book on *Intelligent Life in the Universe*. The contours of his career were set: he would search for extraterrestrial life and its corollary, extraterrestrial intelligence.

He needed only a suitable vehicle, and this the planetary program provided. On several occasions Sagan marveled at the providential timing of his ambitions and the age in which he lived. "Had I been born fifty years earlier, I could have pursued none of these activities. They were all then figments of the speculative imagination. Had I been born fifty years later I also could not have been involved in these efforts, except possibly the search for extraterrestrial life." The coevolution of his career and planetary discovery had the same kind of synchronicity of technology, politics, and astronomical alignments that had made the Grand Tour possible. The planets met their publicist.[17]

In 1971 he became a full professor at Cornell, and a year later began directing its Laboratory for Planetary Studies. Then *Other Worlds* and, with Jerome Agel, *The Cosmic Connection: An Extraterrestrial Perspective* appeared in 1973, a year before Mariner 4 arrived at Mars and three years before the Viking landing, which promised a perfect synthesis of Sagan's obsessions. The year Voyager launched, he published a popular science account of earthly intelligence, *The Dragons of Eden: Speculations on the Evolution of Human Intelligence*; it won a Pulitzer. The next year saw into print his book *Murmurs of Earth*, about the golden records that the Voyagers carried , and another collection, also on the evolution of intelligence, *Broca's Brain*. By then Voyager was well into its rhythm of encounters.

Its project scientist, Edward Stone, early appreciated that planetary exploration was "accessible to the public" in ways many sciences and topics were not. It could be presented in nontechnical terms, it contained a narrative arc inherent within the act of exploring, and it could be used to "communicate to the public" not only what such missions learned but "why scientists do what they do, and why it's interesting to be a scientist." But if JPL's press conferences could present the play-by-play account of Voyagers' mission, it was Sagan who provided the color commentary. It was Sagan, a member of the Voyager imaging team, who appeared regularly as a guest on late-night television (*The Tonight Show*), who was ideally positioned to exploit those encounters and explain them to the layman. As a writer, he did for Voyager what Eric Burgess had done for Pioneer. As a commentator, he made Voyager's saga into a romance.[18]

From the beginning he had grasped the power of the Grand Tour. It was Sagan who had most promoted the golden records, who persuaded NASA to have Voyager 1 take the emblematic photo of Earth and the Moon at the start and the planetary family portrait at its farewell, and who brought the thrill of discovery to the public. Between the time the Voyagers encountered Jupiter and the time they engaged Saturn, while Sagan joined Bruce Murray and Louis Friedman in founding The Planetary Society, he hosted the most spectacularly successful documentary series ever aired by PBS, what he modestly called *Cosmos: A Personal Journey*.

The 1980 TV series was seen by two hundred million viewers; the resulting book became the bestselling science book in English; both won awards. An episode dealing with "Traveler's Tales" featured Voyager at Jupiter. The Voyager spacecraft, Sagan declared, were "the lineal descendants of those sailing-ship voyages of exploration, and of the scientific and speculative tradition of Christiaan Huygens." The one had visited new worlds on Earth, the other wondered about possible "celestial worlds discovered," or "worlds in the planets." The Voyager spacecraft, Sagan insisted, were "caravels bound for the stars."[19]

The Voyagers could carry the public enthusiasm kindled by *Cosmos* throughout the rest of their mission: they were, in a sense, the apex of planetary exploration. But it is equally true that Sagan needed them to carry his message, which they did both physically in their golden records and symbolically by their trek beyond the solar system. Despite Sagan's enormous popularity, NASA veered from the trajectory he preferred for it. The Search for Extraterrestrial Intelligence programs were shut down; the space shuttle throttled planetary exploration; and for a dozen years there was only Voyager to supplement reruns of *Cosmos*. In 1996, as the Voyager twins pushed on toward termination shock, Sagan wrote a sequel to *Cosmos*, *Pale Blue Dot: A Vision of the Human Future in Space*, which opens with Voyager 1's family portrait and showcases the Grand Tour.

It was always Mars, however, that Sagan most obsessed over, and he imagined the colonization and terraforming of the planet as the critical task facing humanity as it moved from exploration to settlement. It was Mars that would reveal the extraterrestrial life that would make more plausible the reality of extraterrestrial intelligence. That discovery, Sagan assumed, would be epiphanous for Earth—"messianic" would not be too strong a term. In this he resembled Arthur C. Clarke, whom he complemented and in many ways succeeded, even writing a novel, *Contact*, whose premise is a connection with an alien intelligence.

Carl Sagan had his own psychic yearnings for an intelligent Other. He presented a curious, seemingly vulnerable, alloy of the skeptical and the wishful, ever pushing his desires before the public even as he countered them with the demands of a disciplined rationalism. Not infrequently he bonded those two sentiments together with double negatives that seemed simultaneously to allow and deny what he wanted. He was a believer in aliens, yet a critic of UFOs; he celebrated an unbounded human spirit, yet skewered religions. Famously agnostic, accepting a Spinoza-like reverence before a lawful universe, he nonetheless effectively populated those galaxies with godlike creatures—not the God of the great monotheisms, but local deities like Clarke's Overlords or the residents of a galactic Olympus or stellar Asgard. His belief that they would be benign and that encounter would be a messianic first-coming drew little support from history; but this was belief, not reason.

It was also literature. An author of Sagan's skills surely appreciated that supposing intelligent Others made encounter possible, and with encounter, something like a traditional narrative. The ideal way to convey information, even science, was through story. But what kind of story could exist if there was no Other to engage, if we talked only to ourselves? What kind of traveler's tale could be told if the traveler never returned? What kind of vision quest could be related if the hero kept going? Assuming contact allowed for an aesthetic closure as well as a thematic one.

Popular writing succeeds because it relies on popular genres. But if the Third Age changed some of the fundamentals of exploring, and especially of encountering, it perhaps required a different style of expression as well. Yet modernism, while full of self-referential paradoxes, spoke in a voice that was hardly attuned to a popular audience. Rather, it preferred to speak in ironic tones to an avant-garde about itself. The trick to expanding that realm was to package new experiences in old forms, and this is what Carl Sagan did with spectacular aplomb. As early Christian missionaries baptized pagan sites into churches, so Sagan rationalized the prophecies of Arthur C. Clarke by immersing them in a more rigorous science.

For his achievements as a popularizer he was often shunned by space scientists, who tended to equate "Saganizing" with shilling. The same group that would have denied Voyager eyes, a camera, would also have denied it a voice, the golden record; and they were prepared to punish anyone who promoted values beyond those most precious to the guild or the prescribed venues by which it expressed itself. Paradoxically, no space scientist was better known than Sagan, and none was so ostracized by his own community. Whether or not his golden records might someday connect with aliens, his books and TV appearances certainly connected in his own time with an audience for whom the subjects he addressed were often instinctively alien.

---

What makes a strength also makes a weakness. One is a distortion of the norm, and typically comes at the cost of neglecting some other trait. So it is with great popularizers.

*Cosmos* had the scope of an electronic epic of exploration. But it is less the work of a bard than of a propagandist. It is more animated by the vision of a Richard Hakluyt than that of a Luiz Vaz de Camões. It has no Old Man of Belem, or Moses who leads his people to a Promised Land but is himself denied entry. It shuns ultimate tragedy for a promised utopianism. (Unlike Camões, who traveled to the Indies and lost an eye fighting there, Sagan traveled in an imaginary spaceship, the whimsically named *Dandelion*.) *Contact* comes close to putting Carl Sagan's celebrated double negatives into novelistic (and very modernist) form, a contact with Others that ends with a journey to self. *Cosmos* will likely survive as one of the era's grand romances. It is unlikely to survive as its *Os Lusíadas*.

*Cosmos* and *Contact* are a long way from the Mars novels of Edgar Rice Burroughs, which Sagan openly confesses were the spark for his extraplanetary imagination. The chrysalis of a rigorous scientific education metamorphosed those boyish fantasies into an enduring adult passion that bonded with a hard-wrought intelligence, and granted a way to project in technological garb his longings for worlds to come. Yet origins can matter as much as futures. By looking to Homer and Virgil, Camões lapses into an imitative style that can seem contrived to modern eyes, and which leaves the poem fettered to archaic forms and an exalted language that edges into bombast; yet that structural quotation also endows the text with an aura of classical gravitas. Originating from the Tarzan-gone-to-Mars pulp fiction of Burroughs, Sagan's vision retains a juvenile joyfulness that suffuses it with an attractive zest, but perhaps never achieves the intellectual heft it needs to transcend its own circumstances and times. That may seem a harsh judgment, and perhaps an unfairly ironic one, but it may not be unearned for one whose view of the future glazed over his understanding of the past.

What endures are the tales told along the way. No one told the Voyagers' better.

# 25. Voyager's Record

As the Voyagers passed through termination shock, interest turned again to another of their instruments, this one prerecorded. Each spacecraft held a gold-plated copper record full of pictures, sounds, and voices of Earth. Of all the artifacts the Voyagers carried, this was the one the public most readily appreciated, and may well become the mission's signature memento.

Like many features of the Grand Tour, it built on a Pioneer prototype. The concept originated with Eric Burgess, a journalist then writing for *The Christian Science Monitor*, who eventually approached Carl Sagan, who along with Frank Drake had just designed a message suitable for communicating with aliens. They modified their ideas into a plaque, which Sagan's current wife, Linda Salzman Sagan, converted into etchings.[20]

The outcome was a collage of symbolic representations: a silhouette of Pioneer, the binary equivalent of decimal eight, the hyperfine transition of neutral hydrogen, the position of the Sun relative to fourteen pulsars and the center of the galaxy, the planets of the solar system and their binary relative distances. And there was one realistic representation, this of the male and female of the species responsible for the creation of the entity. That they were nude was a cause for some hysteria, and equal hilarity. The prospect of anyone finding the plaque was infinitesimally tiny, and of their interpreting it still more remote. But most observers, less addled by aliens and committed to cosmic liaisons, understood that the real message was aimed at Earth.

In October 1974, as Pioneer 11 began its observation phase for Jupiter, Voyager project manager John Casani coyly noted as a "concern" that there was "no plan for sending a message to our extra solar system neighbors," and recommended that the group "coordinate with Barney Oliver" and "send a message." His suggestion lay fallow until December 1976, when he contacted Sagan, who quickly organized a small committee of like-minded partisans that included Philip Morrison, Frank Drake, A. G. W. Cameron, Leslie Orgel, B. M. Oliver, and Stephen Toulmin, and "because some science-fiction writers" had been "thinking about such problems longer than most of the rest of us," he also solicited comments from Isaac Asimov, Robert Heinlein, and Arthur C. Clarke. Drake, a keen advocate of SETI and director of the National Astronomy and Ionosphere Center, suggested a long-playing phonograph record as the ideal medium. That set in motion a scramble to fill the record.[21]

The group had learned some lessons from the public outcry that had attended the Pioneer plaques. This time it sought out diversity—of peoples, of media, of senses. This plaque would display art to convey emotion, not simply exhibit mathematical and physical notations to show cleverness; the record had to "touch the heart as well as the mind." And this time NASA would not tolerate any nudity. A close working team that included Jon Lomberg, Ann Druyan, Linda Sagan, Timothy Ferris, Frank Drake, and, of course, Sagan gathered. Other consultants, such as musicologist Alan Lomax, were recruited to assist in specialty areas. In the end the record housed 118 photographs, 90 minutes of music, an "evolutionary audio essay" on "The Sounds of Earth," a roster of the U.S. senators and representatives with oversight for NASA, and greetings in 55 languages, along with statements from President Jimmy Carter and UN secretary-general Kurt Waldheim, and a "song" from a humpback whale.[22]

The group appreciated that every choice had two audiences, and that the audience on Earth was the most critical. "Great care was taken with all the musical selections," Sagan intoned, "in an attempt to be as fair and representative as possible in terms of geographical, ethnic and cultural distribution, style of music, and the connection with other selected pieces." Music ranged from Bach and Mozart to Chuck Berry singing "Johnnie B. Goode" and Surashri Kesarbai Kerkar performing the raga "Jaat Kahan Ho." The languages of the greetings were spoken by 87 percent of Earth's human population. Voyager's was an ecumenical vision, a musical and photo essay on the Family of Man. Whether or not an alien interceptor might understand it, humans on Earth could, and to that end the outcome was a proselytizing text.[23]

The record itself was made of gold-coated copper, and hence both imperishable and nonmagnetic. To bolster its capacity it was machined to play at $16\,^{2}/_{3}$ revolutions per minute instead of the more typical $33\,^{1}/_{3}$. In place of a manufacturer label, it had a photo of Earth with the caption "United States of America, Planet Earth." An aluminum cover shielded its delicate grooves from the micrometeorites that were expected to cluster between Earth and Jupiter and through the Oort cloud. (Calculations suggested that impacts might degrade some 4 percent of the record's surface by the time Voyager had traveled a hundred million years, or a sixth the distance to the center of the galaxy.) In order to date its conception, the cover was electroplated with uranium-238, whose regular decay makes it a radioactive clock with a half-life of 4.51 billion years, roughly the estimated life of the Earth.[24]

It was that other audience, however, that proved sticky. To help an alien intelligence (or "recipient") make sense of its find, the aluminum cover had etched onto it visual instructions for using the stylus, safely tucked behind, to play the record. A diagram demonstrated how to decode the resulting signals into pictures. And completing the instruction manual were two images: one, a pulsar map that showed Earth's exact location relative to fourteen pulsars, and the other, "two circles" depicting "a drawing of the hydrogen atom in its two lowest states, with a connecting line and digit 1 to indicate that the time interval associated with the transition from one state to the other is to be used as the fundamental time scale, both for the time given on the cover and in the decoded pictures." With these clues, a recipient could presumably play Earth.[25]

Even partisans admit the odds of discovery belong in the realm of infinite numbers, and if the record is discovered, it is questionable that the intelligent recipient might be able to decode it, or even recognize it as a message. The technology of recording has evolved like a fast-mutating virus during Voyager's long traverse. By the time Voyager reached Jupiter and Saturn, vinyl phonograph records were overtaken by magnetic tapes; by the time it reached Uranus and Neptune, tapes were fast fading before CDs; by the time it reached the heliosheath, CDs were passé compared with digital drives and iPods. The phonograph was hopelessly archaic just as the golden record reached the edge of the solar system–in technology years, barely beyond cuneiform tablets. It is doubtful that, presented with the record, even earthlings unfamiliar with phonographs could operate it.

Worse, a recipient was unlikely to decode it, and could never decipher the many languages it held. As Linda Sagan noted, there was no Rosetta Stone on board, "let alone a pocket dictionary." The recipients would know them as sound waves, assuming they had senses for sound. Carl Sagan confessed that "the most likely situation" was that "no human language will be remotely intelligible to an extraterrestrial auditor" without a previously provided "primer." And Frank Drake told the story of an experiment with a group of elite specialists in SETI, the self-styled Order of the Dolphin, who proved unable to decipher a test message. Even earthlings could make no sense of an earthling code. Still, desire begat belief. Science would provide an intergalactic language. That faith, at least, provided a context for the experiment.[26]

In the end, what advocates wanted was recognition that a cosmic Other existed. Voyager was a way of validating their belief. The deeper political agendas had to deal with harmony on Earth and support for SETI. The former value was widely recognized, and applauded. Even with Voyager as a totem, however, and with Sagan, then the most widely recognized spokesman for science, as a publicist, the latter faltered. What survived was an older variant of discovery: the received image of the discoverer. Or, in the eyes of critics, what emerged was another exercise in narcissism: *they* would want to know about *us*. (A historical note: Tom Wolfe's essay on the "Me Generation" was published a few months before the Voyager record team assembled.)

In *Murmurs of Earth*, Sagan concluded his reckoning of the Voyager golden record by imagining its encounter with "the planetary system of AC + 79 3888." The "inhabitants will of course be deeply interested in the Sun, their nearest star, and in its retinue of planets. What an astonishing finding the Voyager record, this gift from the skies, would then represent!" They would "wonder about us." They would appreciate that we were beings with "a positive passion for the future." They would be the significant Other who would affirm what we wish to believe is best about ourselves.[27]

As Voyager's recorders appreciated, the heritage of such efforts—exchanges, memorials, and hopeful first contacts—reached much further back than the Pioneer plaques. A meeting between peoples, especially when neither could speak the other's language, was awkward and could easily become violent. Accordingly, for example, when sailing down new shorelines along the coast of Africa, the usual custom was to leave goods on the beach, which the natives would take, leaving something of their own in exchange. Hard goods and hands-off gestures could substitute for greetings and common protocols. Having established one another's presence and interest, the process could scale up.

But such proceedings could not advance much further without a common language, which is why interpreters loom so large in successful exploration. The process of discovery was mostly a process of transferring knowledge from those living in the discovered lands to those who were discovering them; explorers needed interpreters as much as they needed guides and pilots; and this, too, they sought to systematize. As the Portuguese probed south, they brought Africans to Portugal, often as slaves, to learn Portuguese, and then shipped them out to contact new peoples. Bartolomeu Dias thus took two Africans with him south in the hopes that they could communicate the purpose of the expedition with the new peoples they met; they couldn't. Columbus, on his first voyage, brought an Arab interpreter, Luis de Torres, who found his linguistic skills worthless in the Antilles. Instead, Columbus took a direct approach. "As soon as I came to the Indies, at the first island I discovered I seized some natives, intending them to inquire and inform me about things in these parts. These men soon understood us, and we them, either by speech or signs and they were very useful to us." The Admiral of the Ocean Sea had tapped into a complex maritime civilization, for which discovery meant translating local lore into the scholarship of Europe. He kept his informants throughout the voyage, adding "one inhabitant of each country" to take to Castile in order "to give an account of its nature and products." Unfortunately, the group did what most indigenous peoples did: they told the newcomers what the newcomers wanted to hear, a misconstruction magnified because Europeans heard what they wanted the locals to say.[28]

As contact proliferated, interpreters arose from happenstance as much as from calculation. Survivors of shipwrecks adopted into local tribes, youthful captives, children of mixed parental cultures, slaves, and of course those simply gifted in learning tongues were all at times critical, particularly where some general language had evolved—such as hand-signing on the Great Plains or quechua along the Amazon or pidgin in New Guinea—by which basic communication was possible across a large region. Such interlocutors often became interpreters of culture and politics, not simply translators. Think La Malinche, the consort of Cortés as he moved toward Tenochtitlán; Sacajawea, as she helped Lewis and Clark thread the cultural watershed of the Missouri River and cross the Rockies; Oahu Jack, a Hawaiian, who served the Wilkes Expedition as interpreter through Polynesia and Melanesia; Tupaia, the Tahitian who accompanied Capt. James Cook and named and gave rough directions to seventy-four islands within a few days' sail. Revealingly, when Samuel Johnson and James Boswell heard Captain Cook's account of his travels, they doubted his rendering among the natives since he himself "didn't speak the local languages." They appreciated that indigenous lore could be understood only through indigenous tongues.[29]

The ideal was an interpreter who could also guide, or serve as a pilot. Where none was present, explorers were inclined to seize one, as they did translators or captives. Da Gama crossed the Indian Ocean from Malindi to Calicut by commandeering a pilot (perhaps the leading navigator of the day, Ibn Majid). At Valparaiso, Sir Francis Drake seized John Griego, a Greek pilot in the service of Spain, to navigate the *Golden Hind* to Lima. At California, after vanquishing the *Santa Anna*, Thomas Cavendish took "two young lads born in Japan, which could both write and read their own language," "3 boys born in the isle of Manila," and a "Spaniard, which was a very good pilot until the islands of Ladrones" [Guam], who knew the route between Acapulco and Manila. In 1610 Samuel de Champlain sent a young Frenchman, Etienne Brûlé, to live with the Iroquois and learn their language, and so created the progenitor of a species of frontiersman, the coureur de bois, who could broker between societies. Similar classes of mixed-race, multilingual frontiersmen sprang up everywhere. They were especially prominent around Portuguese settlements in South America and Africa, where as *bandeirantes* and freebooters they not only interlocuted but destabilized whole regions.[30]

But with the desire to learn, there was also a desire for a tangible record of the experience through souvenirs and deposited records. Portuguese *marinheiros*, for example, deposited cruciform stone *padrões* like beach flotsam around Africa. Buffeted by a storm on his return, Columbus wrote a brief account of his journey on parchment, wrapped it in waxed cloth, and stuffed it into a cask, which he committed to the sea. Lewis and Clark erected a memorial post at Fort Clatsop when they reached the Pacific. At the tidal delta of the Bella Coola River, Alexander Mackenzie painted onto stone with vermilion and grease an inscription that identified himself, the date, and his origin, "from Canada, by Land." The sorties of the Franklin Expedition left cairns with notices in six languages, along with instructions to finders to return them to Britain. Robert Peary heaped up a pyramid of ice at the North Pole before taking a photo; Roald Amundsen left a tent and flag at the South Pole.[31]

The creators of the Voyager golden record struggled to describe its significance; their singular point of agreement was the need to say something. As B. M. Oliver observed, "There is only an infinitesimal chance that the plaque will ever be seen by a single extraterrestrial, but it will certainly be seen by billions of terrestrials." There is no need to analogize, however. The urge to leave a mark and to take mementoes would seem hardwired into human travelers. The Voyagers' golden record had a long pedigree of memorials, territorial markers, messages, self-promotions, and simple graffiti. It tapped a yearning by humanity to immortalize a record of itself, and in truth the golden records buried in the vacuum of space will long survive the copies placed with the president, the Library of Congress, and the Smithsonian. They will certainly outlive their creators and may endure to the end of time itself.[32]

Whatever the bid for immortality, the public trope behind the golden record was contact with aliens. On this score, there is little evidence from past exploration to suggest that the reactions imagined by Sagan among both recipient and giver would occur. Yet it is significant that he made his appeal through the classic formula of an encounter. For that to work he had to have an Other, and granted his predisposition for an Other, he had to tell the story through the old devices. The golden record was a means to both ends.

Certainly what expeditions carried as exchange goods and memorials were always a cameo not only of what an exploring society sought from its enterprise but how it wanted to see itself—its best goods, its finest technology, its boldest emissaries. It would plant the cross, raise monuments for trigonometric surveys, deposit political tokens. In return, explorers would be received with awe and delight by the discovered Others. Almost always, however, practice proved otherwise. Explorers were troubled, and often troublesome; indigenes had their own ambitions, and rather than use encounter to transcend their society, they strove to bring the discoverers—their goods, their weapons, and their symbolic presence—into their existing world to more mundane ends.

Grand gestures such as expensive exploring expeditions plead for the ideal and the hopeful: Voyager needed to carry some kind of symbolic talisman against the fate of loss, forgetfulness, extinction, and oblivion. Almost all observers have accepted the golden record as a reasonable totem of Earth, one that however constrained, by its own time, place, values, and personalities, has tried to speak to enduring virtues of hope, perseverance, wisdom, humility, and generosity. If Voyager needed the pretext of an alien Other to so present itself, then that, too, they conceded. If aliens did not exist, it seems, we would have to invent them.

# 26. Voyager's Returns

For all its modifications of that tradition, the Voyager mission tapped into a heritage of exploration—that was its cultural power. But there was always to the Grand Tour a quality that went beyond normal expeditioning, a sense of the providential or, for the more literary, perhaps of the mythical. The aura would have embarrassed hardheaded engineers intent on ensuring that solders would not shake loose during launch and electronics could survive immersion in Jovian radiation. Yet it was there, a tradition that predates exploration and in fact a heritage that Western exploration itself taps into.

Myth, as Joseph Campbell has argued, is "the secret opening through which the inexhaustible energies of the cosmos pour into human cultural manifestation." Its primary purpose is "to supply the symbols that carry the human spirit forward." Myths offer a kind of transcendence, a transfiguration or transport beyond death.[33]

The endlessly mutable myths themselves seem to emerge out of a grand monomyth, the "universal mythological formula of the adventure of the hero," so fecund in humanity's multicultural imagination as to display "a thousand faces." At its core are three rites of passage that prescribe a beginning, middle, and end, or specifically, phases of separation, initiation, and return. In Campbell's summary: "A hero ventures forth from the world of common day into a region of supernatural wonder: fabulous forces are there encountered and a decisive victory is won: the hero comes back from this mysterious adventure with the power to bestow boons on his fellow man." That is not a bad formula for the classic tale of exploration.[34]

The journey has multiple stages, latent in the archetype, not all of which are present in each variant. There is a call to adventure—an appeal to depart from the pale of the mundane into a special realm of marvel. The call is sometimes refused, and the refusal must be overcome. There are helpers who outfit the hero with special devices. Then comes the initiation, the crossing of a formidable and seemingly impermeable threshold. Within the wonder world, the hero undergoes test after test, which he must overcome and by which he is measured. In popular tales these tests are physical, or the outcome of cleverness; in more religious contexts, they are moral. At last he reaches his goal. Paradoxically, that moment of climax cannot be the end, for he must return and bestow his boon—his acquired knowledge, his newfound tools—to society. He must recross the threshold, often after further struggle. Upon his return skeptics may challenge his accounts, rivals question his deeds, an apathetic public ignore his elixirs and enlightenment; and he must relearn how to live within this society. Once done, however, he ends as a celebrated hero or in apotheosis.[35]

The formula fuses journeying with seeking, and places both within a particular cultural setting that emphasizes the disruptive power of new knowledge and the terrible task of resocializing the knower. This broad formula is what the West institutionalized during the Great Voyages into its classic scenario for geographic exploration, an undertaking that goes beyond simple scientific inquiry, beyond routine touring, and beyond untethered restlessness. The journey to unknown lands and seas amalgamates them all. Time and again, exploration has echoed those erstwhile themes.

It is there in the call to go: explorers ardent with desire, possessed by visions or driven by personal demons, looking only for an outlet. The simple mythic structure, however, overlooks the history of the many called who declined to answer. The U.S. Exploring Expedition could not locate a naval officer willing to command it. Capt. Robert FitzRoy could not find a companion to accompany him on the HMS *Beagle* until an intermediate eventually proposed Charles Darwin. When a complementary expedition by the HMS *Rattlesnake* to Australia and New Guinea was proposed, Beaufort wanted the veteran surveyor-captain Alexander Vidal to lead it. But Vidal's wife had died, leaving him with a young family to raise. Much as he craved another expedition, the prospects of being away for four years were too much. The needs of family proved the greater call. He stayed—surely, the better choice. Instead, Capt. Owen Stanley assumed command, and took with him Thomas Huxley as assistant-surgeon-cum-naturalist.[36]

The explorer's return could be awkward, too. Columbus was arrested at the Azores, and after his third voyage he was later clapped in irons in Spain. Magellan, dead in the Philippines, was denounced, and credit for the return of the *Victoria* bestowed on a Spaniard, Juan Sebastián del Cano, as the Spanish state seized the official record to doctor the story. (It was only after Antonio Pigafetta, having made a duplicate of his journal, published a full chronicle, that the world appreciated Magellan's successes.) James Bruce was ridiculed as a liar for his "discovery" of the sources of the Nile. Henry Stanley was shunned as a fraud, then scorned as a "workhouse brat," and then condemned as a freebooting killer. Robert Peary returned not to accolades in a shower of ticker tape but to dismissal, since Frederick Cook had some months earlier announced his own trek to the pole and fraudulently seized the glory. Upon its completion, the voyage of the *Rattlesnake* did for Huxley what the voyage of the *Beagle* did for Darwin; but Captain Stanley died at sea. Meriwether Lewis and William Clark were lionized after their trek; but Lewis never published his journal, and ended a suicide.

It often proved difficult for both explorer and society to reabsorb each other. Instead what endured were the exploration's gathered artifacts and printed record.

The written account, too, had its formula, and this evolved along with other aspects of exploration. The earliest records were ship's logs or journals, or letters from expedition leaders to the Crown; they were often impounded and hoarded in the Treasury as state secrets. The modern version appeared in the Second Age when, not coincidentally, the novel arose as a literary genre and Romantic histories adopted a shared narrative structure. The ship's chronicle assumed a more organic form; at its core was the personal narrative, a literature that enjoyed enormous popularity during the Second Age.

Publicists, too, recognized the literary possibilities, along with their political potential. They appreciated how the personal narrative might communicate with the literate public and could get a story into the larger culture in useful ways, and knew that few explorers could also write adequately. The British Admiralty thus hired a professional writer, John Hawkesworth, to prepare James Cook's journal for publication, and commissioned him to do the same with the records of Cook's British predecessors, John Byron, Philip Carteret, and Samuel Wallis. Hawkesworth did what he was told: he effectively transfigured notes and chronologies into a master narrative. Meanwhile, Georg Forster, one of the naturalists on Cook's second voyage, further pioneered the genre with his *A Voyage Round the World*, and Alexander von Humboldt, whom Forster had inspired, helped devise the continental version with his *Personal Narrative of a Journey to the Equinoctial Regions of the New Continent*.[37]

That set the pattern of recounting: the personal narrative was the nuclear core around which details and scientific studies clustered like electrons. The success of an expedition as a cultural event was often determined by its ability to penetrate the imagination, and this meant a combination of art and literature. Many of the classic accounts had ghostwriters. John Charles Fremont relied on his wife, Bessie, to turn his adventuring into narrative and on her father, Thomas Hart Benton, senator from Missouri, into politics. Charles Wilkes refused the Navy Department's wish to hire a professional writer, and churned out five volumes that only sporadically overcame his turgid prose and crabbed personality. Particularly where subscriptions or book sales financed expeditions, literary skills mattered. Both books and lectures built on a narrative pith in which the explorer assumed the role of hero, answering the call to adventure, overcoming perils, and returning to proclaim his triumph and bestow its meaning to an admiring society. That great coda to the Second Age, Robert Scott's epistles, written as he was dying on the Ross Ice Shelf, cast his vanquished expedition in exactly that light, and made even failure fit the formula for triumph.

In the Third Age the genre split. Part went to manned exploration, where it merged with journalism. Norman Mailer would pose as Aquarius and ponder the Apollo Moon landing; Tom Wolfe's paean to the Mercury astronauts, *The Right Stuff*, appeared a decade later; and participants wrote their own personal narratives, such as Michael Collins's *Carrying the Fire*. But the larger part went into fiction and popular culture, where space travel was easy, adventure challenging, and aliens abundant. Hollywood and TV refilled the old bottles with Klingons, Martians, Overlords. In 1968, as Apollo 8 filmed the earthrise over the Moon, *2001: A Space Odyssey* brought Arthur C. Clarke to the big screen; *Star Trek*—the voyage of the *Beagle* outfitted with warp drive—discovered new worlds weekly, before also heading into the deep space of Hollywood with biennial launches; and *Star Wars* grafted high-tech special effects onto the rootstock of ancient myth.

The deeper split was between fiction and nonfiction. Without human protagonists or their alien proxies, the genre sank into formula fiction. The robots remained snugly within nonfiction, particularly journalism. They bonded insecurely to classic tropes of the quest.

Yet the saga of Voyager seemed rife for just such treatment. However much its engineers might scorn anthropomorphism, however much its scientists distrusted and dismissed the allusive, the metaphoric, and the mythical, however much politics might confound and conflate a literary imagination with simple publicity, Voyager is an artifice of human hand and heart as fully as any poem, and its journey is a narrative of adventure and aspiration that eerily echoes the template of ancient myths. Its trajectory has the arc of a hero's quest.

All the parts are there. The call to adventure: the beckoning to a unique journey, one set by the rare alignment of planets beyond mundane Earth. The initial, political refusal: the ensuing struggle to reinstate and accept, if reluctantly or provisionally, the Grand Tour. The threshold: launch, with their separate shaky initiations. The perils, overcome with the help of clever engineers. The tests—the planetary encounters—successfully met. Each adventure propelled it to another, beyond the horizon, until the explorer had journeyed beyond not merely Earth but all the earthly worlds that inhabit the solar system and then passed through a second veil, the transcendence of termination shock. Voyager's ascent to the stars seems an apotheosis.

Here the narrative arc falls apart. The hero won't return—won't return not from refusal, as often happens in the quest narrative and must be overcome, but by design. With a human actor, this would challenge the premise of exploration. As Hakluyt noted, "All this great labour would be lost, all these charges spent in vain, if in the end our travelers might not be able to return again, and bring safely home into their own native country that wealth and riches, which they in foreign regions with adventure of goods, and danger of their lives have sought for." With a robot as actor, the issue is not lost knowledge but the rent fabric of the mythological form. In Campbell's words, "The adventurer still must return with his life-transmuting trophy. The full round, the norm of the monomyth, requires that the hero shall now begin the labor of bringing the runes of wisdom, the Golden Fleece, or his sleeping princess, back into the kingdom of humanity, where the boon may redound to the renewing of the community, the nation, the planet, or the ten thousand worlds." Or, one might add, the lost megabytes of data from beyond the reach of Earth.[38]

The return is often more forceful than the departure. The fountainhead of Western civilization's adventure epics, *The Odyssey*, is the consuming story of a journey home. The grand gesture of the First Age was a circumnavigation—a voyage beyond that by design would end where it began. The Second Age sent its emissaries out and expected them to come back. Then the Third Age dispatched parties into ice, abyss, and space, and the hero quest crossed another threshold.

Until Voyager. Voyager would not return. It could send back data; it would not itself return. The "cosmogenic cycle" would remain if not broken then recast in a modernist vogue.

---

Of course plenty of explorers have failed to return, and rescue missions have from time to time enormously expanded the range of discovery. The two-decade search for the Franklin expedition effectively mapped northern Canada and the North American stretch of the Arctic Ocean. The search for the missing Burke and Wills expedition quickly fleshed out the geography of interior Australia. The search for the "lost" Livingstone launched the most famous career in African exploration. But not returning was never part of the design. The return was fundamental: it may be said that, as with narrative, the character of the ending dictated the beginning.

Clearly, Voyager has "returned" a cornucopia of images and data, and sent back (as mythic heroes do) a seemingly special lore, "a certain baffling inconsistency between the wisdom brought forth from the deep, and the prudence usually found to be effective in the light world." But the protagonist hero won't return; the narrative arc remains unfinished; the cosmogenic cycle, incomplete. Up to a point, Voyager can evade this issue: it continues to function, still probing outward, still meeting tests. But when its power burns out and its transmissions fall silent, the narrative can no longer pretend that the story fits the old forms.[39]

The Voyagers—reluctant modernists to the end—offer one solution with their golden records: they carry the return destination with them. The records avoid the dilemma, too, by the fiction that the adventure will persist and that Voyager will meet an Other, and that Other will presumably return the information to Earth (as *Star Trek I* imagined). Sagan's belief in a cosmic connection, that we are at base "star stuff," provides another closure. In this version, humanity's earthly existence is the outward journey, and spaceflight is but a return to our ultimate origin, the cosmos. Yet both visions continue to imagine, as technological romance does, the old genres filled with new adventures. They exchange an exploration narrative for an emigration one, part of a larger task of colonizing. They remake the manifest destinies of the Second Age into a destined manifest to the stars. What Joseph Campbell called the "hero with a thousand faces" could, it seems, apply equally to the genre as it morphs with the times.

---

Voyager's visionaries looked to the future. They thought of the Voyagers as instruments, and assumed their journey was simply another incremental moment in what would prove to be an irresistible expansion over the solar system, and beyond. The Grand Tour mission would be followed by more and better missions. They could not know that Voyager would culminate a golden age, that its trek would be unique, that it might require a distinctive narrative. Even the most culturally sensitive such as Sagan looked only outward and forward. They imagined, in Emerson's phrasing, a continued succession of "new lands, new men, new ideas."

Yet the Voyagers looked back as well as ahead. Repeatedly, their most stunning images were those taken when they turned around to review what they had passed, from the volcanoes on Io to the family portrait of the solar system. So, too, much of the Voyager mission's cultural power resided less in its future fruits than in its ancient roots as a quest narrative. The Voyagers were what they were because they looked both ways. It is then a lost opportunity, perhaps, that the two Voyagers were not dispatched one to each world, one to the great beyond of interstellar space and one to the home planet.

That thought never occurred to mission designers. The mission was a voyage of scientific discovery. It could continue only toward the new. Without astronauts the spacecraft, as instrument, had no necessity to return. The reason for return lies not in science but in the logic of the quest narrative. The character of Voyager's journey made the spacecraft more than an automated lab. The power of its mission lay in its trek, and that deserved a suitable narrative, which pointed to a quest.

Oddly, return trajectories were the original task of the JPL project that had led to Minovitch's recognition of gravity propulsion, and hence to the Grand Tour. If there was no exact narrative solution, neither was there more than an approximation of the restricted three-body problem, but that didn't stop the mission. Perhaps given its scientific tasks and engineering constraints, no trajectory would have been possible that could have allowed one of the spacecraft to be pulled back, cometlike, into an orbit around the Sun, and then around Earth. If some mechanism did exist, the journey would likely take decades or, more probably, centuries, or more plausibly still, millennia; and the spacecraft would die long before it arrived. But what a true time capsule it would have made. What a narrative of Return.

The quirks of robotic exploration might seem to have created a mechanical divide that the old formulas cannot bridge. Yet perhaps Voyager requires a change in the genre of the same sort that the Third Age has demanded in our understanding of exploration. It may be that Voyager has caught the *gran volta* of modernism, a literary doubling of the Cape in which the self relates to Other, and that it may well require a literary form, a narrative arc, that more resembles a Klein bottle than an arch.

The Voyager saga may have instinctively found a way around that omission, much as its engineers have repeatedly found software patches to work around broken hardware. That cultural patch lies in the dual plaques, gold and aluminum, that the Voyagers carry. The golden records that Sagan celebrated meant they carry their home with them. The aluminum plates that JPL quietly affixed make them a hero of 5,400 faces.

# Beyond Tomorrow

Yet all experience is an arch wherethro'
Gleams that untravelled world whose margin fades
For ever and for ever when I move.

**—Alfred Tennyson, "Ulysses"**

The untold want, by life and land ne'er granted,
Now, Voyager, sail thou forth, to seek and find.

**—Walt Whitman, *Leaves of Grass***

The passage through the heliosheath will be Voyagers' last measurable event. It begins with termination shock, the final veil before the sanctum sanctorum. It will end with bow shock, beyond which there can be no further encounter. There will be only void. Yet the Voyagers will sail on—*ad astra*, to the stars—without foreseeable cessation. *In saecula saeculorum.*

It is hard to imagine a nonending, like the empty echo of the Marabar Caves. Yet over time the Voyagers' power will fail, their transmissions cease. The need remains, nonetheless, to project a continuation, for even as their trajectories diverge ever wider, their histories converge, shrinking like their declining power into a common emptiness of inertia.

Because of its close targeting past Titan, Voyager 1 sails at 35 degrees northward from the ecliptic at a rate of roughly 3.5 AU a year, and because it was programmed to sweep close to Triton, Voyager 2 arcs some 48 degrees southward of the ecliptic at a rate of 3.1 AU. In 40,272 years Voyager 1 will be within 1.64 light-years of the star AC+79 3888, and 100,000 years later, within 2.35 light-years of the star DM+25 3719. Voyager 2 will pass by objects and stars with more familiar names. In some 26,000 years it will reach the

Oort cloud of ice-comets; in another 20,319 years, Proxima Centauri, then only 3.21 light-years distant, and some 310 years later, Alpha Centauri, a scant 3.47 light-years away. In a further 300,000 years it will cruise 4.32 light-years from Sirius.[40]

The Voyagers will approach their final rite of passage amid a void that no human could endure, a vacuum of meaning no less than of geography. It would seem that their trek must become a journey of movement without purpose and a narrative without events.

They will continue to coast. They won't decompose. They won't rust. They won't break down or wear out from use. With their motive power gone and their equipment switched off, they will no longer be even machines, but rather metallic statues or gangly headstones. They will move out of sheer inertia, slowed or quickened by the soft geography of the galaxy, the friction of interstellar gases, and the pull of distant gravitational fields. Light-years from now they may crash into a solid body, or vanish into a gaseous vortex. They may be captured by a planet or a star, or join the vast swirl of the Milky Way, another particle amid trillions. They will pass worlds unbounded, without a scan, a nod, or a message. They will simply persist, an endurance without sentience, record, or end, passing the black holes of space and the dark matter of time.

But while this may be the Voyagers' fate, it won't be their story.

The Third Age is yet young. It may yet last for another century or more.

Like the others it has its distinctive traits and its ideal explorer. The Great Voyages had their questing mariners and indomitable conquistadors—their pilot-admirals like Columbus and Magellan, their great captains like Cortés and Coronado, and their maritime warriors like Albuquerque. The Second Age had its far-ranging naturalists and peripatetic natural philosophers with their unquenchable curiosity, their balky instruments, and their personal narratives. It could boast of La Condamine and Humboldt, Darwin and Wallace, and ended with the unyielding wills of professional explorers such as Stanley and Amundsen. The Third Age has its machines, some staffed by people, most piloted remotely or granted some slack as semiautonomous robots. It has the *Trieste* and the *Alvin, Jason,* and *ABE,* the Autonomous Benthic Explorer. It has Mariner 2, Viking, and Voyager.

Among these new explorers Voyager may serve as synecdoche, as a testimony and a defining gesture. The Grand Tour offered the greatest possible traverse; it trekked farther and longer, saw more for the first time, spanned the entire geographic realm of the solar system, and climaxed a self-proclaimed golden age. Among that Earth-launched constellation of travelers Voyager offers the boldest silhouettes: it contrasts most starkly with the all-too-human explorers of the Second Age and the human-crewed capsules of Apollo, Cosmos, Skylab, and the fatally flawed space shuttle. The Third Age could proceed only with people and machines in sync, but as Voyager has demonstrated, people do not have to be *in* the machines. If Voyager does not perhaps offer the fullest synthesis of the age—for it did not send probes to the planets or land itself on discovered moons—if it has traded a breadth of reconnaissance for intensity of inquiry, it has coasted past the hard geography of new worlds and has cruised through the soft geography of interplanetary space as nothing else.

And if, unlike the human hero of myth, it has not apotheosized, it has become iconic and as immortal a monument as civilization might build. It will outlast the pyramids, coliseums, and the *Mona Lisa.* It will outlast its rival realms, the ice sheets and the abyssal plains, which will rise and fall under the push and pull of tectonic and climatic tides. It might perhaps outlast Earth.

Those associated with Voyager sensed from the beginning that it had a special destiny.

The allure of the Grand Tour suggested an almost mythological birth, as though the heavens had foreordained it: it could not be denied, it had to happen. And the Voyagers were lucky. Through perils and glitches and malfunctioning parts, they survived, they defied odds and manufacturing warranties, they endured to the end.

Voyager surprised, and kept surprising, and when the planetary program stalled shortly after launch, Voyager surprised everyone with its power to keep a grander vision alive. Ellis Miner noted, with almost impossible understatement, "There were more discoveries made on Voyager than I expect to ever see made on any single mission. It is probably the most successful mission ever done and likely ever to be done." As Voyager 2 approached Neptune, Dick Laeser, mission director and later project manager, said simply, "I have no desire to do much else except to ride this thing all the way out into interstellar space."[41]

Most participants found it difficult, in discussing Voyager, to avoid both the trite and the self-laudatory, and they stumbled; for them, the Voyagers' continuing trek was the spacecrafts' own narrative and statement. Professional pundits have tended to follow Carl Sagan's lead and stress both the canonical power of the Voyagers' images and the enforced vision that Earth is merely a speck in the cosmos, of scant significance, that Voyager's power lies in its capacity to humble and to offer hope of some future redemption, perhaps with interstellar contact.

Few believed that. They sensed, if they could not give voice to their sentiments, that there was little about Voyager that spoke of a mandatory meekness or frailty. Perhaps if viewed from the stellar Olympus of Clarke's Overlords or the cool mathematics of Sagan's cosmic intelligence, Earth seems trite and its inhabitants boorishly arrogant. But viewed from Earth, the Voyagers look positively Promethean, expressions of an indomitable moment when humanity returned fire to the heavens. Their power was not to humble but to inspire. Its observers recognized that Voyager's like would not come again. Norm Haynes, project manager for the Neptune encounter, explained in terms many participants recycled, "It wasn't an once-in-a-lifetime experience. It was a one-time experience." It did what couldn't be repeated. Voyager's saga was, as Edward Stone put it, injecting new juice into an old cliché, "the journey of a lifetime." The lifetime was humanity's.[42]

There are those for whom the quest for newer worlds must point outward, for whom it means the discovery and occupation of distant places. And there are those for whom the quest points inward, for whom it means the rediscovery of Earth, or further, a deeper discovery of the human heart. But all can look to Voyager with awe, pride, and faith that whatever we are, we will endure, and all can savor its journey as an expression of a shared longing that newer and better worlds are indeed possible, and that Earth might be among them.

# Afterword

*Voyager* is not the text I set out to write. What intrigued me about the Voyager mission—apart from its sheer audacity and awe—was its long history, which is to say, a lengthy and complex narrative that I thought might braid with a general chronicle of geographic discovery by Western civilization. I hoped that I might use a stream of commentaries, drawn from the exploring past, to shepherd the story, much as Voyager's overseers used course corrections to keep it on trajectory, such that each narrative could reinforce the other. Voyager could carry the grand narrative of Western exploration, the grand narrative of Western exploration, could propel Voyager, and their collective story could have a common tempo throughout.

I couldn't make it work. One or the other had to be the primary vehicle, and I chose Voyager. That left the Great Ages of Discovery as a commentator but not a co-chronicle; and without the prospects for sustained counterpoint, I opted to replace a chattering stream of recalibrating observations with fewer but larger set pieces that could highlight particular themes of relevance to both. Accordingly, I shifted the early chapters from a continuous narrative into a format more traditional and analytical. This meant a lot of text to get Voyager to launch and a big payload to carry upward. What results is an interpretive history whose internal rhythms mimic those that led to Voyager's launch and journey.

The Great Ages will have to wait for a full-spectrum history of their own. They enter this text as an organizing principle that allows for comparisons and contrasts, that is, for context. The conceit has itself a context, however, that may be worth explicating. I wish to acknowledge my intellectual debts for its evolution.

The idea of parsing the grand sweep of exploration by Western civilization into eras derives from a 1974 graduate seminar on nineteenth-century America in which William Goetzmann read a paper that argued for a "second great age of discovery." It took no great leap of imagination to see the latter half of the twentieth century as part of a third great age. I exploited Goetzmann's phrase, if not wholly his idea, in a paper read at an AAAS meeting in 1976 and subsequently published under the title "From the Grand Canyon to the Marianas Trench: The Earth Sciences after Darwin," and then in my 1976 doctoral dissertation, written under Goetzmann's general supervision, about an American geologist and explorer, later published as *Grove Karl Gilbert* (1980). Our separate lines of inquiry converged in 1986 with his book *New Lands, New Men: America and the Second Great Age of Discovery*, and my book *The Ice: A Journey to Antarctica*.

Over the succeeding years I have tinkered with and refined the idea in a handful of papers read to conferences and occasionally published, used it as an organizing device for a course I taught on exploration history, and relied on it as an informing conceit for *How the Canyon Became Grand* (1998). It was embedded in the text I delivered on "The Future of Exploration" at the Sarton Memorial Lecture at the AAAS meeting (2002), and in "Seeking Newer Worlds," delivered at a workshop co-sponsored by NASA and the National Air and Space Museum and later published in *Critical Issues in the History of Spaceflight* (2006). By then I had decided I ought to apply these ideas directly to the space program, with Voyager as my preferred vehicle. As before, the three ages get refracted through a particular subject. In the past, these were places; here, an expedition. Even Voyager, however, has only so much narrative thrust: it cannot lift the entire Third Age, and it carries it outward to space when so much of the era will explore the depths of the oceanic abyss. But it is a start.

Several people have assisted in this project, which has too often resembled the tempo of the Voyager mission, full of frenzied activity and long voids. I would like to note particularly Julie Cooper, archivist at JPL; David Fries at NASA's History Office; Stan Seibert, who gave up scarce discretionary time to help a mathematically challenged friend check some calculations; and Lydia Pyne, who offered comments on a draft. A special thanks goes to Wendy Wolf for defibrillating parts of texts and ideas that threatened to sink into reverie, self-absorption, or obscurantism and for reminding me that the narrative really is the message. While the text, unlike the Voyager spacecrafts, cannot carry an aluminum plaque with all their names, it carries their presence.

All were thrusters, helping to stabilize and point. The primary propulsion has come as always from Sonja, who looks ever heavenward, yet always manages to have her feet planted squarely on Earth.

Brittlebush Valley, August 2009

# Appendix

## CHRONOLOGY OF MAJOR LUNAR AND PLANETARY MISSIONS (LAUNCH DATES FOR SUCCESSES ONLY)

1957

**USSR:** Sputnik 1

**USSR:** Sputnik 2

1958

**USA:** Explorer 1

**USA:** Vanguard 1

1959

**USA:** Pioneer 4—lunar flyby

**USSR:** Luna 2—lunar impact

**USSR:** Luna 3—lunar flyby

1962

**USA:** Ranger 4—lunar impact

**USA:** Mariner 2—Venus flyby

1964

**USA:** Mariner 4—Mars flyby

1965

**USA:** Rangers 8, 9—lunar impact

**USSR:** Luna 5—lunar soft landing

1966

**USSR:** Lunas 9 and 13—lunar landers

**USSR:** Lunas 10 through 12—lunar orbiters

**USA:** Surveyor 1—lunar lander

**USA:** Lunar Orbiter 1, 2

1967

**USA:** Lunar Orbiter 3 through 5

**USSR:** Venera 4—Venus probe

**USA:** Surveyors 3 through 6—lunar lander
**USA:** Mariner 5—Venus flyby

1968

**USA:** Surveyor 7—lunar lander
**USSR:** Luna 14—lunar orbiter
**USSR:** Zonds 5 and 6—lunar orbit and return
**USA:** Apollo 8—lunar orbit and return

1969

**USSR:** Venera 5 and 6—Venus probes
**USA:** Mariner 6 and 7—Mars flybys
**USA:** Apollo 10—lunar orbit and return
**USA:** Apollo 11—lunar landing
**USSR:** Zond 7—lunar flyby and return
**USA:** Apollo 12—lunar landing

1970

**USA:** Apollo 13—aborted lunar landing
**USSR:** Venera 7—Venus lander
**USSR:** Luna 16—lunar sample return
**USSR:** Zond 8—lunar flyby and return
**USSR:** Luna 17/Lunokhod 1—lunar rover

1971

**USA:** Apollos 14 and 15—lunar landings
**USSR:** Mars 2 and 3—Mars orbiters and landers
**USA:** Mariner 9—Mars orbiter
**USSR:** Luna 19—lunar orbiter

1972

**USSR:** Luna 20—lunar sample return
**USA:** Pioneer 10—Jupiter flyby
**USSR:** Venera 8—Venus probe
**USA:** Apollos 16 and 17—lunar landings

1973

**USSR:** Luna 21/Lunokhod 2—lunar rover
**USA:** Pioneer 11—Jupiter/Saturn flyby
**USA:** Skylab
**USSR:** Mars 4 through 7—Mars flybys, orbiters, landers
**USA:** Mariner 10—Venus and Mercury flyby

1974

**USSR:** Luna 22—lunar orbiter

1975

**USSR:** Veneras 9 and 10—Venus orbiters and landers
**USA:** Vikings 1 and 2—Mars orbiters and landers

1977

**USA:** Voyagers 1 and 2—Grand Tour

1978

**USA:** Pioneer Venuses 1 and 2—Venus orbiter and probes
**USA:** ISEE-3/ICE—Comet flybys

**USSR:** Veneras 11 and 12—Venus orbiters and landers

1981

**USSR:** Veneras 13 and 14—Venus orbiters and landers

1983

**USSR:** Veneras 15 and 16—Venus orbiters

1984

**USSR:** Vegas 1 and 2—Venus landers and Comet Halley flyby

1989

**USA:** Magellan—Venus orbiter

**USA:** Galileo—Jupiter orbiter and probe

**USA:** Hubble space telescope

**USA/ESA:** Ulysses—Jupiter flyby and solar orbiter

1994

**USA:** Clementine—lunar orbiter, attempted asteroid flyby

1996

**USA:** NEAR—asteroid Eros orbiter

**USA:** Mars Global Surveyor—Mars Orbiter

**USA:** Mars Pathfinder—Mars lander and rover

1997

**USA:** Cassini—Saturn orbiter

1998

**USA:** Lunar Prospector—lunar orbiter

**USA:** Deep Space 1—asteroid and comet flyby

1999

**USA:** Stardust—Comet Coma sample return

—

Source: NASA Planetary Exploration Timeline

## STATUS OF VOYAGERS (AUGUST 2009)

| | VOYAGER 1 | VOYAGER 2 |
|---|---|---|
| **Distance from Earth:** | 16,387,000,000 km | 13,252,000,000 km |
| **Total distance traveled:** | 20,992,000,000 km | 20,004,000,000 km |
| **Velocity (to Earth):** | 131,389 km/hr | 107,374 km/hr |
| **Round-trip light time:** | 30 hr, 21 min, 14 sec | 24 hr, 24 min, 18 sec |
| **Propellant remaining:** | 26.63 kg | 28.28 kg |
| **Electrical output:** | 277.4 watts | 278.7 watts |

Source: JPL Voyager Weekly Operations Report http://voyager.jpl.nasa.gov/mission/weekly-reports/index.htm

## THE GRAND TOUR AND ITS ENCOUNTERS

FIGURE 1 THE GRAND TOUR

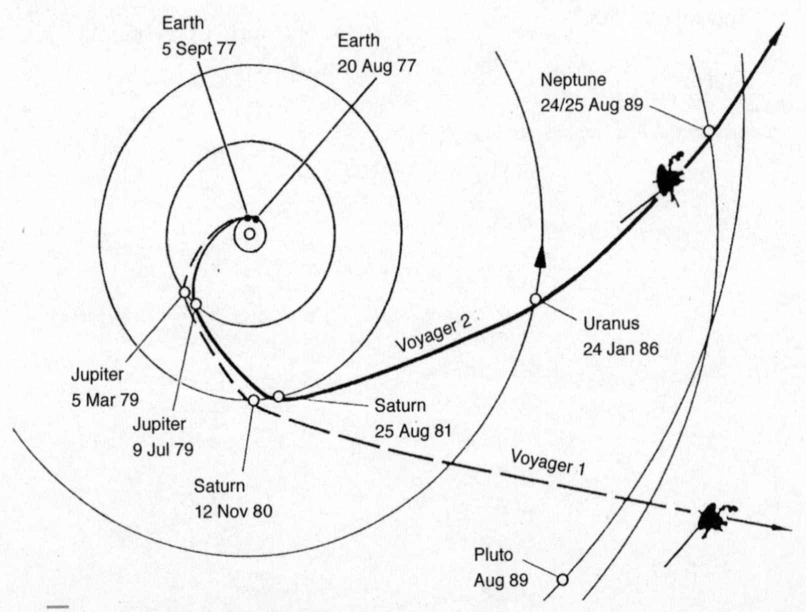


Source: *Voyager Neptune Travel Guide*

FIGURE 2 ENCOUNTER: VOYAGER 1 AT JUPITER

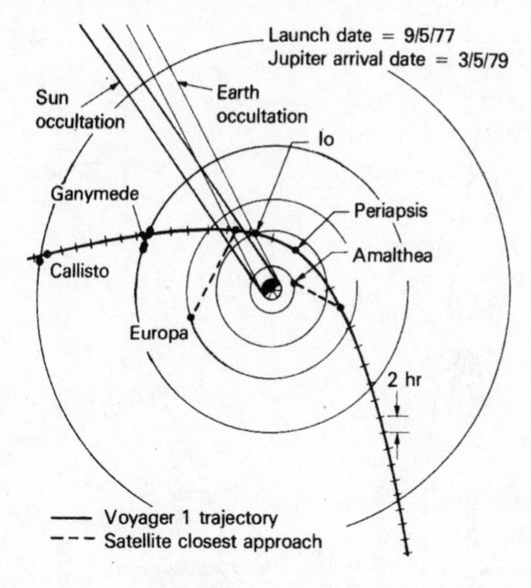


Source: NASA

FIGURE 3 ENCOUNTER: VOYAGER 2 AT JUPITER

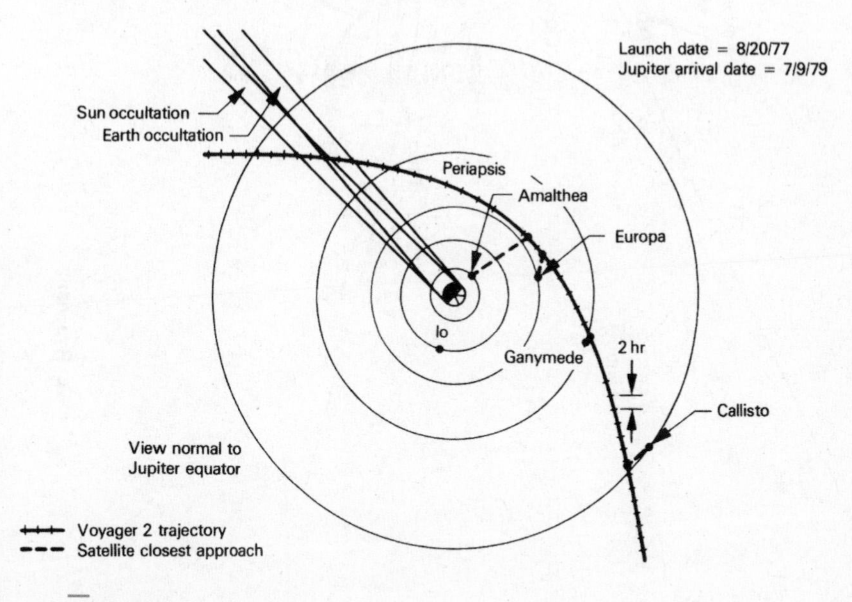


Source: NASA

FIGURE 4 ENCOUNTER: VOYAGER 1 AT SATURN

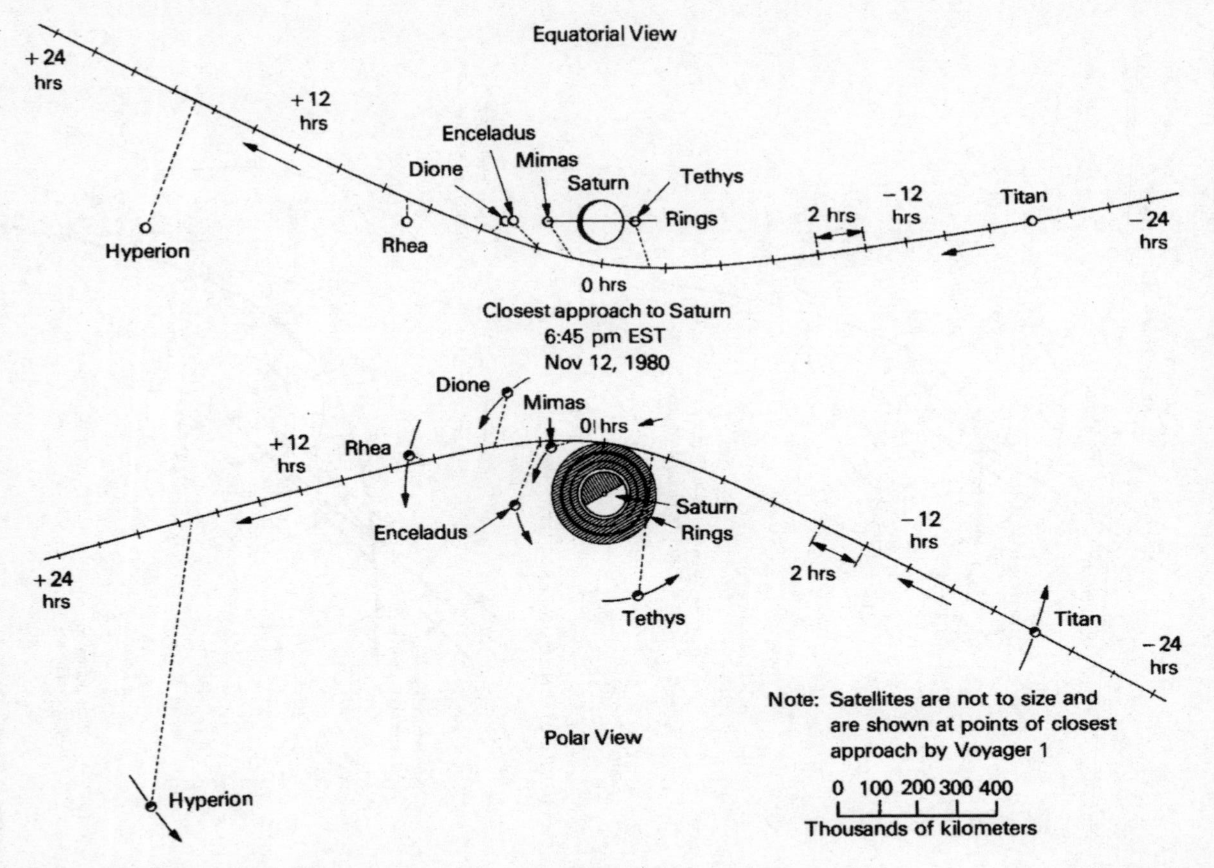


Source: NASA

FIGURE 5 ENCOUNTER: VOYAGER 2 AT SATURN

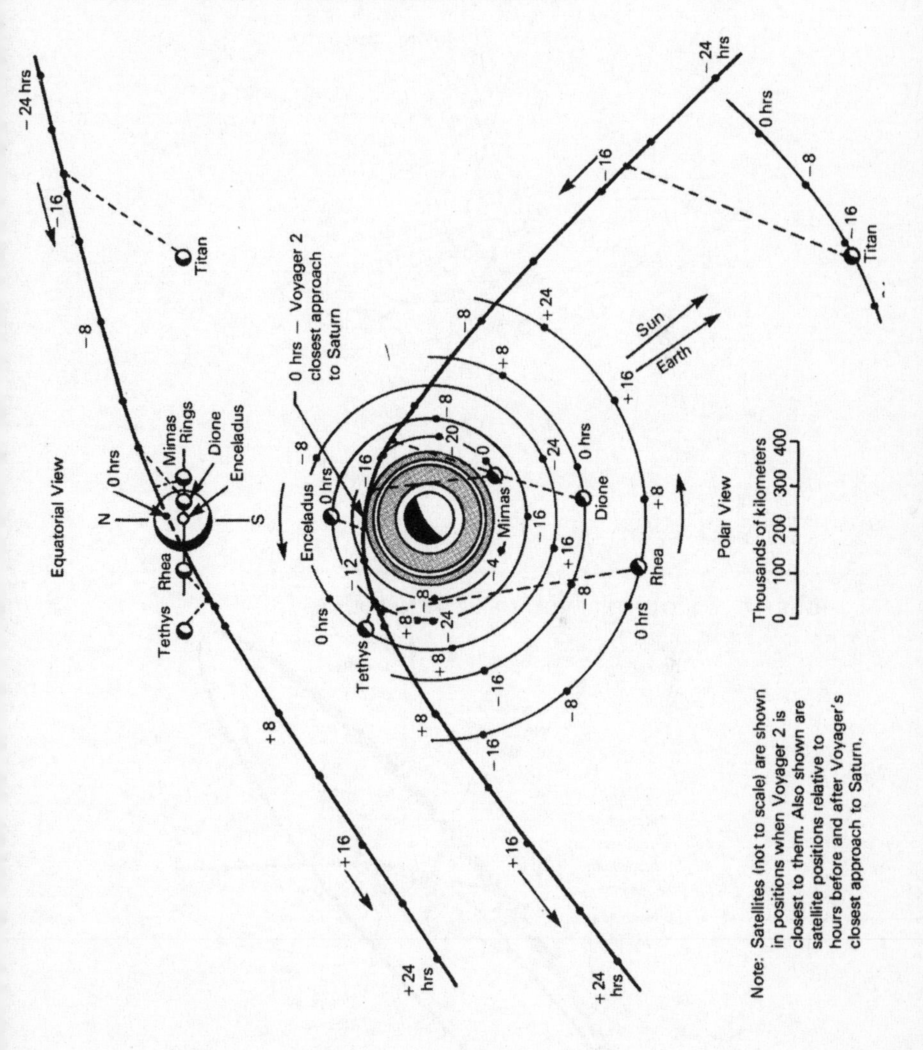


Source: NASA

FIGURE 6 ENCOUNTER: VOYAGER 2 AT URANUS

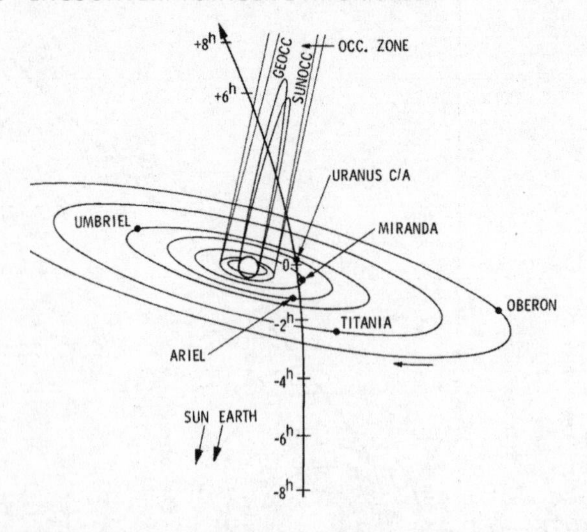


Source: NASA

FIGURE 7 ENCOUNTER: VOYAGER 2 AT NEPTUNE

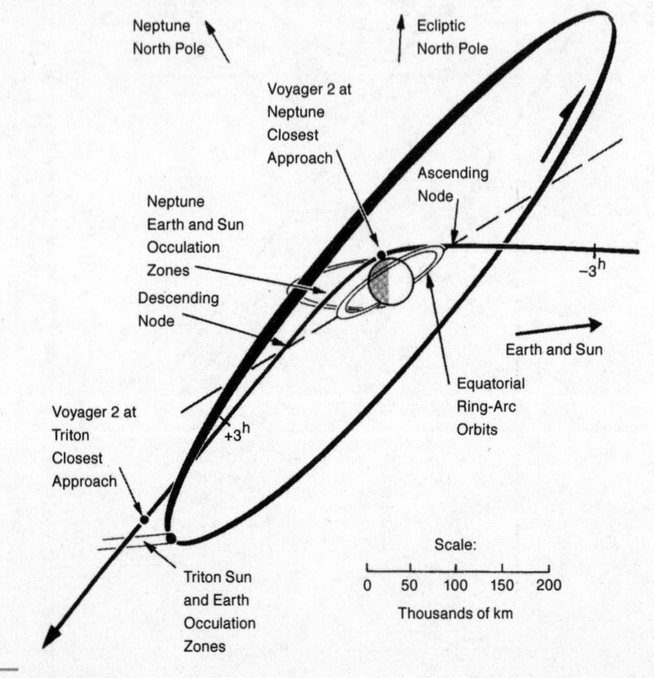


Source: NASA

FIGURE 8 VOYAGER INTERSTELLAR MISSION AND MAP OF THE FOUR FAR-TRAVELER SPACECRAFT LEAVING THE SOLAR SYSTEM

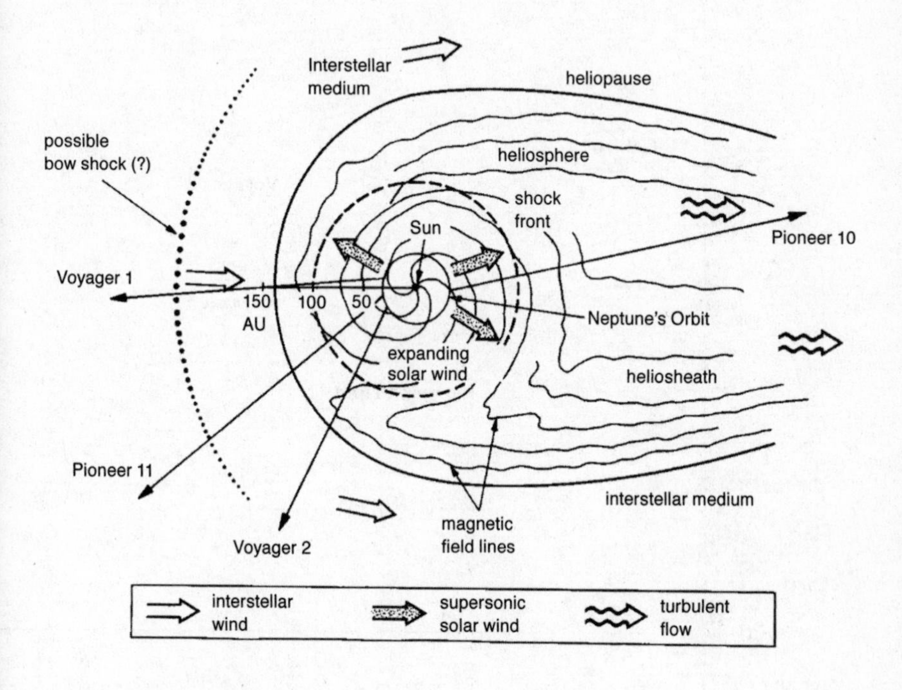


Source: *Voyager Neptune Travel Guide*

FIGURE 9 EXPLORING THE SOLAR SYSTEM: A SPACE-TIME CONTINUUM

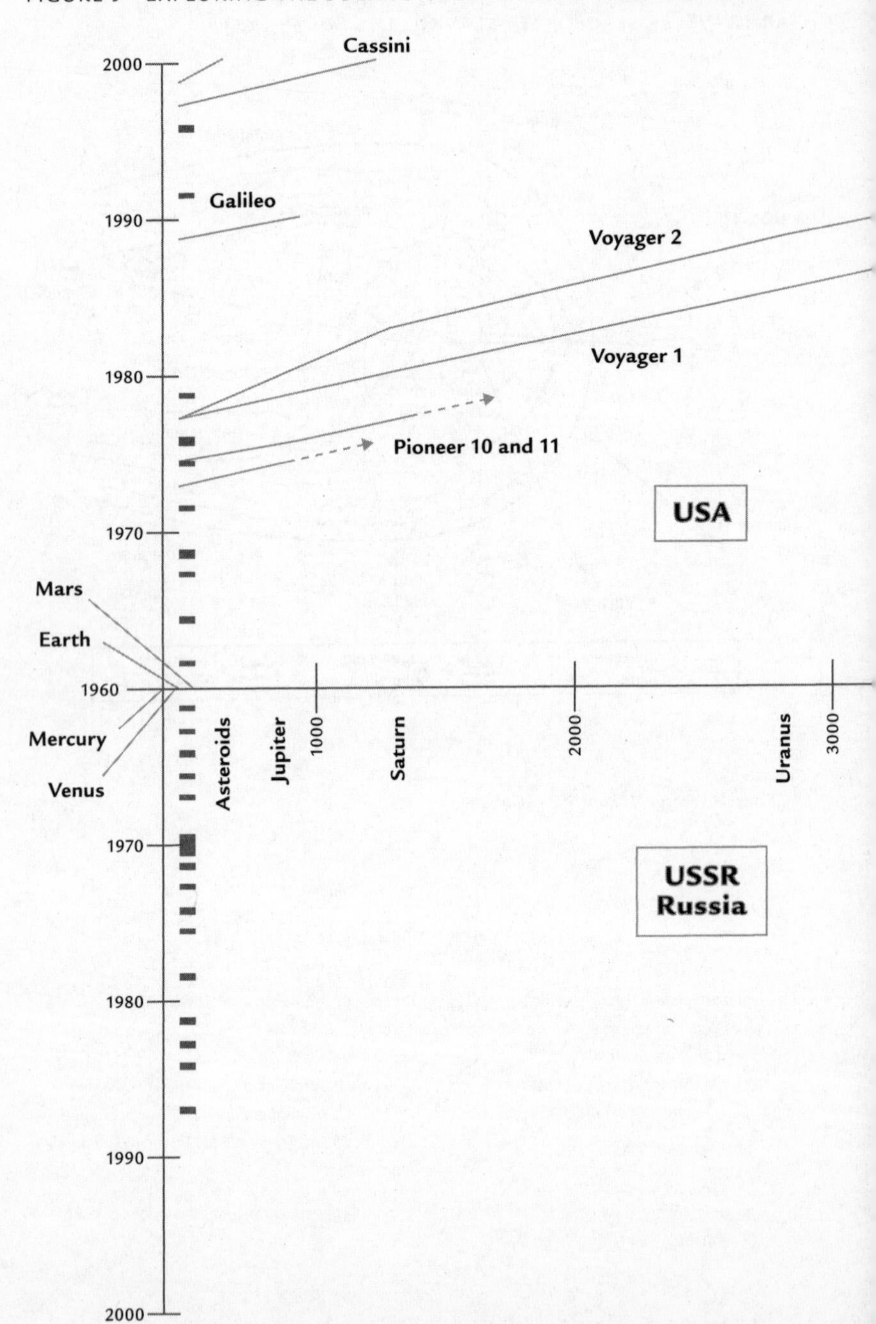

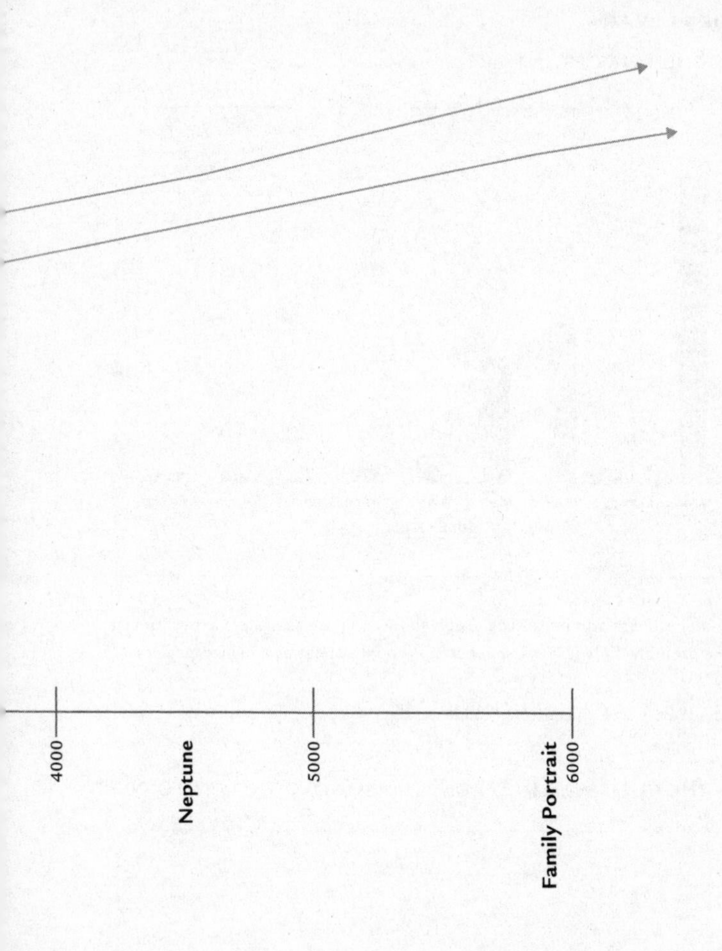


The horizontal axis locates the planets by distance from the Sun (million km). The vertical axis gives years of launches and encounters for the United States (top) and the USSR (bottom). On this scale missions to the inner planets appear as mere hachures. It's easy to see how the immense trek of the Voyagers dominated distance, and given the long hiatus after launch, the history of planetary exploration by NASA. While the Pioneers continued their outward trek as well, their missions officially ended with Jupiter and Saturn. The Voyagers had their missions redefined as they progressed.

Note both the dominating position of Voyager spatially by encountering the outer planets and its domination of American launches for a decade.

—

Data source: NASA, Planetary Exploration Timeline

## THE COLDEST WAR

FIGURE 10 COLD WAR FRONTIERS

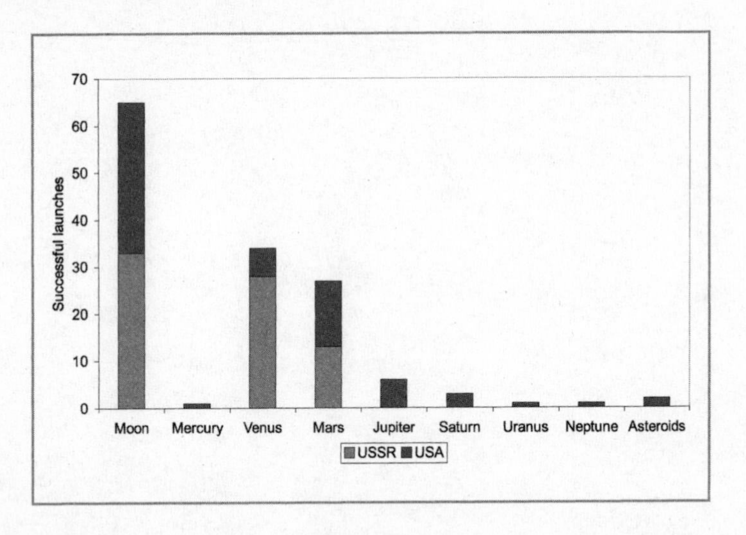


Note how the inner planets dominate. The graph shows only successful launches, or else Mars—the scene for many failed Soviet spacecraft—would appear as a higher target.

Source: NASA Planetary Exploration Timeline

FIGURE 11 THE COLD WAR IN SPACE: A CHRONOLOGY (1957 TO 2000).

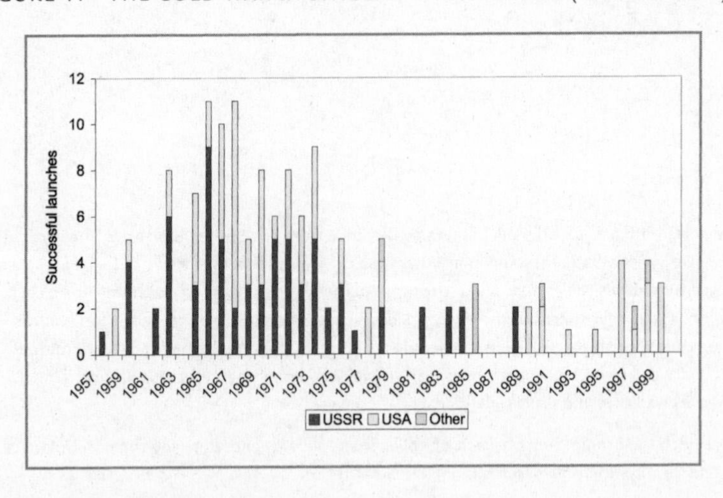


Source: NASA Planetary Exploration Timeline

# Notes

## MISSION STATEMENT: VOYAGER OF DISCOVERY

1. See, for example, Asif A. Siddiqi, *Deep Space Chronicle: A Chronology of Deep Space and Planetary Probes 1958–2000*. Monographs in Aerospace History, no. 24. NASA SP-2002-4524 (Washington, D.C.: NASA, 2002), and Brian Harvey, *Russia in Space: The Failed Frontier?* (Chichester, UK: Springer Praxis, 2001).

## PART 1: THE BEGINNING OF BEYOND: JOURNEY OF AN IDEA

### CHAPTER 1. ESCAPE VELOCITY

1. Bruce Murray, *Journey into Space: The First Three Decades of Space Exploration* (New York: W. W. Norton, 1989), p. 15.
2. On shifting rationales, see Jacob Darwin Hamblin, *Oceanographers and the Cold War: Disciples of Marine Science* (Seattle: University of Washington Press, 2005); the introduction is a useful summary of motives, so similar to those for space.
3. An excellent study of this convergence is available in Walter A. McDougall, *The Heavens and the Earth: A Political History of the Space Age* (New York: Basic Books, 1985).

4. Expression "exploration's nation" taken from William H. Goetzmann, "Exploration's Nation," pp. 11–36, in Daniel J. Boorstin, ed., *American Civilization* (New York: McGraw-Hill, 1972).
5. Letter, W. H. Pickering to Dr. Thomas O. Paine, July 1, 1969, JPL Archives 214, no. 66.
6. Ibid.
7. Ibid. A thorough survey of the uses of exploration as justification for the space program is available in Roger D. Launius, "The Historical Dimension of Space Exploration: Reflections and Possibilities," *Space Policy* 16 (2000): 23–38.
8. Wernher von Braun, Frederick I. Ordway III, and Dave Dooling, *Space Travel: A History*, 4th ed., rev. (New York: Harper and Row, 1975), p. 281.
9. Ibid.

## CHAPTER 2. GRAND TOUR

10. There are many accounts of NASA's origins; see Roger E. Bilstein, *Orders of Magnitude: A History of the NACA and NASA, 1915–1990*. NASA SP-4406 (Washington, D.C.: NASA, 1989), and Homer E. Newell, *Beyond the Atmosphere: Early Years of Space Science*, NASA History Series, NASA SP-4211 (Washington, D.C.: NASA, 1980); exchanges between George Kistiakowsky (presidential science advisor) and Lloyd Berkner, p. 124. For an excellent political history of the space age, see McDougall, *The Heavens and the Earth*. For the perspective of JPL, see Clayton R. Koppes, *JPL and the American Space Program: A History of the Jet Propulsion Laboratory* (New Haven, Conn.: Yale University Press, 1982).
11. Quote from Rocket and Satellite Research Panel in Appendix D, Newell, *Beyond the Atmosphere*, p. 427.
12. Craig B. Watt, "The Road to the Deep Space Network," *IEEE Spectrum* (April 1993): 50. JPL quotes from Henry C. Dethloff and Ronald A. Schorn, *Voyager's Grand Tour: To the Outer Planets and Beyond* (Washington, DC: Smithsonian Books, 2003), pp. 10–12, and Koppes, *JPL*, pp. 90–95.
13. Dethloff and Schorn, *Voyager's Grand Tour*, pp. 14–16; Koppes, *JPL*, pp. 102–5.
14. Lack of interest explained in Oran W. Nicks, *Far Travelers: The Exploring Machines*, NASA SP-480 (Washington, D.C.: NASA, 1985), p. 14.
15. On S-1, Dethloff and Schorn, *Voyager's Grand Tour*, p. 19; Nicks, *Far Travelers*, p. 101.
16. Nicks, *Far Travelers*, p. 17.

17. Dethloff and Schorn, *Voyager's Grand Tour*, pp. 21–22. G. W. Morgenthaler and R. G. Morra, eds., "Unmanned Exploration of the Solar System," *Advances in Astronomical Sciences* 19 (1965); Maxwell W. Hunter II, "Unmanned Scientific Exploration Throughout the Solar System," *Space Science Reviews* 6 (1975): 601–54.
18. For a concise account of Voyager Mars, see Nicks, *Far Travelers*, pp. 170–74.
19. An excellent digest of events is available in J. K. Davies, "A Brief History of the Voyager Project: The End of the Beginning," *Spaceflight* 23, no. 5 (March 1981): 35–41, although the article's value is compromised by sloppy citations.
20. See the 1963 summary, M. A. Minovitch, "The Determination and Characteristics of Ballistic Interplanetary Trajectories under the Influence of Multiple Planetary Attractions," Technical Report No. 32-464, JPL, Oct. 31, 1963. For claims that others were also closely tracking the concepts, see M. W. Hunter II, "Unmanned Scientific Exploration Throughout the Solar System," *Space Science Reviews* 6 (1967): 601–54.
21. Flandro has given several accounts of how he arrived at the Grand Tour scheme, and of the connection (or not) to Minovitch: Gary Flandro, in David W. Swift, *Voyager Tales: Personal Views of the Grand Tour* (Reston, Va.: American Institute of Aeronautics and Astronautics, 1997), pp. 62–70; Dr. Gary A. Flandro, "Discovery of the Grand Tour Voyager Mission Profile," pp. 95–98, in Mark Littman, *Planets Beyond: Discovering the Outer Solar System* (Mineola, N.Y.: Dover Publications, 2004; rev. ed.). See also Tony Reichhardt, "Gravity's Overdrive," *Air and Space/Smithsonian* 8, no. 6 (1994): 77.
22. See Flandro in Littman, *Planets Beyond*, pp. 95–96.
23. Ibid., p. 97. G. A. Flandro, "Utilization of Energy Derived from the Gravitational Field of Jupiter for Reducing Flight Time to the Outer Solar System," *JPL NASA SPS* IV (1965): 37–35. Published version: G. A. Flandro, "Fast Reconnaissance Missions to the Outer Solar System Utilizing Energy Derived from the Gravitational Field of Jupiter," *Astronautica Acta* 12, no. 4 (1966): 329–37. See also Minovitch's update, "Utilizing large planetary perturbations for the design of deep space, solar probe, and out-of-ecliptic trajectories," Technical Report No. 32-849, JPL (December 15, 1965).
24. Flandro, in Littman, *Planets Beyond*, p. 97. Homer Joe Stewart, "New Possibilities for Solar System Exploration," *Astronautics and Aeronautics*

4 (Dec. 1966): 26–31; James Long, "To the Outer Planets," *Astronautics and Aeronautics* 7 (June 1969): 32–48. Murray quoted in Dethloff and Schorn, *Voyager's Grand Tour*, pp. 44–45.

25. Long, "To the Outer Planets," 7, quote on p. 47; William Pickering, "Grand Tour," *American Scientist* 58, no. 2 (March–April 1970): 148–55; E. M. Repic, "Outer-Planet Exploration Missions" (Part B of NAS8-24975), Final Report, Vol. 1, Summary. Space Division, North American Rockwell, S.D., 70-32-1 (January 1970). Note, however, that the NAS Space Science Board, in its report "Planetary Exploration 1968–1975," did not boost missions to the outer planets, clinging to Venus and Mars, with a flyby of Mercury (NAS-NRC, July 1968). Note: See JPL Archives, 122; item no. 2 is the basis for the popular article distilled by Long (1969). Some critical memos trace the very interesting thinking (and the complexity of choices among alternatives) that led to the Grand Tour. See, in particular, Fred H. Felberg to F. E. Goddard et al., Interoffice Memorandum OPP 68-102, May 23, 1968, Subject: Review of AST Candidates, JPL 230, no. 28; and Letter, W. H. Pickering to D. P. Hearth, March 2, 1970, JPL 214, no. 74.
26. See Long, "To the Outer Planets," pp. 33–34. An excellent, balanced survey of the evolution of the Grand Tour is available in Andrew J. Butica, "Voyager: The Grand Tour of Big Science," chapter 11, in Pamela E. Mack, ed., *From Engineering Science to Big Science*. NASA SP-4219 (Washington, D.C.: 1998). See also David Rubashkin, "Who Killed Grand Tour?" ms., NASA History Office, Historical Reference Collection.
27. Several sources for funding: Dethloff and Schorn, *Voyager's Grand Tour*; J. K. Davies, "A Brief History of the Voyager Project," *Spaceflight* 23, no. 3 (March 1981): 35–41; Bruce Murray, *Journey into Space*, pp. 138–42.
28. Schurmeier quoted in Dethloff and Schorn, *Voyager's Grand Tour*," p. 45.
29. A good account available in Butica, "Voyager: The Grand Tour of Big Science." Quotes from Ellis D. Miner, *Uranus: The Planet, Rings, and Satellites* (New York: Ellis Horwood, 1990), pp. 101–2. See "The Grand Tour," draft manuscript, 11-1-69, JPL Archives 214, no. 70, for a simplified version of the early conception, and Advanced Planetary Mission Study Team, "Grand Tour Definition Study, Vol. 1. Mission Analysis, March 15, 1971," JPL Archives 122, no. 1, for the fuller, final draft.
30. National Academy of Sciences, "The Outer Solar System: A Program for Exploration," report of a study by the Space Science Board, June 1969 (Washington, D.C.: NAS, 1969).

31. There are several published accounts of these events, but Butica, "Voyager: The Grand Tour of Big Science," is a very handy synopsis.
32. I follow Dethloff and Schorn, *Voyager's Grand Tour*, pp. 52–57; quote on budget, p. 56.
33. Space Science Board, "The Outer Solar System: A Program for Exploration," p. 5.
34. Fletcher quotes from Fred H. Felberg to Memorandum to Record, Interoffice Memorandum OPP 72-45, February 24, 1972, JPL Archives 230, no. 51.
35. Quotes from preface, which contains a good summary of preceding studies, NAS-NRC Space Science Board, "Priorities for Space Research, 1971–1980: Report of a Study on Space Science and Earth Observations Priorities" (Washington, D.C.: National Academy of Sciences, 1971), pp. 17–18. "Satellite imaging" quote from NAS-NRC Space Science Board, "Outer Planets Exploration 1972–1985" (Washington, D.C.: National Academy of Sciences, 1971).
36. As always, the details are more complicated than the inherited explanations imply. Fletcher spoke to JPL in February 1972 and offered his own thoughts. The Grand Tour, as originally conceived, did not have support from either OMB or the White House, but a stripped-down version to Jupiter and Saturn for 1977 did, and would likely be considered in the next budgetary round. He also listed two "lessons" from his time as NASA administrator: that any specific decision makes enemies but few friends, and that there was, in reality, little occasion to trade dollars from one NASA program to another; each had to be justified on its own. Source: Fred H. Felberg to Memorandum to Record, Interoffice Memorandum OPP 72-45, February 24, 1972, JPL Archives 230, no. 51.
37. The Mariner study commenced immediately after the SSB report; see F. H. Felberg, Interoffice Memorandum OPP 71-117, "Mariner Outer Planets Missions Study," JPL Archives 230, no. 47. For a good chronology of the revival, see Dethloff and Schorn, *Voyager's Grand Tour*, pp. 60–64.
38. Pickering quote: F. H. Felberg to R. J. Parks, Interoffice Memorandum OPP 71-124, September 16, 1971, JPL Archives 230, no. 47.
39. See Harris M. Schurmeier to MJS Review Board, November 6, 1974, JPL Archives, Flight Collections, Folio 54.
40. Dethloff and Schorn, *Voyager's Grand Tour*, p. 88, for Final Spacecraft; Casani quote, p. 106.
41. On the reconstituted design, see Roger D. Bourke et al., "Mariner Jupiter/Saturn 1977: The Mission Frame," *Astronautics and Aeronautics*

(Nov. 1972), pp. 42–49. For a detailed internal review of final trajectory trade-offs, see Letter, J. R. Casani to R. A. Mills, "MJS77 Flight Trajectory Selection and Rationale," June 22, 1976, JPL Archives 36, no. 60.

42. Dethloff and Schorn, *Voyager's Grand Tour,* pp. 106–7. An excellent summary of the Voyager preparations as viewed by an insider is Miner, *Uranus,* chapter 6; quotes from p. 105.
43. Dethloff and Schorn, *Voyager's Grand Tour,* p. 105.
44. The story of naming preserved in oral interviews, the results of two published as: Charles Kohlhase, in Swift, *Voyager Tales,* p. 85, and Dethloff and Schorn, *Voyager's Grand Tour,* pp. 105–7. See also Nicks, *Far Travelers,* pp. 170–73, which is better on the antecedent history. On final choice and NASA approval, see J. R. Casani, Interoffice Memo, MJS77-JRC-77-63, Subject: Project Name, March 4, 1977, NASA History Office, Historical File 005566.
45. Malyn Newitt, *A History of Portuguese Overseas Expansion, 1400–1668* (New York: Routledge, 2005), p. 58.

CHAPTER 3. GREAT AGES OF DISCOVERY

46. This analysis of three ages of discovery is an adaptation of "Seeking Newer Worlds: An Historical Context for Space Exploration," pp. 8–35, in Steven J. Dick and Roger D. Launius, eds., *Critical Issues in the History of Spaceflight,* NASA SP-2006-4702 (Washington, D.C.: NASA, 2006). As the common title suggests, this entire book before you is an attempt to extend and elaborate the arguments of that essay, and by focusing on the Voyager mission to find a more focused mode of expression for the ideas within it. I note here, as in the afterword, that the idea of a Second Age of Discovery belongs to William H. Goetzmann, who developed it most fully in *New Lands, New Men: America and the Second Great Age of Discovery* (New York: Viking, 1986).
47. I have relied on that doyen of the founding age of discovery, J. H. Parry. Among his many works are three that serve especially as syntheses: *The Establishment of the European Hegemony, 1415–1715,* 3rd ed., rev. (New York: Harper and Row, 1966); *The Discovery of the Sea* (Berkeley: University of California Press, 1981); and *The Age of Reconnaissance: Discovery, Exploration and Settlement, 1450–1650* (New York: Praeger, 1969).
48. Parry, *The Discovery of the Sea*, op cit.
49. Harry Wolf, *The Transits of Venus: A Study of Eighteenth-Century Science* (Princeton, N.J.: Princeton University Press, 1959), p. 83.

50. See William H. Goetzmann, "Exploration's Nation: The Role of Discovery in American History," in Daniel J. Boorstin, ed., *American Civilization: A Portrait from the Twentieth Century* (New York: McGraw-Hill, 1972).
51. As a useful way to summarize this explosion, see the flawed but indispensable, J. N. L. Baker, *A History of Geographical Discovery and Exploration* (New York: Cooper Square Publishers, 1967).
52. J. Tuzo Wilson, *I.G.Y: The Year of the New Moons* (New York: Alfred Knopf, 1961), p. 324. Chapman characterized IGY's mission as learning "more about the fluid envelope of our planet–the atmosphere and oceans–over all the earth and at all heights and depths." But this was academic abstraction. Wilson came closer to the mark when he observed that it was the yet-unvisited places that mattered, that IGY proposed a planetary inventory as conceived by geophysicists. The founding geophysicists fretted most over the outer boundary of Earth, which is where they most wanted IGY to go. Chapman, in *IGY Annals* 1 (Jan. 28, 1957): 3.
53. Wilson, *I.G.Y*, pp. 275, 320, 324, 219–25 passim.

## CHAPTER 4. VOYAGER

54. The best summary is Raymond L. Heacock, "The Voyager Spacecraft," Institution of Mechanical Engineers, *Proceedings 1980*, 194, no. 28 (1980). Digests are available in most books on Voyager; for example, Ellis D. Miner, *Uranus: The Planet, Rings and Satellites* (New York: Ellis Horwood, 1990), pp. 119–32. See also Nicks, *Far Travelers*, pp. 15–17.
55. Mark Wolverton, *The Depths of Space: The Story of the Pioneer Planetary Probes* (Washington, D.C.: Joseph Henry Press, 2004), pp. 58–61.
56. S. E. Morison says "c. 55 feet" in *The Great Explorers: The European Discovery of America* (New York: Oxford University Press, 1978), p. 385. A more detailed study by Eugene Lyon recommends sixty-seven feet; "*Niña*: Ship of Discovery," in Jerald T. Milanich and Susan Milbrath, eds., *First Encounters: Spanish Explorations in the Caribbean and the United States, 1492–1570* (Gainesville: University of Florida Press, 1989).
57. Gilbert data: Arvid M. Johnson and David D. Pollard, transcribers, "Part of the Field Notes of Grove Karl Gilbert for the period 20 June 1875 to 24 November 1876 taken during his study of the Henry Mountains and Areas to the West," School of Earth Sciences, Stanford University (1977), pp. 87–88, for list of items carried. On HMS *Challenger*, see the published deck plans (available online at www.19thcenturyscience.org/HMSC/HMSC-INDEX/Deck-Plans.html; accessed October 13, 2008).

58. Nicks, *Far Travelers*, p. 71. On dimensions, see Heacock, "Voyager Spacecraft."
59. Nicks, *Far Travelers*, pp. 80–82.
60. A good digest of robotic needs is found in Roger D. Launius and Howard E. McCurdy, *Robots in Space: Technology, Evolution, and Interplanetary Travel* (Baltimore, Md.: Johns Hopkins University Press, 2008), pp. 140–45.
61. Casani, in Swift, *Voyager Tales*, pp. 118–19. Statistics: Voyager Mission Planning Office Staff, *The Voyager Uranus Travel Guide* JPL D-2580 (August 15, 1985), p. 107.
62. Casani, in Swift, *Voyager Tales*, p. 119.
63. Raymond Heacock, in Swift, *Voyager Tales*, p. 152; Nicks, *Far Travelers*, p. 80.
64. Raymond Heacock, in Swift, *Voyager Tales*, p. 152.
65. Quote from Morison, *The Great Explorers*, p. 563.
66. Quote from William H. Goetzmann, *Exploration and Empire: The Explorer and Scientist in the Winning of the American West* (New York: Alfred A. Knopf, 1967), p. 58.
67. Casani, in Swift, *Voyager Tales*, p. 120.
68. Good summary of early history in James E. Tomayko, *Computers in Spaceflight: The NASA Experience*. NASA Contractor Report 182505 (March 1988). But an excellent distillation is available in Raymond L. Heacock, "The Voyager Spacecraft," *The Institution of Mechanical Engineers, Proceedings 1980* 194, no. 28 (1980): 221–22.
69. Tomayko, *Computers in Spaceflight*, chapter 6; Miner, *Uranus*, p. 118.
70. Quotes from Ben Evans, with David M. Harland, *NASA's Voyager Missions: Exploring the Outer Solar System and Beyond* (Chichester, UK: Springer Praxis, 2004), p. 67.
71. Nicks, *Far Travelers*, p. 253.
72. Antonio Pigafetta, *Magellan's Voyage: A Narrative Account of the First Circumnavigation*, trans. and ed., R. A. Skelton (New York: Dover Publications, 1994), p. 57.
73. See J. A. Parry, *The Discovery of the Sea* (Berkeley: University of California Press, 1981).

## CHAPTER 5. LAUNCH

74. Murray quoted in Swift, *Voyager Tales*, p. 215.
75. Account of ship technology from the inestimable J. H. Parry, *The Age of Reconnaissance*, chapter 3, and *The Discovery of the Sea*, chapter 1.

76. Rocket history from Wernher von Braun and Frederick I. Ordway III, *Rocket's Red Glare* (Garden City, N.Y.: Anchor Press, 1976), and Wernher von Braun, Frederick I. Ordway III, and Dave Dooling, *Space Travel: A History* (New York: Harper and Row, 1985; 4th ed).
77. On pinnaces and longboats, see Morison, *The Great Explorers*, p. 16.
78. Cape Canaveral information from NASA, "Kennedy Space Center Story," www.nasa.gov/centers/kennedy/about/history/story/ch1_prt.htm; accessed August 30, 2007.
79. On Franklin's dove, see Owen Beattie and John Geiger, *Frozen in Time: The Fate of the Franklin Expedition* (Toronto: Greystone Books, 1987), p. 16. On the storms around Voyager, see Bruce Murray, *Journey into Space: The First Thirty Years in Space* (New York: W. W. Norton, 1989), p. 143.

## PART 2: BEYOND THE SUNSET: JOURNEY ACROSS THE SOLAR SYSTEM

### CHAPTER 6. NEW MOON

1. On Voyager malfunctions, I rely on several accounts: J. E. Davies, "A Brief History of the Voyager Project: Part 2," *Spaceflight* 23, no. 5 (May 1981): 71–72; Bruce Murray, *Journey into Space: The First Thirty Years of Space Exploration* (New York: W. W. Norton, 1989), pp. 143–47. A good summary exists in Miner, *Uranus,* pp. 108–12. For Voyager 2, see Swift, *Voyager Tales*, especially John Casani, pp. 119–20; "Anxiety attack" and "Superautonomous" in Murray, *Journey into Space*, pp. 146–47.
2. Miner, *Uranus,* pp. 108–12. A good journalistic account of Voyager woes, beginning with the faulty Centaur rocket, is available in Joel Davis, *Flyby: The Interplanetary Odyssey of Voyager 2* (New York: Atheneum, 1987), pp. 42–46. For detailed accounts of malfunctions, see Davies, "Brief History," and Heacock, "The Voyager Spacecraft."
3. Malyn Newitt, *A History of Portuguese Overseas Expansion, 1400–1668* (New York: Routledge, 2005), pp. 15–16; Donald Rayfield, *The Dream of Lhasa: The Life of Nikolay Przhevalsky: Explorer of Central Asia* (Columbus: Ohio University Press, 1976), p. 69.
4. Best summary is in Goetzmann, *New Lands, New Men*, chapter 6; number of expeditions, p. 266.
5. Quote from Fergus Fleming, *Barrow's Boys* (New York: Grove Press, 1998), p. 1.

6. See Lisle A. Rose, *Assault Against Eternity: Richard E. Byrd and the Exploration of Antarctica, 1946–47* (Annapolis, Md.: Naval Institute Press, 1980).
7. An excellent institutional history that addresses this issue directly is Peter J. Westwick, *Into the Black: JPL and the American Space Program 1976–2004* (New Haven, Conn.: Yale University Press, 2007); a distilled synopsis is available on pp. 309–13.
8. Nicks, *Far Travelers*, pp. 154–55.

CHAPTER 7. CRUISE

9. Sieur de Champlain, "Treatise on Seamanship and the Duty of a Good Seaman," Appendix II, in Samuel Eliot Morison, *Samuel de Champlain: Father of New France* (Boston: Atlantic Monthly Press, 1972).
10. Ibid.
11. My analysis follows the criticisms summarized in NASA's review of JPL oversight of Voyager, reproduced in Dethloff and Schorn, *Voyager's Grand Tour*, pp. 141–42. The fullest account of the April 1978 communications breakdown is in Miner and Wessen, *Neptune*, pp. 76–77.
12. Quote from Clayton R. Koppes, *JPL and the American Space Program: A History of the Jet Propulsion Laboratory* (New Haven, Conn.: Yale University Press, 1982), p. ix. My analysis follows closely this excellent study.
13. Again, I follow Koppes, *JPL*, closely, pp. ix–x.
14. Craig B. Watt, "The Road to the Deep Space Network," *IEEE Spectrum* (April 1993): 50. Nicks, *Far Travelers*, p. 17. Other JPL quotes from Koppes, *JPL*, pp. 90–95.
15. Pickering quoted in Koppes, *JPL*, p. 112.
16. This is a primary theme of two outstanding institutional histories: Koppes, *JPL*, op. cit., and Peter J. Westwick, *Into the Black: JPL and the American Space Program 1976–2004* (New Haven, Conn.: Yale University Press, 2007). It is present in all of the published accounts of Voyager, although the liveliest and most informative is Murray, *Journey into Space*.
17. There are several accounts of the Great Surveys. I follow Goetzmann, *Exploration and Empire*, chapters 12–16. The other prime contenders are Wallace Stegner, *Beyond the Hundredth Meridian* (Boston: Houghton Mifflin, 1953), which hedges into Powell hagiography, and Richard A. Bartlett, *Great Surveys of the American West* (Norman: University of Oklahoma Press, 1962).
18. Quote from Goetzmann, *Exploration and Empire*, p. 589.

CHAPTER 8. MISSING MARS

19. See Carl O. Sauer, *The Early Spanish Main* (Berkeley: University of California Press, 1966), pp. 46, 138; and Morison, *The Great Explorers*, pp. 390–91.
20. Camões, *The Lusíads*, 9:44 and 8:98. Díaz in Bernal Díaz, *The Conquest of New Spain*, trans. by J. M. Cohen (New York: Penguin Books, 1963), p. 274.
21. Richard Hakluyt, *Voyages and Discoveries: The Principal Navigations, Voyages, Traffiques, and Discoveries of the English Nation*, ed. and abridged by Jack Beeching (New York: Penguin, 1973), p. 60.
22. Columbus quote from Cecil Jane, trans. and ed., *The Four Voyages of Columbus: A History in Eight Documents* (New York: Dover Publications, 1988), p. 12. The Spanish phrasing reads: "esta es para desear, e, v [ista], es para nunca dexar."
23. Felipe Fernández-Armesto, *The Pathfinders: A Global History of Exploration* (New York: W. W. Norton, 2006), p. 131; for a better take on Henry's career overall, see Peter Russell, *Prince Henry "The Navigator": A Life* (New Haven, Conn.: Yale University Press, 2000). Morison, *The Great Explorers*, pp. 390–91.
24. Columbus, in Morison, *The Great Explorers*, pp. 390–91, and quotes from p. xvi. Fernández-Armesto, *The Pathfinders*, p. 145–47; Pigafetta, *Magellan's Voyage*, p. 37.
25. Fernández-Armesto, *The Pathfinders*, p. 145.
26. Orders quoted in Newby, *The Rand McNally World Atlas of Exploration*, p. 134.
27. Best summary of purposes and practices of the American exploration throughout the century is Goeztmann, *Exploration and Empire*; Ashley quote, p. 105; King Survey quote from p. 437.
28. Apsley Cherry-Garrard, *The Worst Journey in the World: Antarctic 1910–1913* (New York: Penguin Books, 1970, reprint), pp. 642–43.
29. Ibid., p. 643.
30. John Parker, *Books to Build an Empire* (Amsterdam: N. Israel, 1965), p. 39.
31. Ibid., p. 102.
32. Hakluyt, *Voyages and Discoveries*, p. 31.
33. Ibid., p. 38.
34. Morison, *The Great Explorers*, p. 127. See Richard H. Grove, *Green Imperialism: Colonial Expansion, Tropical Island Edens, and the Origins of Environmentalism, 1600–1860* (Cambridge, UK: Cambridge University Press, 1995), p. 249.

35. This paragraph is a close paraphrase and partial quote from Goetzmann, *New Lands, New Men*, pp. 229–30.
36. For a distillation of his thoughts, see "Which Way Is Up?" pp. 124–33, in Arthur C. Clarke, *The Challenge of the Spaceship* (New York: Ballantine, 1961).
37. Clarke, *Exploration of Space* (1951), pp. 183–85.
38. Ibid., pp. 186–87, 194–95.
39. Kurt Vonnegut Jr., *The Sirens of Titan* (New York: Dell, 1959), p. 30.
40. *Lusíads,* 1:27, 2:45.
41. Ibid., 4:94–95.
42. The Old Man's soliloquy runs from 4:94 to 4:104.
43. Daniel Defoe, *Robinson Crusoe* (New York: Charles Scribner's Sons, 1920), pp. 2, 4, 368.
44. Swift, *Gulliver's Travels*, pp. 359–60.
45. Bertrand H. Bronson, ed., *Samuel Johnson: Rasselas, Poems, Selected Prose* (New York: Holt, Rinehart and Winston, Inc., 1958), p. 511.
46. "Of these wishes that they had formed they well knew that none could be obtained": Bronson, ed., *Samuel Johnson*, p. 612.
47. James Boswell, *The Life of Samuel Johnson* (Garden City, N.Y.: Doubleday and Co., Inc., 1946), p. 355.
48. Amitai Etzioni, *The Moon-Doggle* (Garden City, N.Y.: Doubleday and Co., Inc., 1964), pp. ix, 111.
49. The statistic comes from an outstanding general survey of motives in Roger Launius, "Compelling Rationales for Space Flight: History and the Search for Relevance," pp. 37–71, in Steven J. Dick and Roger D. Launius, eds., *Critical Issues in the History of Spaceflight*. NASA SP-2006-4702 (Washington, D.C.: 2006). Fernández-Armesto, *Pathfinders*, p. 399. Two other useful articles on popular support (and funding) are Roger D. Launius, "Public Opinion Polls and Perceptions of U.S. Human Spaceflight," *SpacePolicy* 19 (2003): 163–75, and Howard E. McCurdy, "The Cost of Space Flight," *Space Policy* 10, no. 4 (1994): 277–89.
50. The weight of literature on Mars could sink the International Space Station. A concise summary of the major events, at least for the American program, is Thor Hogan, *Mars Wars: The Rise and Fall of the Space Exploration Initiative*, NASA SP-2007-4410 (Washington, D.C.: NASA, 2007), pp. 1–21. "Algorithm" quote from Thomas O. Paine, in foreword of Wernher von Braun, *The Mars Project* (Springfield: University of Illinois Press, 1991; reprint of 1953 edition), p. vii.
51. Ray Bradbury et al., *Mars and the Mind of Man* (New York: Harper and Row, 1973).

52. The Saturn and Jupiter events were recorded and are available in poor-quality DVDs from JPL.
53. Roger D. Bourke et al., "Mariner Jupiter/Saturn 1977: The Mission Frame," *Astronautics and Aeronautics* (Nov. 1972): 42–49.

## CHAPTER 9. CRUISE

54. Miner, *Uranus*, pp. 110–11. A complete account of the episode is also available in Miner and Wessen, *Neptune*, pp. 76–77.
55. For a succinct summary of what Voyager navigation involved, see Robert Cesarone interview, in Swift, *Voyager Tales*, pp. 262–73. A good account of adjustments is available in J. K. Davies, "A Brief History of the Voyager Project," *Spaceflight* 23, no. 5 (May 1981): 72–73.
56. Major sources: Lloyd A. Brown, *The Story of Maps* (New York: Dover Publications, 1977); Parry, *The Discovery of the Sea*, and *The Age of Reconnaissance*; and for a concise distillation, Samuel Eliot Morison, *The Great Explorers*, pp. 26–32.
57. The description follows Parry, *Discovery of the Sea*, pp. 147–50.
58. Morison, *Champlain*, pp. 256, 267.
59. For the specific makeup of the NAV group and its techniques, see Miner, *Uranus*, pp. 117–18. The best general accounts of navigation in principle and as practiced for Voyager are Robert John Cesarone, pp. 261–73, and Charles E. Kohlhase, pp. 83–100, in Swift, *Voyager Tales*; and Nicks, *Far Travelers*, esp. pp. 22–24, 29–30, 41, 56.
60. I found Nicks's explanation in *Far Travelers* the most enlightening, and have followed his accounts, although most deal with earlier spacecraft and handle Voyager especially for the complicated maneuvers around Saturn.

## CHAPTER 10. ENCOUNTER: ASTEROID BELT

61. The Pioneer bibliography is appropriately long. A good summary is Mark Wolverton, *The Depths of Space: The Story of the Pioneer Planetary Probes* (Washington, D.C.: Joseph Henry Press, 2004). Two overlapping but lively surveys by some of the principal participants and the project's best-known journalist are available in Richard O. Fimmel, William Swindell, and Eric Burgess, *Pioneer Odyssey*, NASA SP-396 (NASA, 1977), and an updated version, in Fimmel, et al., *Pioneer*, op. cit.
62. Quote by John Naugle, in Wolverton, *Depths of Space*, p. 140, which also distills Pioneer's achievements on p. 4.

63. See Fimmel et al., *Pioneer*, pp. 84–85, with a more detailed explanation of the imaging system on pp. 251–52.
64. The story of the choice of routes is told in all the Pioneer accounts. The most elaborate is in Wolverton, *Depths of Space*, pp. 140–49. But see also, Robert S. Kraemer, *Beyond the Moon: A Golden Age of Planetary Exploration 1971–1978* (Washington, D.C.: Smithsonian Institution, 2000), pp. 77–78.
65. See Fimmel et al., *Pioneer*, pp. 248–50. Burgess offers a more detailed account in Eric Burgess, *Far Encounter: The Neptune System* (New York: Columbia University Press, 1991), pp. 57–58. Wolverton, *Depths of Space*, pp. 184–85.
66. Wolverton gives an eloquent eulogy in *Depths of Space*, pp. 199–209 and 224–25.
67. Kraemer, *Beyond the Moon*, pp. 71–72; Wolverton, *The Depths of Space*, pp. 95, 98. A good description of the asteroid belt problems, anticipated and real, is in Richard O. Fimmel, James Van Allen, and Eric Burgess, *Pioneer: First to Jupiter, Saturn, and Beyond*. NASA SP-446 (Washington, D.C.: NASA, 1980), pp. 91–92.
68. Wolverton, *Depths of Space*, pp. 97–98.
69. Kraemer, *Beyond the Moon*, pp. 73–74; Wolverton, *Depths of Space*, pp. 112–15.

## CHAPTER 11. CRUISE

70. Quoted in Bruce Mazlish, ed., *The Railroad and the Space Program: An Exploration in Historical Analogy* (Boston, Mass.: MIT Press, 1965), p. 4.
71. Ibid.

## CHAPTER 12. ENCOUNTER: JUPITER

72. By far the best account for encounter is David Morrison and Jane Samz, *Voyage to Jupiter*, NASA-SP-439 (Washington, D.C.: NASA, 1980); for above details, see p. 56. I have followed this text closely, largely doing what any collage does, namely, deface the original to make a new whole.
73. Ibid., pp. 58, 60.
74. Ibid., pp. 60, 63.
75. Ibid., pp. 74–86; Stone quote on p. 86.
76. Ibid., pp. 74–86; Callisto quote quote on p. 86.
77. Ibid., p. 75.
78. Ibid., Soderblom quote on p. 67.
79. Ibid., pp. 86–91.

80. Alexander von Humboldt, *Personal Narrative of a Journey to the Equinoctial Regions of the New Continent*, abridged and translated by Jason Wilson (London: Penguin Books, 1995), p. 26. Darwin quoted in Jason Wilson, introduction to Humboldt, *Personal Narrative*, p. xxxvi. Charles Darwin, *The Voyage of the Beagle*, ed. Leonard Engel (Garden City, N.Y.: Doubleday, 1962), pp. 2–3.
81. I again follow, in this entire section, Morrison and Samz, *Voyage to Jupiter*, pp. 97–99.
82. Ibid., pp. 101–2.
83. Soderblom quote from ibid., p. 106; Mutch quote, ibid., p. 108.
84. Cited in ibid., pp. 114–15.
85. Felipe Fernández-Armesto, *The Canary Islands After the Conquest: The Making of a Colonial Society in the Early Sixteenth Century* (Oxford: Clarendon Press, 1982), p. 205, fn 6. For a general consideration of the Canaries within early European expansion, see Fernández-Armesto, *Pathfinders*, pp. 122–51.
86. Fernández-Armesto, *The Canary Islands After the Conquest*, p. 15; on Sancho Panza: from Fernández-Armesto, *Pathfinders*, p. 145.
87. See Columbus, *Four Voyages*, p. 12.
88. For an introduction to the tropic island as Eden and the role of traveling naturalists generally, see Grove, *Green Imperialism*.
89. Thomas More, *Utopia*, trans. Paul Turner (Baltimore, Md.: Penguin Books, 1965), pp. 38, 42.
90. An excellent survey of space-based utopias is available in DeWitt Douglas Kilgore, *Astrofuturism: Science, Race, and Visions of Utopia in Space* (Philadelphia: University of Pennsylvania Press, 2003), which includes an insightful chapter on O'Neill. I have also found very useful the excellent summary in Roger D. Launius and Howard E. McCurdy, *Robots in Space: Technology, Evolution, and Interplanetary Travel* (Baltimore, Md.: Johns Hopkins University Press, 2008), chapter 2.

CHAPTER 13. CRUISE

91. Sources: Morison, *The Great Explorers*, pp. 353–54; Fernández-Armesto, *Pathfinders*, pp. 174–78; and for an especially lively account, Alfred W. Crosby, *Ecological Imperialism: The Biological Expansion of Europe, 900–1900* (Cambridge, UK: Cambridge University Press, 1986), pp. 113–19.
92. Charles Kohlhase, ed., *The Voyager Neptune Travel Guide*. JPL Publication 89-24 (Pasadena, Calif.: JPL, 1989), pp. 103–4.

93. I rely on several accounts, listed in these notes. Of particular value is Tony Reichhardt, "Gravity's Overdrive," *Air and Space/Smithsonian* 8, no. 6 (1994): 72–78. Details of Minovitch's work are difficult to extract from JPL sources, ostensibly because they are covered by privacy guidelines governing personnel disputes, but likely because of threatened lawsuits. Some of the major documents are posted on Minovitch's Web site (www.gravityassist.com). The responses of his JPL supervisors are unavailable. On JPL's inability to furnish details, I quote Julie Cooper, archivist, who helped as much as she could: "The correspondence about Minovitch was denied clearance on the grounds that it is personnel related—like correspondence about any dispute between an employer and employee. Our clearance procedures at JPL have changed since the documents were processed back in the early 1990s, so they weren't separated from the rest of the collection, as discreet records should be. So, the correspondence won't be available to any other researchers, and you won't be able to use those documents again." E-mail to author, Oct. 6, 2008.
94. M. A. Minovitch, "A Method for Determining Interplanetary Free-Fall Reconnaissance Trajectories," JPL Technical Memo no. 312-130 (August 23, 1961). Earlier memos that summer had inquired into conic trajectories of various kinds.
95. Figures from Reichhardt, "Gravity's Overdrive," p. 76. Inner versus outer planet contributions from G. A. Flandro, "Fast Reconnaissance Missions to the Outer Solar System Utilizing Energy Derived from the Gravitational Field of Jupiter," *Astronautica Acta* 12, no. 4 (1966): 334.
96. For the evolution of publications, see M. A. Minovitch, "The Determination and Characteristics of Ballistic Interplanetary Trajectories Under the Influence of Multiple Planetary Attractions," JPL Technical Report no. 32-464 (1963); M. A. Minovitch, "Utilizing Large Planetary Perturbations for the Design of Deep-Space, Solar-Probe, and Out-of-Ecliptic Trajectories," JPL Technical Memo no. 312-514 (1965). For a general survey of efforts, see Richard L. Dowling et al., "The Origin of Gravity-Propelled Interplanetary Space Travel," 41st Congress of the International Astronautical Federation, IAA-90-630 (1990), pp. 1–19. For a parallel project, outside JPL, see Maxwell W. Hunter II, "Unmanned Scientific Exploration Throughout the Solar System," *Space Science Reviews* 6 (1967): 601.
97. Cutting quote: Joe Cutting to Mike Minovitch, interoffice memo, Jan. 21, 1964, JPL, p. 2 (accessible at www.gravityassist.com).
98. Reichhardt, "Gravity's Overdrive," pp. 76–77.

99. I follow Tony Reichhardt, "Gravity's Overdrive," pp. 72–78; Kohlhase, ed. *The Voyager Neptune Travel Guide*, pp. 103–9. Quote from Minovitch Web site home page, www.gravityassist.com; accessed September 1, 2007.
100. Richard L. Dowling et al., "The Origin of Gravity-Propelled Interplanetary Space Travel," *41st Congress of the International Astronautical Federation* (1990), IAA-90-630, p. 16.
101. On the limitations of Minovitch's methods, see Gary Flandro in Littman, *Planets Beyond*, p. 97.
102. See Robert Merton, "Singletons and Multiples in Scientific Discovery," *Proceedings of the American Philosophical Society*, vol. 105, no. 5 (October 1971), pp. 470–86. Einstein's general theory of relativity is an exception.
103. Account from Morison, *The Great Explorers*, p. 400.
104. I follow William H. Goetzmann, *Exploration and Empire*, pp. 117–20, 276–78.

## CHAPTER 14. ENCOUNTER: SATURN

105. My account of the Voyagers' encounter follows closely the excellent narrative in David Morrison, *Voyages to Saturn*, NASA SP-451 (Washington, D.C.: NASA, 1982). He describes the coincidence of the Great Conjunction on page 50.
106. Ibid., pp. 50–56.
107. Ibid., pp. 56–61.
108. Ibid., pp. 62, 65.
109. Ibid., pp. 67–68, 94.
110. Quote from Bradford Smith, in Morrison, *Voyages to Saturn*, p. 69, and for Pioneer 11, from James Van Allen, in Wolverton, *The Depths of Space*, pp. 150, 154.
111. Morrison, *Voyages to Saturn*, pp. 70–76.
112. Details from ibid., pp. 78–85. Quotes: Bradford Smith, p. 84; Larry Soderblom and Morrison, p. 85.
113. Stone quoted in Morrison, *Voyages to Saturn*, p. 86.
114. Ibid., pp. 59, 67, 70–71.
115. Ibid., pp. 89–90.
116. Summary of findings from ibid., p. 93.
117. The event was advertised as a "planetary festival" that would extend the three days before encounter and would include symphony orchestras, lectures, educational activities, and an invitational black-tie dinner at $150 per plate to which members of Congress would also be invited. This would be the last planetary encounter for years, but acceptance

rates had "traditionally run low." See Memo, Lynne Murphy to Terry Finn, July 15, 1981, NASA Historical Reference Collection, File 005580.

118. Summary of session in Morrison, *Voyages to Saturn*, p. 94. The session was not transcribed or published, but was filmed for possible TV broadcast. The tape is available through JPL.
119. Bruce Murray and Louis D. Friedman, "Our Founders: The Planetary Society Founders' Statement," www.planetary.org/about/founders, accessed December 23, 2007.
120. For a synopsis, see William Sinclair, "The African Association of 1788," *African Affairs* 1, no. 1 (1901): 145–50.
121. Again I follow the chronology of events laid out in Morrison, *Voyages to Saturn*, pp. 96–135.
122. Quote from ibid., p. 100.
123. Ibid., pp. 117–21.
124. Ibid., pp. 119–21.
125. Quote from ibid., p. 123.
126. Quoted in ibid., p. 127.
127. Quotes from ibid., p. 133.
128. My historical analysis is a mere gloss of George R. Stewart's classic, *Names on the Land: An Historical Account of Placenaming in the United States*, 4th ed. (San Francisco: Lexikos, 1982); quotes from pp. 11–12. Columbus quote from J. M. Cohen, ed. and trans., *The Four Voyages of Christopher Columbus* (London: Penguin Books, 1969), p. 2.
129. Stewart, *Names on the Land*, pp. 23, 315.
130. Charge quoted from USGS Board on Geographic Names Web site: http://geonames.usgs.gov, accessed December 30, 2007.
131. For the procedure, see "How Names Are Approved," USGS Astrogeology Research Program, http://planetarynames.wr.usgs.gov/approved.html, accessed December 31, 2007.
132. "Sources of Planetary Names," USGS Astrogeology Research Program: http://planetarynames.wr.usgs.gov/jsp/append4.jsp, accessed December 30, 2007.
133. This synopsis distills a wonderful, dialogue-rich passage in Henry S. F. Cooper, Jr., *Imaging Saturn: The Voyager Flights to Saturn* (New York: Holt, Rinehart and Winston, 1982), pp. 99–103.
134. Incident reported in Nicks, *Far Travelers*, p. 76. Quote on "no geology" from Bruce Murray, in Swift, *Voyager Tales*, p. 217. End quote from Dethloff and Schorn, *Voyager's Grand Tour*, p. 145.
135. Nicks, *Far Travelers*, p. 121.

136. On Voyager 1 photo: Ben Evans with David M. Harland, *NASA's*, caption for color plate 1.
137. Best account is Miner, *Uranus*, pp. 153–54.
138. Ibid., pp. 141–43.
139. Ibid., pp. 156–57. The critical official documents are found in JPL, "Voyager: Project Plan, Part 2: Voyager Uranus/Interstellar Mission," PD 618-5, Part 2 (July 22, 1981), JPL Archives. The option had long been retained; see, for example, Decision Paper, AD/Deputy Administrator to S/Associate Administrator for Space Science, October 29, 1975, Selection of a Uranus Mission Option, in which "one spacecraft is launched on a trajectory that retains the Uranus targeting option."
140. The most useful document is JPL, "Voyager: Project Plan, Part 2"; quotes from p. 3-1. For a brief assessment of what the reduced staffing meant, see Ellis Miner, in Swift, *Voyager Tales*, p. 309.

CHAPTER 15. CRUISE

141. For background on the DSN, see Craig B. Waff, "The Road to the Deep Space Network," *IEEE Spectrum* (April 1993): 50–57, and "The Struggle for the Outer Planets," *Astronomy* 17, no. 9 (Sept. 1989): 33–52, and for a good general survey, Douglas J. Mudgway, *Uplink-Downlink: A History of the Deep Space Network 1957–1997.* NASA SP-2001-4227 (Washington, D.C.: NASA, 2001). Also useful for contemporary status, Roger Ludwig and Jim Taylor, "Voyager Telecommunications," JPL DESCANSO, Deep Space Communications and Navigation Systems, Design and Performance Summary Series (n.d.). On the role of Pioneer, see Wolverton, *The Depths of Space*, pp. 184–85.
142. See Mudgway, *Uplink-Downlink*, pp. xliv–xlvi; Nicks, *Far Travelers*, pp. 157–58.
143. JPL, *The Voyager Uranus Travel Guide*, p. 24.
144. I follow Miner, *Uranus*, p. 178.
145. Ibid., pp. 178–80. Quote on watch battery: Ben Evans with David M. Harland, *NASA's*, p. 172.
146. Miner, *Uranus*, pp. 180–81; Douglas J. Mudgway, *Uplink-Downlink*, pp. 193–96.
147. Mudgway, *Uplink-Downlink*, p. 200.
148. Ibid., p. 201.
149. Ibid., pp. 202–3.
150. Miner, *Uranus*, p. 182.
151. Mudgway, *Uplink-Downlink*, p. 203.

152. Miner, *Uranus*, pp. 177–86.
153. Jay T. Bergstralh, ed., *Uranus and Neptune: Proceedings of a Workshop Held in Pasadena, California, February 6–8, 1984*, NASA Conference Publication 2330 (Washington, D.C.: NASA, 1984).
154. I follow Miner, *Uranus*, p. 186. Quote from JPL, *The Voyager Uranus Travel Guide*, pp. 9, 6; on ring discovery, p. 17.
155. Miner, *Uranus*, pp. 186–89. For a more accessible review, see JPL, *Voyager Uranus Travel Guide*, pp. 19–24, and for a copy of the master timeline, p. 77.
156. Miner, *Uranus*, pp. 288–89.

CHAPTER 16. ENCOUNTER: URANUS

157. JPL, "Voyager. Project Plan: Part 2," pp. 3–4, 5–6. See also Miner, *Uranus*, p. 195.
158. Miner, *Uranus*, pp. 195–96.
159. Ibid., p. 196.
160. JPL, *The Voyager Uranus Travel Guide*, Voyager Project Document no. 618-150 (Pasadena, Calif.: Caltech, 1985), p. 6. The projected record of what Voyager 2 might find is in *The Voyager Uranus Travel Guide*, pp. 79–94, and the actual record of results is in Miner, *Uranus*, pp. 190–93, 195–96.
161. Miner, *Uranus*, pp. 297–98.
162. Ibid., pp. 198–200, 203. For an interesting comparison, see how deeds matched expectations in JPL, *Voyager Uranus Travel Guide*, and JPL, "Voyager. Project Plan. Part 2: Voyager Uranus/Interstellar Mission."
163. Miner, *Uranus*, pp. 201–202, 207; quote from p. 202.
164. Ibid., pp. 210–78. For the best quasi-popular summary, see the special issue of *Science* 233 (July 4, 1986): 39–107; E. D. Miner and E. C. Stone, "Voyager at Uranus," *Journal of the British Interplanetary Society* 41 (1988): 49–62.
165. Miner, *Uranus*, pp. 282–84; names on pp. 282–83; descriptions from pp. 285–319, and quote from p. 319. For fuller technical account, see also published articles, especially the special issue of *Science*.
166. Von Braun story in Wolverton, *The Depths of Space*, p. 127.
167. James Van Allen, "Space Science, Space Technology, and the Space Station," *Scientific American* 254 (January 1986): 32–39, and "Myths and Realities of Space Flight," *Science* 232 (May 30, 1986): 1075–76.
168. JPL efforts to avoid conflict and Beggs's response reported by Charles Kohlhase in *Voyager Tales*, p. 96.
169. Miner, *Uranus*, pp. 207–8.

170. Murray, *Journey into Space*, p. 235. The contest among competing space interests has a long trail of literature. An interesting survey of public opinion is Roger D. Launius, "Public Opinion Polls and Perceptions of US Human Spaceflight," *Space Policy* 19 (2003): 163–75. For particulars cited in text, see the following: for Apollo's oversize profile, see Etzioni, *The Moon-Doggle*, pp. 13–14; for Skylab competition, see Morrison and Samz, *Voyage to Jupiter*, p. 101; Kraemer, *Beyond the Moon*, pp. xvi–xviii.
171. Miner and Wessen, *Neptune*, pp. 265–67.
172. Robert D. Ballard, *The Eternal Darkness: A Personal History of Deep-Sea Exploration* (Princeton, N.J.: Princeton University Press, 2000), p. 227.
173. See Nicks, *Far Travelers*, esp. pp. 190, 245–48, 250; quote from p. 245.
174. Statistics from Newby, *The Rand McNally World Atlas of Exploration*, p. 135.
175. Quote from Tim Jeal, *Stanley: The Impossible Life of Africa's Greatest Explorer* (New Haven, Conn.: Yale University Press, 2007), p. 111.
176. Cherry-Garrard, *The Worst Journey in the World*, pp. 606, 613, 642.
177. Nicks, *Far Travelers*, p. 254.
178. The story is told with fascinating detail and empathy in Connolly and Anderson, *First Contact*. For a rawer version, see Leahy and Crain, *The Land That Time Forgot*.
179. Columbus quote from Morison, *The Great Explorers*, p. 403. Díaz, *The Conquest of New Spain*, pp. 59, 64–65.
180. Diaz, *Conquest*, pp. 60–61.
181. Quotes from Jeal, *Stanley*, pp. 358–59.
182. For an interesting survey, see James G. Bellingham and Kanna Rajan, "Robotics in Remote and Hostile Environments," *Science* 318 (Nov. 16, 2007): 1098–1102.
183. Stone quote in Swift, *Voyager Tales*, p. 57. Nicks quote from Nicks, *Far Travelers*, p. 32.
184. Carl Sagan, *Cosmos* (New York: Random House, 1980), p. 125; Nicks, *Far Travelers*, p. 32; Murray, quoted in Swift, *Voyager Tales*, pp. 210–11.
185. Miner, *Uranus*, pp. 202–3.
186. Ibid., pp. 206–7.

## CHAPTER 17. CRUISE

187. On sources of Neptune information at the time, see Miner and Wessen, *Neptune*, pp. 1–62.
188. My sources for this phase of the journey are Miner and Wessen, *Neptune*, with Miner quote from p. 147, and "better spacecraft" quote, p. 157;

Kohlhase, ed., *The Voyager Neptune Travel Guide*; and Evans with Harland, *NASA's*. On staffing: see Ellis Miner, p. 309, and Charles Kohlhase, p.89, in Swift, *Voyager Tales*.

189. Quote from Miner and Wessen, *Neptune*, p. 147. The phrase refers to a Larry McMurtry novel of that name, released as a Hollywood movie in 1971.
190. Jay T. Bergstralh, ed., *Uranus and Neptune*. On staffing: see Ellis Miner, p. 309, and Charles Kohlhase, p. 89, in Swift, *Voyager Tales*. Quotes from Evans with Harland, *NASA's*, pp. 204, 206, and Kohlhase, *Voyager Neptune*, pp. 123–29, 136 (this book is probably the best single source). Transmission numbers from Miner and Wessen, *Neptune*, p. 147.
191. Miner and Wessen, *Neptune*, pp. 147–49; Kohlhase, *Voyager Neptune*, pp. 78–80; Evans with Harland, *NASA's*, pp. 202–4.
192. Miner and Wessen, *Neptune*, pp. 158–60.
193. Evans with Harland, *NASA's*, pp. 204–6. Best summary is in Miner and Wessen, *Neptune*, pp. 158–66.
194. Kohlhase, *Voyager Neptune*, p. 72.
195. Best account is in Evans with Harland, *NASA's*, p. 206. Eight years reference and quote: from Kohlhase, *Voyager Neptune*, p. 68. Miner and Wessen, *Neptune*, p. 167; Evans with Harland, *NASA's*, p. 206.
196. See Miner and Wessen, *Neptune*, pp. 154–56. See also Kohlhase, *Voyager Neptune*, pp. 73–74, 89.
197. Miner and Wessen, *Neptune*, pp. 155–56.
198. Best summary in Kohlhase, *Voyager Neptune*, pp. 77–78.
199. For partial listing of stresses, Dethloff and Schorn, *Voyager's Grand Tour*, p. 141.
200. Quoted by Charles Kohlhase, in Swift, *Voyager Tales*, p. 88. Miner quote from Swift, *Voyager Tales*, p. 311.
201. On Humboldt, see Helmut de Terra, *Humboldt: The Life and Times of Alexander von Humboldt 1769–1859* (New York: Octagon Books, 1979); on the Institut d'Egypte, see Nina Burleigh, *Mirage* (New York: HarperCollins, 2007); on the *Challenger* expedition, see Richard Corfield, *The Silent Landscape: The Scientific Voyage of the HMS Challenger* (Washington, D.C.: Joseph Henry Press, 2003).
202. There are many accounts available; a good popular version that grants special attention to Steller is Corey Ford, *Where the Sea Breaks Its Back* (Anchorage: Alaska Northwest Books, 1966).
203. Numbers from Eric Linklater, *The Voyage of the Challenger* (Garden City, N.Y.: Doubleday, 1972), p. 270, and quotes from p. 24. For a good com-

parison with modern oceanographic discoveries, see Corfield, *The Silent Landscape*, op. cit.

204. The plaque briefly flashed onto the screen before the spacecraft blasts off appears to identify it as one of the Pioneers.
205. The story of oceanic research has become a growth industry. I found particularly helpful as a compendium of recent discoveries Tony Koslow, *The Silent Deep: The Discovery, Ecology, and Conservation of the Deep Sea* (Chicago: University of Chicago Press, 2007). For a detailed examination of one career as it evolved through the broader field, see Henry Menard, *The Ocean of Truth: A Personal History of Global Tectonics* (Princeton, N.J.: Princeton University Press, 1986), and for the larger research context, Jacob Darwin Hamblin, *Oceanographers and the Cold War: Disciples of Marine Science* (Seattle: University of Washington Press, 2005).
206. See Ballard, *The Eternal Darkness*, p. 5.
207. The *Challenger* story has been oft-told. Koslow, *Silent Deep*, op. cit, provides a useful sketch. For accounts both more thorough and more popular, see, respectively, Corfield, *The Silent Landscape*, and Linklater, *The Voyage of the Challenger.*
208. Ballard, *The Eternal Darkness*, pp. 225–31, and in "The Explorers," *The Universe Beneath the Sea* (Beverly Hills, Calif.: Oxford Television Co., 1999).

## CHAPTER 18. ENCOUNTER: NEPTUNE

209. My account closely follows Miner and Wessen, *Neptune*, pp. 174–76. For the official documentation, see JPL, "Voyager. Project Plan: Part 3: Voyager Neptune/Interstellar Mission. Revision A," JPL Archives 44, no. 48.
210. Miner and Wessen, *Neptune*, pp. 174–76; Kohlhase quote and estimates from Kohlhase, ed., *Voyager Neptune Travel Guide*, p. 89.
211. While most encounters were given in local time at JPL, this one was recorded officially in UTC (coordinated universal time), which results in 3:56 UTC on August 25, or 11:56 EDT August 24. The usual date for the encounter is thus listed as August 25, 1989. See E. C. Stone and E. D. Miner, "The Voyager 2 Encounter with the Neptunian System," *Science* 246 (December 15, 1989): 1418.
212. I follow Miner and Wessen, *Neptune*, pp. 176–81.
213. Ibid., p. 181.
214. Ibid.
215. Ibid., pp. 181–82.

216. Information and paraphrases from Miner and Wessen, *Neptune*, pp. 255–61.
217. Source: ibid., p. 259.
218. Wilson, *I.G.Y*, pp. 162–63.
219. Rachel Carson, *The Sea Around Us*, rev. ed. (New York: New American Library, 1961), pp. 196, 123.
220. Wilson, *I.G.Y*, p. 245. For an overview of exploring marine geology, see H. William Menard, "Very Like a Spear," in Cecil Schneer, ed., *Two Hundred Years of Geology in America: Proceedings of the New Hampshire Bicentennial Conference on the History of Geology* (Hanover, N.H.: University Press of New England, 1979), p. 21. Menard published a fuller, book-length personal survey of the revolution with *The Ocean of Truth: A Personal History of Global Tectonics* (Princeton, N.J.: Princeton University Press, 1986).
221. Wilson, *I.G.Y*, p. 190.
222. Useful summary of field in Dethloff and Schorn, *Voyager's Grand Tour*, pp. 146–53; Kuiper identification on p. 147. On the USAF series, see G. P. Kuiper, ed., *The Solar System*, 4 vols. (Chicago: University of Chicago Press, 1953–62). Also: Harold C. Urey, *The Planets: Their Origin and Development* (New Haven, Conn.: Yale University Press, 1952).
223. Robert Baker, *Astronomy*, 8th ed. (New York: D. Van Nostrand Co., 1964), p. 209. G. P. Kuiper, ed., *The Earth as a Planet*, Vol. 2: *The Solar System* (Chicago: University of Chicago Press, 1954), p. v.
224. On Pioneer 6, Wolverton, *The Depths of Space*, p. 126. On Voyager: JPL, *The Voyager Uranus Travel Guide*, p. 106; Bruce Murray, Michael Malin, and Ronald Greeley, *Earthlike Planets* (San Francisco: W. H. Freeman, 1981), p. xi. Any modern text will demonstrate the changes, but for a handy digest, see David J. Stevenson, "Planetary Science: A Space Odyssey," *Science* 287, no. 5455 (Feb. 11, 2000): 997–1005.
225. Nicks, *Far Travelers*, p. 219; Miner and Wessen, *Neptune*, p. xix; Van Allen quoted in Wolverton, *The Depths of Space*, p. 154.
226. Stone quoted in Swift, *Voyager Tales*, p. 38. Soderblom quoted in JPL video *And Then There Was Voyager*. I have confirmed his use of the quote and its original source by personal interview. The expression also appears in a slightly different version by Edward Stone, in Swift, *Voyager Tales*, p. 28.
227. For "legend and myth" quote: Roger Launius, foreword, in Kraemer, *Beyond the Moon*, p. x. Murray quoted in Swift, *Voyager Tales*, p. 215.
228. Murray, in Swift, *Voyager Tales*, p. 214.

229. Careful readers will note the irony of my criticizing one analogy by creating another. So be it.
230. Murray, in Swift, *Voyager Tales*, p. 212; Letter, Bruce Murray to Sir Arthur C. Clarke, April 23, 1980, JPL Archives, 223, no. 45.
231. Murray, in Swift, *Voyager Tales*, pp. 212–13.
232. Ibid., p. 213. On the federal budget, see Roger D. Launius, "Public Opinion Polls and Perceptions of US Human Spaceflight," *Space Policy* 19 (2003): 163–75. Data sources: Web sites for NASA, MLB, and NFL. For publishing and restaurants: "The 24-Billion-Dollar Question," PublishingTrends.com, www.publishingtrends.com/copy/batch0one/3-00-24billion.html (20 June 2008).
233. Murray, in Swift, *Voyager Tales*, pp. 215, 210.

## CHAPTER 19. CRUISE

234. See Janet Vertesi, "'Seeing Like a Rover': Embodied Experience on the Mars Exploration Rover Mission," CHI 2008, April 5–10, 2008, Florence, Italy. I thank Valerie Olson for alerting me to this item and sending me a copy. On symbiosis between explorers and machines, see Nicks, *Far Travelers*, p. 94, and Ballard, *The Eternal Darkness*, pp. 299–311.
235. "NASA investigates virtual space," BBC News: http://news.bbc.co.uk/go/pr/fr/-/2/hi/technology/7195718.stm, published 2008/01/18. I would like to thank Valerie Olson for correspondence early in this project in which she alerted me to the question of virtual exploration.

## CHAPTER 20. LAST LIGHT

236. Several accounts exist. I follow Charlene Anderson, "Voyager's Last View," on the Planetary Society's Web site: www.planetary.org/explor/topics/space_missions/voyager/family_portrait.html, accessed June 23, 2008; and the account given in Carl Sagan, *Pale Blue Dot* (New York: Ballantine, 1994), pp. 1–7; and the detailed recollection of Candice Hansen, in Swift, *Voyager Tales*, pp. 323–25. Constance Holden, "Voyager's Last Light," *Science* 248 (1990): 1308, gives the date as the "evening of candice 13 February 1990." Granted the long process, it spilled over into February 14, which is what has been generally reported.
237. Quote from Sagan, *Pale Blue Dot*, p. 5.
238. For an interesting account of the competing interests, see Candice Hansen, in Swift, *Voyager Tales*, p. 324.

239. Quote from Sagan, *Pale Blue Dot*, p. 5. An interesting UPI story by Rob Stein dates the event to February 13, 1990, a day earlier; see NASA Historical Collection, File 005578.
240. Woude and Hansen quoted in Stein, NASA Historical Reference Collection, File 005578; Stone quoted in John Noble Wilford, "From Voyager 1, a Space Snapshot of 6 Planets," *New York Times*, June 7, 1990.

## PART 3: BEYOND THE UTMOST BOND: JOURNEY TO THE STARS

### CHAPTER 21. VOYAGER INTERSTELLAR MISSION

1. Best summaries: Miner and Wessen, *Neptune*, pp. 182–84, and Kohlhase, ed., *The Voyager Neptune Travel Guide*, pp. 151–54.
2. On costs: JPL, *Voyager Uranus Travel Guide*, pp. 105–6. On prospects for shutting down: Andrew Lawler, "NASA Plans to Turn Off Several Satellites," *Science* 307 (March 11, 2005): 1541; Tony Reichhardt, "NASA's Funding Shortfall Means Journey's End for Voyager Probes," *Nature* 434 (March 10, 2005): 125, which lists the costs as $4.2 million annually. The discrepancy seems to derive from the cost of both spacecraft and data analysis.
3. Sources on heliosheath are many. Good diagram and brief explanation on NASA JPL Web site: www.voyager.jpl/nasa.gov/mission/mission.html, accessed June 22, 2008. For what mission planners expected, see Kohlhase, ed., *The Voyager Neptune Travel Guide*, pp. 151–54, and Miner and Wessen, *Neptune*, pp. 182–84.
4. Christian quoted in "Voyager 2 Proves Solar System Is Squashed," JPL Voyager Web site: www.voyager.jpl.nasa.gov/news/voyager_squashed.html, accessed June 22, 2008.
5. See Kohlhase, *Voyager Neptune Travel Guide*, p. 153.
6. Timetables: Miner and Wessen, *Neptune*, p. 184; Kohlhase, *Voyager Neptune Travel Guide*, p. 155, and 155–67 passim.

### CHAPTER 22. FAR TRAVELERS

7. Sources: Wolverton, *The Depths of Space*; data, p. 225. See also the NASA Pioneer Web site for technical information and dates. Velocity calculations from JPL Web site on Solar System Dynamics, Horizon System: http://ssd.jpl.nasa.gov/?horizons.

8. Calculations from JPL Web site on Solar System Dynamics, Horizon System: http://ssd.jpl.nasa.gov/?horizons. Other information from Wolverton, *Depths of Space*, p. 225, and NASA Pioneer mission Web sites.
9. On the fate of Columbus's ships, see Morison, *The Great Explorers*, on the *Niña*, p. 386. The *Caird* was preserved at Dulwich College, South London, then did a stint at the National Maritime Museum before returning to Dulwich and a career of exhibitions.
10. The best composite account is in Wikipedia; accessed June 25, 2008.
11. Vessel history from ibid.
12. Alicia Chang, "New tasks given to old NASA spacecraft," AP, reported on *Yahoo! News*, July 3, 2007. Plans include redirecting the Deep Impact spacecraft into use as an observatory, although this was part of its two-phase mission.

CHAPTER 23. NEW WORLDS, NEW LAWS

13. Direct quotes and intervening passages (paraphrased) are from Jeal, *Stanley*, p. 225.
14. Best source on Portuguese expansion is Newitt, *A History of Portuguese Overseas Expansion*; for the origins, see p. 41.
15. The story of Arvid Pardo and the "common heritage" concept is a well-told tale, but see George V. Galdorisi and Kevin R. Vienna, *Beyond the Law of the Sea: New Directions for U.S. Oceans Policy* (Westport, Conn.: Praeger, 1997), p. 25 for the observation that the concept, and nearly the language, was in American policy, additional fallout from the Antarctic Treaty.
16. "Treaty on Principles Governing the Activities of States in the Exploration and Use of Outer Space, including the Moon and Other Celestial Bodies," U.S. Department of State; www.state.gov/t/ac/trt/5181.htm, accessed June 26, 2008.

CHAPTER 24. VOYAGER'S VOICE

17. Sagan quoted in Tom Head, ed., *Conversation with Carl Sagan* (Jackson: University of Mississippi Press, 2006), p. xvii.
18. Stone quoted in Swift, *Voyager Tales*, p. 39.
19. Carl Sagan, *Cosmos* (New York: Random House, 1980), pp. 120–21.

CHAPTER 25. VOYAGER'S RECORD

20. See Fimmel et al., *Pioneer*, pp. 248–50. Burgess offers a more detailed account in Burgess, *Far Encounter*, pp. 57–58.

21. Casani quote: October 16–17, 1974, Mariner Jupiter/Saturn 1977, Mission and Systems Design Review, Concern/Action Control Sheet, JPL Archives 36, no. 29. Quotes from Sagan, "For Future Times and Being," in Carl Sagan, et al., *Murmurs of Earth: The Voyager Interstellar Record* (New York: Random House, 1978), p. 11. My account of the golden record derives primarily from this source.
22. Sagan et al., *Murmurs*, preface, pp. 146, 162.
23. Sagan, "For Future Times and Beings," p. 37.
24. Sagan, et al., *Murmurs*, epilogue, p. 234; Sagan, "For Future Times and Beings," p. 37.
25. Sagan, "For Future Times and Beings," p. 37.
26. Linda Sagan, "A Voyager's Greetings," in Sagan et al., *Murmurs*, p. 125; Carl Sagan, "For Future Times and Beings," p. 23; F. D. Drake, "Foundations of the Voyager Record," pp. 47–53.
27. Sagan, *Murmurs*, epilogue, p. 236.
28. Christopher Columbus, *The Four Voyages*, pp. 118, 80.
29. Boswell, *The Life of Samuel Johnson*, p. 355.
30. Hakluyt, *Voyages and Discoveries*: Drake, p. 177; Cavendish, pp. 287–88. On Brûlé, see William H. Goetzmann and Glyndwr Williams, *The Atlas of North American Exploration* (New York: Prentice Hall, 1992), p. 59.
31. The Columbus story comes from Morison, *The Great Explorers*, p. 422.
32. Oliver quoted in Sagan, "For Future Times and Beings," p. 11. On the disposition of the records, see Letter, Robert A. Frosch to S. Dillon Ripley, Dec. 19, 1977, NASA History Office, File 005580. Interestingly, Sagan and Casani were denied copies. One might also note the slightly blasphemous associations possible between the Voyager golden record and Joseph Smith's golden plates, dug up across the Finger Lakes from Sagan's Laboratory for Planetary Studies—both metal tablets that record an era and identify its originators, subsequently buried deeply to preserve that memory, a record to be found ages hence, written in a peculiar language surely unintelligible to its eventual finder, save with a special apparatus by which to interpret it, a message ultimately that speaks to the character of its creators. The Voyager golden records and the gold plates of Moroni—the prophet of the space age channeling the prophet of Mormonism? Would an alien intelligence confronted with the association be a lumper or a splitter?

## CHAPTER 26. VOYAGER'S RETURNS

33. Joseph Campbell, *The Hero with a Thousand Faces,* 2nd ed. (Princeton, N.J.: Princeton University Press, 1968), p. 3.
34. Ibid., pp. 11, 21, 30.
35. Ibid., pp. 245, 38.
36. Jordan Goodman, *The Rattlesnake* (London: Faber and Faber, 2005), pp. 18–19.
37. Stories on Hawkesworth and Forster from Lynne Withey, *Voyages of Discovery: Captain Cook and the Exploration of the Pacific* (Berkeley: University of California Press), pp. 175, 312–13.
38. Campbell, *Hero,* p. 193; Hakluyt, *Voyages and Discoveries,* p. 163.
39. Campbell, *Hero,* p. 217.

## BEYOND TOMORROW

40. I follow the scenarios sketched in Kohlhase, ed., *The Voyager Neptune Travel Guide,* pp. 151–67. Other versions exist on the JPL Voyager Web site, particularly the projected scenario for shutting down instruments.
41. Miner quoted in Douglas Fulmer, "Memories of Voyager," *Ad Astra* (July/August 1993): 48. Laeser quote in Kohlhase, ed., *Voyager Neptune Travel Guide,* p. 205.
42. Haines quoted in Fulmer, "Memories of Voyager," p. 48; Stone, in Swift, *Voyager Tales,* p. 57.

# Sources

V*oyager: Seeking Newer Worlds in the Third Great Age of Discovery* is a study—call it perhaps a meditation—about context. It is not a scholarly reconstruction based on original archives but one that seeks to achieve insights by arranging facts, many of which are common knowledge, and observations, few of which come from unexplored sources. It strives to generate new understanding by emphasizing comparisons, contrasts, and continuities.

Still, some work in archives was useful, and necessary. I have consulted only two such sets of primary documents: the JPL Archives and the NASA Historical Reference Collection at the NASA History Office. Each has guides accessible online. For NASA resources, see also Stephen J. Garber, compiler, *Research in NASA History: A Guide to the NASA History Program* (Washington, D.C.: NASA, June 1997), and the fine line of monographs the NASA History Office has sponsored.

Most of my research derives from published accounts. These fall into three general areas. Some—quite a few—deal specifically with Voyager, and I am happy to defer to them on matters of design, construction, and operations. Some pertain to exploration generally. Some are specific to the era under examination, particularly its institutions. Those works that I found most pertinent I have listed here. Since exploration is among the most heavily chartered historical literature around, I include only major works that I reviewed for several purposes, or to help establish general concepts. Books or articles consulted for a single fact or purpose are included in the notes where and when they are cited.

## BOOKS (AND MAJOR ARTICLES) WHOLLY OR SIGNIFICANTLY ABOUT VOYAGER

Burgess, Eric. *Far Encounter: The Neptune System* (New York: Columbia University Press, 1991).

——. *Uranus and Neptune: The Distant Giants* (New York: Columbia University Press, 1988).

Butrica, Andrew J. "Voyager: The Grand Tour as Big Science." In Pamela E. Mack, ed. *From Engineering to Big Science: The NACA and NASA Collier Trophy Research Project Winners.* NASA SP-4219 (Washington, D.C.: NASA, 1998).

Cooper, Jr., Henry S. F. *Imaging Saturn: The Voyager Flights to Saturn* (New York: Holt, Rinehart and Winston, 1982).

Davies, J. K. "A Brief History of the Voyager Project," Part 1: *Spaceflight* 23 (March 1981); Part 2: *Spaceflight* 23 (March 1981); Part 3: *Spaceflight* 23 (May 1981); Part 4: *Spaceflight* 23 (Aug.–Sept. 1981); Part 5: *Spaceflight* 24 (Feb. 1982); Part 6: *Spaceflight* 24 (June 1982).

Davis, Joel. *Flyby: The Interplanetary Odyssey of Voyager 2* (New York: Atheneum, 1987).

Dethloff, Henry C., and Ronald A. Schorn. *Voyager's Grand Tour: To the Outer Planets and Beyond* (Washington, D.C.: Smithsonian Institution Press, 2003).

Evans, Ben, with David M. Harland. *NASA's Voyager Missions* (Chichester, UK: Springer Praxis, 2004).

Heacock, Raymond L. "The Voyager Spacecraft," Institution of Mechanical Engineers, *Proceedings 1980,* 194, no. 28 (1980).

JPL. *Voyager: The Grandest Tour: The Mission to the Outer Planets.* NASA CR-197708 (Pasadena, Calif.: JPL, 1991).

Kohlhase, Charles, ed. *The Voyager Neptune Travel Guide.* JPL Publication 89-24 (Pasadena, Calif.: JPL, 1989).

Kraemer, Robert S. *Beyond the Moon: A Golden Age of Planetary Exploration 1971–1978* (Washington, D.C.: Smithsonian Institution Press, 2000).

Littmann, Mark. *Planets Beyond: Discovering the Outer Solar System* (Mineola, N.Y.: Dover Publications, 2004; reprint of 1990 ed.).

Miner, Ellis D. *Uranus: The Planet, Rings, and Satellites* (Chichester, UK: Ellis Horwood, 1990).

Miner, Ellis D., and Randi R. Wessen. *Neptune: The Planet, Rings, and Satellites* (Chichester, UK: Springer Praxis, 2002).

Morrison, David. *Voyages to Saturn.* NASA SP-451 (Washington, D.C.: NASA, 1982).

Morrison, David, and Jan Samz. *Voyage to Jupiter.* NASA SP-439 (Washington, D.C.: NASA, 1980).

Murray, Bruce. *Journey into Space: The First Thirty Years of Space Exploration* (New York: W. W. Norton, 1989).

Sagan, Carl. *Pale Blue Dot* (New York: Ballantine, 1994).

Sagan, Carl, et al. *Murmurs of Earth: The Voyager Interstellar Record* (New York: Random House, 1978).

Stone, E. C. "Voyager Mission: Encounters with Saturn," *J. Geophys. Res.* 88, no. A11 (Nov. 1983): 8639–42.

——. "The Voyager Mission Through the Jupiter Encounters," *J. Geophys. Res.* 86, no. A10 (Sept. 1981): 8123–24.

Stone, E. C., and A. L. Lane. "Voyager 1 Encounter with the Jovian System," *Science* 204, no. 4396 (June 1, 1979): 945–48.

——. "Voyager 2 Encounter with the Jovian System," *Science* 206, no. 4421 (Nov. 23, 1979): 925–27.

Stone, E. C., and E. D. Miner. "Voyager 1 Encounter with the Saturnian System," *Science* 212, no. 4491 (Apr. 10, 1981): 159–63.

——. "Voyager 2 Encounter with the Saturnian System," *Science* 215, no. 4532 (Jan. 29, 1982): 499–504.

——. "Voyager 2 Encounter with the Uranian System," *Science* 233 (1986): 39–43.

——. "Voyager 2 Encounter with the Neptunian System," *Science* 246, no. 4936 (Dec. 15, 1989): 1417–21.

Swift, David W. *Voyager Tales: Personal Views of the Grand Tour* (Reston, Va.: AIAA, 1997).

Voyager Mission Planning Office Staff. *Voyager Uranus Travel Guide.* JPL D-2580 (Pasadena, Calif.: JPL, August 1985).

## BOOKS ABOUT EXPLORATION, INCLUDING SPACE EXPLORATION

*Atlas of Exploration* (New York: Oxford University Press, 1997).

Baker, J. N. L. *A History of Geographical Discovery and Exploration* (New York: Cooper Square Publishers, 1967).

Ballard, Robert D. *The Eternal Darkness: A Personal History of Deep-Sea Exploration* (Princeton, N.J.: Princeton University Press, 2000).

Broad, William J. *The Universe Below: Discovering the Secrets of the Deep Sea* (New York: Simon and Schuster, 1997).

Burrows, William. *Exploring Space: Voyages in the Solar System and Beyond* (New York: Random House, 1990).

——. *This New Ocean: The Story of the First Space Age* (New York: Random House, 1998).

Carson, Rachel. *The Sea Around Us* (New York: Oxford University Press, 1951; rev., 1961 ed.).

Clarke, Arthur C. *The Exploration of Space* (New York: Harper and Brothers, 1951).

Dick, Steven J., and Roger D. Launius, eds. *Critical Issues in the History of Spaceflight*. NASA SP-2006-4702 (Washington, D.C.: NASA, 2006).

Fernández-Armesto, Felipe. *Pathfinders: A Global History of Exploration* (New York: W. W. Norton, 2006).

——, ed. *The Times Atlas of World Exploration* (New York: HarperCollins, 1991).

Fimmel, Richard O., James Van Allen, and Eric Burgess. *Pioneer: First to Jupiter, Saturn, and Beyond*. NASA SP-446 (Washington, D.C.: NASA, 1980).

Foerstner, Abigail. *James Van Allen: The First Eight Billion Miles* (Iowa City: University of Iowa Press, 2007).

Goetzmann, William H. *Exploration and Empire* (New York: Knopf, 1966).

——. *New Lands, New Men: The United States and the Second Great Age of Discovery* (New York: Viking, 1986).

Hakluyt, Richard. *Voyages and Discoveries: The Principal Navigations, Voyages, Traffiques and Discoveries of the English Nation*. Ed. and abridged by Jack Beeching (New York: Penguin, 1973).

Harvey, Brian. *Russia in Space: The Failed Frontier?* (Chichester, UK: Springer Praxis, 2001).

——. *Russian Planetary Exploration: History, Development, Legacy and Prospects* (Chichester, UK: Springer Praxis, 2007).

Kilgore, DeWit Douglas. *Astrofuturism: Science, Race, and Visions of Utopia in Space* (Philadelphia: University of Pennsylvania Press, 2003).

Koppes, Clayton R. *JPL and the American Space Program* (New Haven, Conn.: Yale University Press, 1982).

Koslow, Tony. *The Silent Deep: The Discovery, Ecology, and Conservation of the Deep Sea* (Chicago: University of Chicago Press, 2007).

Launius, Roger D. *Frontiers of Space Exploration*, 2nd ed. (Westport, Conn.: Greenwood Press, 2004).

Launius, Roger D., and Howard E. McCurdy. *Imagining Space: Achievements, Predictions, Possibilities 1950–2050* (San Francisco: Chronicle Books, 2001).

——. *Robots in Space: Technology, Evolution, and Interplanetary Travel* (Baltimore, Md.: Johns Hopkins University Press, 2008).

Lewis, Richard S. *From Vinland to Mars: A Thousand Years of Exploration* (New York: Quadrangle Books, 1976).

Logsdon, John M., ed., with Amy Paige Snyder, Roger D. Launius, Stephen J. Garber, and Regan Anne Newport. *Exploring the Unknown: Selected Documents in the History of the U.S. Civil Space Program, Volume IV: Exploring the Cosmos.* NASA SP-4407 (Washington, D.C.: NASA, 2001).

Logsdon, John M., ed., with Dwayne A. Day, and Roger D. Launius. *Exploring the Unknown: Selected Documents in the History of the U.S. Civil Space Program, Volume II: External Relationships.* NASA SP-4407 (Washington, D.C.: NASA, 1996).

Logsdon, John M., ed., with Linda J. Lear, Jannelle Warren Findley, Ray A. Williamson, and Dwayne A. Day. *Exploring the Unknown: Selected Documents in the History of the U.S. Civil Space Program, Volume I: Organizing for Exploration.* NASA SP-4407 (Washington, D.C.: NASA, 1996).

Logsdon, John M., ed., with Ray A. Williamson, Roger D. Launius, Russell J. Acker, Stephen J. Garber, and Jonathan L. Friedman. *Exploring the Unknown: Selected Documents in the History of the U.S. Civil Space Program, Volume II: External Relationships.* NASA SP-4407 (Washington, D.C.: NASA, 1999).

Logsdon, John M., ed., with Roger D. Launius, David H. Onkst, and Stephen J. Garber. *Exploring the Unknown: Selected Documents in the History of the U.S. Civil Space Program, Volume III: Using Space.* NASA SP-4407 (Washington, D.C.: NASA, 1998).

Logsdon, John M., ed., with Stephen J. Garber, Roger D. Launius, and Ray A. Williamson. *Exploring the Unknown: Selected Documents in the History of the U.S. Civil Space Program, Volume VI: Space and Earth Science.* NASA SP-4407 (Washington, D.C.: NASA, 2004).

McDougall, Walter A. *The Heavens and the Earth: A Political History of the Space Age* (New York: Basic Books, 1985).

Menard, Henry. *Islands* (New York: Scientific American, 1986).

——. *The Ocean of Truth: A Personal History of Global Tectonics* (Princeton, N.J.: Princeton University Press, 1986).

Morison, Samuel Eliot. *The Great Explorers: The European Discovery of America* (New York: Oxford University Press, 1978).

Mudgway, Douglas J. *Uplink-Downlink: A History of the Deep Space Network 1957–1997.* NASA SP-2001-4227 (Washington, D.C.: NASA, 2001).

Newell, Homer E. *Beyond the Atmosphere: Early Years of Space Science.* NASA SP-4211 (Washington, D.C.: NASA, 1980).

Newitt, Malyn. *A History of Portuguese Overseas Expansion, 1400–1668* (New York: Routledge, 2005).

Nicks, Oran. *Far Travelers: The Exploring Machines.* NASA SP-480 (Washington, D.C.: NASA, 1985).

Parry, J. H. *The Age of Reconnaissance: Discovery, Exploration, and Settlement, 1450–1650* (New York: Praeger, 1969).

——. *The Discovery of the Sea* (Berkeley: University of California Press, 1981).

Pyne, Stephen J. *How the Canyon Became Grand: A Brief History* (New York: Viking, 1998).

——. *The Ice: A Journey to Antarctica* (Iowa City: University of Iowa Press, 1986).

——. "Seeking Newer Worlds: An Historical Context for Space Exploration," pp. 7–36. In Steven J. Dick and Roger D. Launius, eds., *Critical Issues in the History of Spaceflight* (Washington, D.C.: NASA, 2006).

Roland, Alex, ed. *A Spacefaring People: Perspectives on Early Space Flight* (Washington, D.C.: Government Printing Office, 1985).

Scammel, G. V. *The World Encompassed: The First European Maritime Empires c. 800–1650* (Berkeley: University of California Press, 1981).

Siddiqi, Asif A. *Deep Space Chronicle: A Chronology of Deep Space and Planetary Probes 1958–2000.* Monographs in Aerospace History, No. 24. NASA SP-2002-4524 (Washington, D.C.: NASA, 2002).

Westwick, Peter J. *Into the Black: JPL and the American Space Program 1976–2004* (New Haven, Conn.: Yale University Press, 2007).

Wilson, J. Tuzo. *I.G.Y: The Year of the New Moons* (New York: Alfred Knopf, 1961).

Wolfe, Harry. *The Transits of Venus* (Princeton, N.J.: Princeton University Press, 1959).

Wolverton, Mark. *The Depths of Space: The Story of the Pioneer Planetary Probes* (Washington, D.C.: Joseph Henry Press, 2004).

## USEFUL WEB SITES (ALL ACCESSED NOVEMBER 1, 2008)

### JPL

JPL maintains two sites that are particularly pertinent, one on Voyager itself and one on solar system dynamics (Horizons) by which it is possible to calculate the ephemeris for both spacecraft.
On Voyager: http://voyager.jpl.nasa.gov/index.html
Horizons: http://ssd.jpl.nasa.gov/?horizons

### NASA

NASA has several useful sites on Voyager and on historical research.
For Voyager: http://www.nasa.gov/mission_pages/voyager/index.html
and http://nssdc.gsfc.nasa.gov/planetary/voyager.html
For historical publications, archives, and services, see:
http://history.nasa.gov/series95.html

### THE PLANETARY SOCIETY

See: http://www.planetary.org/explore/topics/space_missions/voyager/

### USGS ASTROGEOLOGY RESEARCH PROGRAM

See: http://astrogeology.usgs.gov/Missions/Voyager/

# Index